Hans Hermes

Introduction to Mathematical Logic

Translated from German
by Diana Schmidt

New York · Heidelberg · Berlin 1973

Dr. HANS HERMES
o. Professor an der Albert-Ludwigs-Universität
Mathematisches Institut
Abteilung für mathematische Logik
und Grundlagen der Mathematik
78 Freiburg i. Br.
Hermann-Herder-Straße 10

Dr. DIANA SCHMIDT
6368 Bad Vilbel-Heilsberg
Schlesienring 4

By permission of the publishers B. G. Teubner, Stuttgart, only authorized English translation of the German original edition „Einführung in die mathematische Logik, 2. Aufl.“, which appeared in the series „Mathematische Leitfäden“, edited by Professor G. Köthe.

AMS Subject Classifications (1970): 02-01, 02B05, 02B10

ISBN 0-387-05819-2 Springer-Verlag New York - Heidelberg - Berlin
ISBN 3-540-05819-2 Springer-Verlag Berlin - Heidelberg - New York

Library of Congress Catalog Card Number 72-76760

Preface

This book grew out of lectures. It is intended as an introduction to classical two-valued predicate logic. The restriction to classical logic is not meant to imply that this logic is intrinsically better than other, non-classical logics; however, classical logic is a good introduction to logic because of its simplicity, and a good basis for applications because it is the foundation of classical mathematics, and thus of the exact sciences which are based on it.

The book is meant primarily for mathematics students who are already acquainted with some of the fundamental concepts of mathematics, such as that of a group. It should help the reader to see for himself the advantages of a formalisation. The step from the everyday language to a formalised language, which usually creates difficulties, is discussed and practised thoroughly. The analysis of the way in which basic mathematical structures are approached in mathematics leads in a natural way to the semantic notion of consequence.

One of the substantial achievements of modern logic has been to show that the notion of consequence can be replaced by a provably equivalent notion of derivability which is defined by means of a calculus. Today we know of many calculi which have this property. Some of these calculi are characterised by particular elegance and symmetry, which, however, is bought at a price: The process of operating with them bears little relation to the practices which mathematicians have adopted through years of experience in proving. In order that the mathematician's experience should not be useless when it comes to deriving by means of a calculus, calculi have been developed which, to varying degrees, mirror the traditional mathematical methods. The calculus which is treated here, a form of sequent calculus, is of this sort. For technical reasons, we initially consider only negation, conjunction and generalisation, in order to keep the theoretical treatment as short as possible. However, the other logical connectives are treated later on.

In order to make the significance of the completeness theorem clear, second-order logic is also discussed briefly, using Peano's axiom system as a starting-point.

Using model-theoretic methods, we prove a satisfiability theorem of A. Robinson; and, with the aid of this, Beth's definability theorem and Craig's interpolation lemma.

I am greatly indebted to Heinz-Dieter Ebbinghaus, Walter Oberschelp, Dieter Rödding, Dieter Titgemeyer and Rainer Titgemeyer for their valuable help.

Finally, my thanks are due to Diana Schmidt for a most careful translation of this book, which was first published in 1963 in German.

Freiburg i. Br., Summer 1972 Hans Hermes

Table of Contents

I. Introduction

In this chapter we start from facts which are more-or-less familiar to every mathematician. It is a characteristic of mathematics that every statement (other than the axioms themselves) has to be established by an argument of a nature which is essentially different from that of the experiments and experiences on which scientists base their theories. By analysing mathematical procedure, we arrive at an idea of consequence based on semantic concepts. The attempt to master mathematics by means of algorithms leads to the study of logic calculi, which are used to operate with and manipulate mathematical statements just as elementary arithmetic is used to do this with figures. It is at this stage at the latest that we realize that we can only successfully apply formal logic to mathematics if we formalize the mathematical statements. A formal language suited to this purpose is built up in the second chapter and investigated more closely in the following ones. Thus, the first chapter is of a merely preparatory nature. However, in spite of this it is especially important, because it bridges the gap between traditional and fully formalized mathematics.

§ 1. The task of logic

What is logic? It is relatively easy to give a mathematician a preliminary answer to this question. In preparation for this, we shall say a little about the nature of mathematics. Many people have tried to define the character of mathematics, both as subject matter and as regards methods. The subject matter of mathematics is continually increasing and being enriched by new disciplines, so that, rather than simply enumerating such disciplines, we should have to find a way of describing adequately the contents of mathematical theories. Thus, in the nineteenth century, there was an attempt to characterise mathematics as the science of quantity. Today, however, we have to admit that this description is much too limiting; for example, it does not include basic mathematical ideas like group theory or topology. These are concerned with fundamental structures, to which, in their purest form, the attention of mathematicians has only been drawn since the end of the last century. Mathematics today is, in many of its areas, so dominated by the idea of structures like these that, following Bourbaki, many people would like to conceive of the whole of mathematics as the science of structures.

Thus, the characterisation of mathematics by its subject matter has changed completely and raises the problem that we cannot be certain whether today's views about it will still be accepted tomorrow. However, ideas about the methods of mathematics have remained basically the same for over two thousand years. These methods can be characterised by a catchword: the mathematician is expected to prove the theorems of his science. This requirement distinguishes mathematics clearly from the natural sciences, in which experiments can also be used to establish statements.

If considered purely externally, a proof for a mathematical theorem can be seen to consist of a well-defined way of referring it back to mathematical theorems which were already known to be true, by showing that it is a "consequence" of these known theorems. Of course, the theorems which, in a proof, are supposed to be true must in their turn be referred back to other theorems which are known to be true. This process cannot go on ad infinitum; we must have started somewhere. Now it is conceivable that there could be mathematical theorems which could be proved without appealing to earlier ones. However, in general, a mathematician who is developing a particular theory starts from particular theorems which he lays down without proof as the basis of his theory. In this case, these theorems are called the axioms of the theory. Since Euclid's time, the idea of building up a mathematical discipline axiomatically has been an ideal of mathematicians; certain theorems, the axioms, are assumed without proof. A system of axioms like this then defines a mathematical theory. The theorems of this theory are (apart from the axioms themselves) all those statements which can be proved starting from the axioms. Mathematicians can express in many different ways the fact that a theorem can be proved starting from a given system of axioms. For example, they can say: the theorem can be referred back, or reduced, to the axioms; the theorem can be inferred, derived, or deduced, from the axioms; the theorem is a consequence of the axioms; the theorem follows from the axioms.

The above-mentioned expressions are used more-or-less synonymously in mathematical conversation, but the idea which they describe is not very precisely defined. Later, starting from different points of entry, we shall give two exact definitions of the idea, for which we shall use the words consequence and derivability (see §3 and Chap. III, §1 and 2; also §5 and Chap. IV, §1 and 2). These words will then be used later in the sense in which they have been defined. They indicate possible ways of making the idea, described above in the main text, precise. One of the main aims of the reflections of this book will be to show that, in spite of their different definitions, the ideas of consequence and derivability are equivalent (cf. Chap. V).

In order to indicate that a proof is valid, mathematicians use expressions like hence, thus, therefore, because, since, consequently, as a result of.

We indicated previously that there is a well-defined connection between a mathematical theorem and the theorems to which it is referred back in a proof. This connection carries over to the mathematical theorem and the axioms of the mathematical theory to which it belongs. It is the task of logic[1] to investigate this relationship. Since the above-mentioned connection is obtained through proofs, we can also say that logic has to investigate the nature and purpose of proofs. After the above reflections it is clear that a knowledge of logic is a basic requirement for the understanding of mathematical methods.

In connection with mathematical methods, it is worth noticing that a mathematician learns, during the course of his training, to infer logically, but that in general he does not know what this means, and in particular he obtains no overall idea of the logical rules which he may apply in a proof. This remarkable state of affairs is presumably connected with the fact that it is only a relatively short time since science obtained a clear understanding of the relationships which exist in this field. Logic has indeed been practised since ancient times, but syllogistics, which Aristotle created, is only fragmentary and is insufficient for the needs of mathematics. After many unsuccessfull efforts, for example those of Leibniz, it was only in the middle of the last century that a breakthrough was made. This was pioneered by George Boole, who, among other things, published his work "An investigation of the laws of thought, on which are founded the mathematical theories of logic and probabilities" in 1854, and Gottlob Frege, whose first book, which appeared in 1870, bears the title: "Begriffsschrift, eine der arithmetischen nachgebildete Formelsprache des reinen Denkens." So-called "formal" languages are used to facilitate the formulation of logical rules. (We shall say more about this at the end of this section.) Boole tried to use normal algebraic formulas for this purpose. However, the algebraification which this involved is no longer used (except in specialized investigations). The symbolisation which Frege chose was modelled more on everyday language, but fell down on its unusual two-dimensional notation. In 1900, Guiseppe Peano created, in his "Formulaire de mathématiques" a more natural symbolism, which is, in essence, also the basis of Alfred North Whitehead and Bertrand Russell's epoch-making work "Principia Mathematica" (1910/13). Since then, logic (sometimes also called "logistics") has become an important independent discipline, which bridges the gaps between sciences differing as widely as epistemology, mathematics, the theoretical sciences and linguistics.

In this book we shall be treating classical predicate logic. The relationship of this fundamental part of logic to the totality of all logical theories can be compared with that of the infinitesimal calculus to the whole of analysis. Predicate logic, unlike Aristotle's syllogistic logic, is sufficient for the construction of mathematical theories.

We have just ascertained that logic investigates the connection which exists between the axioms and the theorems of mathematical theories. That is to say, logic deals with the statements of mathematical theories (just as geometry deals with points, lines and planes, and analysis deals with numbers and functions). Since logic talks about the statements of mathematics, it is sometimes also called metamathematics.

Although here we have assigned to logic the task of talking about mathematical theories, this is really an unnecessary restriction. Proofs are conducted and conclusions drawn in other sciences, and even in the deliberations of everyday life. However, such proofs in other fields are usually not of the same "strictness" as mathematical proofs. For

[1] What we here call logic is the same as what is also called formal logic. In philosophy, the word logic is often used with a wider meaning. To denote logic as "the theory of correct thinking" can lead to misunderstandings, for it is not at all clear what is meant by "correct" thinking, if we take thinking to mean a psychophysical process rather than objectivising it.

this reason it is, in general, much harder to analyse them as regards their logical content. As the state of the development of the sciences stands today, mathematics is still the field of application par excellence of logic.

Although the word "statement" can also be understood as something spoken, it is best to imagine the statements of mathematics as written statements. It is also possible to think of a written statement as a name for the "ideal" statement which it represents, which is itself not written down on paper but has an abstract existence in its own right. But we shall only want to admit those "statements in their own right" which we can represent in writing. However, this being the case, we can do without all statements except the written ones from the start.

The advances which the formalisation of mathematics has brought about are well known. However, a glance at the run of mathematics books shows us that only a relatively small proportion of mathematical statements is expressed in formulas; the greater part of them consists of text expressed in everyday language. This is a serious disadvantage when, as in logic, we want to take the mathematical statements themselves as the subject of investigation, for the language of everyday conversation is a living language, and, as such, is capable of changing continually and is, in many ways, not fixed enough. In order to make mathematical statements the subject of a strict theory, we have first of all to strip them of their linguistic garb and to formalize them completely. Thus, we shall start Chapter II by building up the language of predicate logic as a formal language sufficient to represent mathematical statements.

§ 2. Examples of mathematical proofs

In this section we shall give three examples, simple in nature, of mathematical proofs. They will be taken from geometry, arithmetic and group theory respectively. In the next section, we shall use these examples to illustrate and explain the idea of consequence.

2.1 Example from geometry. The axioms of geometry were laid down in their original form by Euclid. After more than two thousand years, it has been discovered that the euclidean axioms are not in fact sufficient for the deduction of all the geometrical theorems which had been claimed to follow from them. This discovery led to a revision of the euclidean system of axioms. The axioms which, today, are usually taken as the basis of geometry are to be found in Hilbert's classic work, "Foundations of geometry". These axioms are divided into several groups. The first group of axioms (the so-called axioms of connection) consists of eight axioms, which (in a formulation which differs inessentially from the original) read as follows:

H1 Two distinct points A, B always lie on a line c.

H2 Any two distinct points A, B do not lie on more than one line c.

H3 For every line c there are at least two points which lie on c. There are three points which do not all lie on one line.

H4 Three points A, B, C not lying on the same line always lie in a plane α.

H5 Any three distinct points A, B, C not lying on the same line do not lie in more than one plane α.

H6 If two points A, B which lie on a line c lie in a plane α, then every point which lies on c lies in α.

H7 If a point A lies in two planes α, β, then there is a point B, distinct from A, which also lies in α and β.

H8 There are at least four distinct points A, B, C, D which do not lie in a plane.

A simple consequence of this system of axioms is the

Theorem. For every two distinct lines c, d on which a point A lies there is a plane α such that every point which lies on c or on d also lies in α.

Proof. By H3, there are at least two distinct points on each line. Therefore there is a point A_1 on c which is distinct from A and a point B_1 on d which is distinct from A. A, A_1 and B_1 do not all lie on one line. We prove this indirectly: if A, A_1 and B_1 lay on a line b, then A and A_1 would lie both on b and on c. Therefore, by H2, we should have $b = c$. Moreover, A and B_1 would lie both on b and on d. Therefore, by H2, we should have $b = d$, and hence $c = d$, contrary to hypothesis. A, A_1 and B_1 are pairwise distinct: Since, by construction, $A \neq A_1$ and $A \neq B_1$, we could have at most $A_1 = B_1$. But then A and A_1 would lie both on c and (since $A_1 = B_1$) on d, so that by H2 we should have $c = d$, which is not the case. Thus we have three distinct points A, A_1, B_1 which do not all lie on one line. Now, by the first part of H4 there is a plane α in which A, A_1, B_1 lie. We shall have completed the proof if we can show that every point on c and every point on d also lies in α. We know that the two distinct points A and A_1 lie on c and in α. Therefore, by H6, every point of c lies in α. Similarly for d: the two distinct points A, B_1 lie on d and in α, therefore, by H6, every point of d also lies in α.

The axioms H2, H3, H4 and H6 were used to prove the theorem.

2.2 Example from arithmetic. Arithmetic can be built up on a system of axioms which was suggested by Peano at the turn of the century.

The system of axioms consists of the following 5 axioms:

P1 0 is a natural number.

P2 If x is a natural number, then the successor of x is always a natural number.

P3 If x is a natural number, then the successor of x is always different from 0.

P4 If x and y are natural numbers and the successor of x is equal to the successor of y, then x = y.

P5 (Axiom of induction). If 0 has an arbitrary property P and if, whenever x has the property P, the successor of x also has the property, then every natural number has the property P.

It is on the axiom of induction P5 that "proofs by induction" are based. These are thus not regarded as pure logical inferences in Peano's construction of arithmetic. As a simple example of an inference from the Peano axioms we shall prove the

Theorem. If x is a natural number, then x is always different from the successor of x.

Proof. We consider the property P which is characterised as follows: A thing x has the property P if x is a natural number and if x is different from the successor of x. We now prove two lemmas:

Lemma 1. 0 has the property P.

Proof of Lemma 1. By P1, 0 is a natural number. By P3, for every natural number x, the successor of x is different from 0. Thus, in particular, the successor of 0 is different from 0.

Lemma 2. If x is a natural number which has the property P, then the successor of x always has the property P.

Proof of Lemma 2. Let x be an arbitrary natural number with the property P. Then, by P2, the successor of x is also a natural number. If the successor of x did not have the property P, then the successor of x would be equal to the successor of the successor of x. Then, by P4, x would be equal to the successor of x, which is not the case since x has the property P. This contradiction shows that the successor of x must have the property P.

Now we can complete the proof of the theorem. By P5 and Lemmas 1 and 2 we prove that every natural number has the property P. Thus, in particular, for every natural number x, the successor of x is different from x.

All the Peano axioms were used to prove the theorem.

2.3 Example from group theory. The axioms of group theory can be found in different forms. One possible form is the following: we consider a two-place operation, which is denoted as a product. The product of x and y is denoted by xy. We require the product to satisfy the following three axioms, put forward by Huntington:

G1 (Associative law) $(xy)z = x(yz)$.

G2 (Existence of a right quotient) For every x and z there is at least one y such that $xy = z$.

G3 (Existence of a left quotient) For every y and z there is at least one x such that $xy = z$.

As an example of an inference from this system of axioms we prove the

Theorem of the uniqueness of the right quotient. For every x and z there is at most one y such that $xy = z$ [2])

Proof. Let $xy_1 = z$ and $xy_2 = z$, i.e.

(1) $xy_1 = xy_2$

We have to show that $y_1 = y_2$. By G3 there is a u such that:

(2) $uy_1 = y_1$.

Again by G3, there is for x and u a v such that

(3) $vx = u$.

Finally, by G2, there is for y_1 and y_2 a w such that $y_1w = y_2$.

Now we infer:

$$\begin{aligned} y_1 &= uy_1 && \text{by (2)} \\ &= (vx)y_1 && \text{by (3)} \\ &= v(xy_1) && \text{by G1} \\ &= v(xy_2) && \text{by (1)} \\ &= (vx)y_2 && \text{by G1} \end{aligned} \qquad \begin{aligned} y_1 &= uy_2 && \text{by (3)} \\ &= u(y_1w) && \text{by (4)} \\ &= (uy_1)w && \text{by G1} \\ &= y_1w && \text{by (2)} \\ &= y_2 && \text{by (4).} \end{aligned}$$

§ 3. The notion of consequence

3.1 Statement of the problem. We express the relationship between the axioms and the theorems of a mathematical theory by saying: The theorems are consequences of the axioms. We want to try to clarify this relationship and to give a definition of the relation of consequence. We shall only determine this idea in a provisional way, in order to provide a foundation for the exact definition which we shall give in the second chapter on the basis of a formal language.

Our considerations of the notion of consequence will, in particular, lead us to reformulate two of the systems of axioms put forward in the last section.

In order to motivate the definition we are seeking, we shall start from the examples of the last section, in which we gave three systems of axioms and, for each system, a theorem which follows from it. We want, at first, to consider only the relation of consequence which holds between the axioms and the given theorems, not the proofs which we gave; i.e. not the methods which led us to recognise the existence of this relation. We shall consider these methods in the next section.

3.2 Groups. If the ideas which are familiar to mathematicians are taken into account, the considerations which are necessary to clarify the ideal of consequence can be carried out most convincingly with reference to the axiom system for group theory (see the third example in the last section). Let us therefore remind ourselves of the concept of a group.

---

[2] Correspondingly, we can clearly also prove the uniqueness of the left quotient.

(Note that it was not necessary to define this concept in the last section, although usually, in mathematics , the concept of a group and the axiom system for group theory are introduced simultaneously. It is, however, also possible to set up an axiom system for group theory in which the word "group" appears. We shall go into this possibility in 3.5).

By a group $\mathfrak{C}$ we mean a domain ω of elements together with a two-place operation Φ which can be carried out within the domain (so that the result of the operation, applied to two elements of the domain given in a particular order, is again an element of the domain) and for which the axioms G1, G2 and G3 are valid.

That the axiom G1 is valid, or holds true, for a group $\mathfrak{C}$ means: if $\mathfrak{x}$, $\mathfrak{y}$, $\mathfrak{z}$ are (not necessarily distinct) elements of ω, then $\Phi(\Phi(\mathfrak{x},\mathfrak{y}),\mathfrak{z})$ is the same as $\Phi(\mathfrak{x},\Phi(\mathfrak{y},\mathfrak{z}))$. That G2 holds true for $\mathfrak{C}$ means that for given (not necessarily distinct) elements $\mathfrak{x}$, $\mathfrak{z}$ of ω there is always at least one element $\mathfrak{y}$ of ω such that $\Phi(\mathfrak{x},\mathfrak{y})$ is the same as $\mathfrak{z}$. The situation is similar for G3.

The domain ω is also called a domain of individuals or an object domain.

3.3 Group-theoretic statements. For the following, we shall use the concept of a group-theoretic statement. In preparation for this, we shall look first at some arbitrary mathematical statements. The axioms H1,...,H8, P1,...,P5, G1,...,G3, as well as the theorems of the last section, are examples of mathematical statements. These statements are valid statements of the particular mathematical disciplines. However, in order for something to be simply a mathematical statement, it is not required to be valid. Thus, the following are also examples of mathematical statements:

(*) There is a natural number x which is equal to the successor of its own successor (arithmetical statement).

(**) xy = yx holds for all x and y (group-theoretic statement).

If we consider these mathematical statements and those given above, we recognise that the components of which the statements are built belong to two different categories:

(1) Components which are specifically concerned with the given theories, and

(2) components which are universal and can appear in the statements of any mathematical theory.

The second category contains logical expressions such as and, not, or, either - or, for all, there exists, is identical with, and also variables. The words (or groups of words) which belong to the first category show to which mathematical theory the given statement belongs. They are called specific notions of the theory. The specific notions are either primitive notions of the theory, i.e. concepts which appear in at least one axiom of the theory, or defined notions. A defined notion is referred back to the primitive notions by means of a definition (sometimes by means of a detour via other notions which have already been defined). The definition of a mathematical notion is required to be such that, with its help, the defined notion can be eliminated from every statement of the theory. Thus, to a certain extent, a definition is only a means of abbreviation, and it can in principle be dispensed with (although in practice a definition can be an invaluable aid). Thus, in our considerations, we can leave aside the defined notions and confine ourselves to the primitive notions of the given theory.

What we have just said about definitions can perhaps be clarified by an example. In geometry, we talk about lines which meet and lines which lie in a plane. These notions are not primitive notions of the theory. This might be open to doubt in the case of the latter. However, a closer inspection reveals that the axioms only mention points (and not lines) which lie in a plane. The uncertainty is finally removed when we notice that Hilbert did indeed define what it means for a line to lie in a plane.

The two notions which have just been mentioned can be introduced by the following definitions:

Definition 1. The lines b and c meet if and only if they are different and there is a point A which lies both on b and on c.

Definition 2. The line c lies in the plane α if and only if every point A which lies on c also lies in α.

Let us consider any mathematical statement which contains notions which have been introduced by definition in this way, e.g. the (valid) statement:

(I) Two lines b, c which meet must lie in a plane α.

In every such statement, the defined notions can be eliminated with the help of their definitions. In this way, statement (I) can be transformed into the statement:

(II) If b and c are different lines, and if there is a point A which lies both on b and on c, then there is a plane α such that every point which lies on b and every point which lies on c also lies in α.

(I) can be regarded as an abbreviation for the statement (II).

Inductive definitions, as they are called, play a special part. For example, on the basis of Peano's system of axioms, addition is usually introduced by an inductive definition, which consists of the following two equations:

$$x + 0 = x$$

$$x + \text{successor of } y = \text{successor of } (x + y).$$

It is not immediately clear whether this definition of addition is in fact an explicit definition, i.e. a definition in the sense of an abbreviation. Thus, for example, we cannot immediately see how we could, with the help of the two equations above, express the equation $x + y = z$ in such a way that the addition symbol would no longer appear in it. It was Dedekind ("The nature and meaning of numbers", 1888) who first showed that every inductive definition can in fact be expressed as an explicit definition. We shall not enlarge on this point.

The primitive notions of geometry, as they appear in the axioms H1,...,H8, are the concepts point, line, plane, lies on, lies in. The primitive notions of Peano's system of axioms are natural number, 0 and successor. The axiom system for group theory which was considered in the last section has only one primitive notion, namely the notion times (expressed by the notation xy for the product of x and y)[3].

By a group-theoretic statement we shall mean a statement in which (if necessary after elimination of other terms by means of definitions) the only specific notion which appears is the primitive notion times of group theory. (We can introduce the concepts of arithmetical and geometrical statements in a similar way.)

The classification of the components of a statement which we have discussed in this section is fundamental; we shall enlarge upon it in § 6 and § 7. As long as we work on the basis of a natural language, we shall always be able, in individual cases, to dispute as to whether a particular word or phrase is a specific or a universal component of a statement. The situation only becomes completely clear when we go over to a formal language.

3.4 Consequences of the axioms for groups. With the aid of the concept of a group we can now describe what it means to say that a group-theoretic statement A follows from (is a consequence of) an axiom system G1, G2, G3 for group theory. Let us consider at first a given group C. As regards the statement A, exactly one of two cases holds: either A holds true for C, or A does not hold true for C. If we assume that A follows from G1, G2 and G3, then we shall expect A to hold true for the group C; for we shall require a reasonable notion of consequence to be such that a statement which follows from the axioms of group theory holds true in every group. This requirement of the notion of consequence is a fundamental one. We can ask ourselves whether the notion of consequence should be restricted by any other requirement. Any such restriction would mean that there would be at least one group-theoretic statement A which did not follow from the axioms of group theory although it held true for every group. However, there is no apparent reason why such a statement should not also be a theorem of group theory. Therefore, we do not want to restrict the notion of consequence any further. If a group theoretic statement holds true for every group, then we want to regard it as a consequence of the axioms G1, G2, G3.

Thus, to sum up, we have the required

Definition. A group-theoretic statement A follows from the axioms G1, G2, G3 of group theory if and only if it holds true for every group.

A proof, based on this definition, that a group-theoretic statement A does not follow from the axioms of group theory can be provided by finding a group C for which A does

[3] There are axiom systems for group theory which contain other primitive notions as well as the notion times. Cf. also 3.5.

not hold. There are non-abelian (i.e. non-communative) groups. This shows that the statement (**) in 3.3 does not follow from the axioms of group theory.

In an attempt to describe in general terms the relationship between a group and the axioms of group theory, we can say: The axiom system G1, G2, G3 contains the primitive notion times. This primitive notion has no fixed meaning, and must therefore be regarded as a word without content as far as the group-theoretic axioms are concerned. However, in any given group C, the primitive notion times is interpreted by assigning the two-place group operation Φ to it. Thus, we can call the group operation Φ belonging to a group C an interpretation of times.

The axiom system G1, G2, G3 contains no other primitive notions which can be interpreted. The domain ω of elements of a group is not the interpretation of a primitive notion in G1, G2, G3, but merely the "foundation" on which the function Φ operates.

One more basic remark on the idea of a group: As we introduced this idea in 3.2, a group consists of a domain (a set) of elements, together with a two-place operation on this domain. However, the word "group" is also often used in a rather looser way, so that it is understood to mean only the domain mentioned above. It is expedient to distinguish these two meanings of the word clearly (which is not always done as much as is desirable in mathematics). If, by a group (in the looser sense of the word) we mean only the set of the elements of a domain, then in general there are different operations on this domain which, together with the domain, form different (and, in general, not even isomorphic) groups (in the original sense of the word) (see exercise 1). In the following, whenever we talk about a group (without further qualification), we shall always mean a group in the original sense.

3.5 Another axiom system for group theory. It is also possible to set up axiom systems for group theory in which the word "group" (to be understood in the looser sense) appears. We want to give an axiom system like this. In order to draw attention to the fact that here "group" is to be understood in the looser sense, we shall write "group*" instead.

G*0 The product operation assigns uniquely an element xy (i.e. x times y) to any two elements x, y of the group*. xy is also an element of the group*.

G*1 For any elements x, y, z of the group*, (xy)z = x(yz).

G*2 For every x in the group* and every z in the group* there is at least one y in the group* such that xy = z.

G*3 For every y in the group* and every z in the group* there is at least one x in the group* such that xy = z.

This axiom system has two primitive notions, group* as well as times. Thus, both these primitive notions must be interpreted in any interpretation. As a basis for the interpretation we need, as before, a domain of individuals ω. The primitive notion times is (as for the axiom system G1, G2, G3) to be interpreted by an operation. The primitive notion group* stands for a property, namely the property which elements of the domain have if and only if they are elements of the group. Thus, group* is to be interpreted by a property over the object domain. A group in the original sense, but now defined by the axiom system G*0, G*1, G*2, G*3, is thus a domain of individuals ω, together with a property $\mathfrak{P}$ of elements of this domain and a two-place operation Φ on ω, provided that the axioms G*0, G*1, G*2, G*3 hold true for the property $\mathfrak{P}$ and the operation Φ. Thus, for example, G*0 means that, to any two elements $\mathfrak{x}$, $\mathfrak{y}$ of ω which have the property $\mathfrak{P}$, the operation Φ assigns an element $\Phi(\mathfrak{x}, \mathfrak{y})$ of the domain such that $\Phi(\mathfrak{x}, \mathfrak{y})$ also has the property $\mathfrak{P}$.

Let us compare the original axiom system G1, G2, G3 for group theory with the axiom system G*0, ..., G*3 which we have just considered. The concept of a group which goes with G1, G2, G3 is such that a group is determined by a domain of individuals ω and an operation Φ, whereas that which goes with G*0,...,G*3 is such that a group is determined by a domain of individuals ω, a property $\mathfrak{P}$ and an operation Φ. It must be agreed that a characterisation of a group by means of ω, Φ, as in the case of the first axiom system, is better suited to our purposes than a characterisation by means of ω, $\mathfrak{P}$, Φ, for with (ω, Φ), every element of ω is also an element of the group, but with $(\omega, \mathfrak{P}, \Phi)$ only those elements of ω which have the property $\mathfrak{P}$ can be regarded as elements of the group in the intended sense. In the consideration of a particular group, it seems an unnecessary luxury to take into consideration at all those elements of ω which do not have the property $\mathfrak{P}$. For this reason, we shall give preference to our original axiomatisation of group theory.

3.6 Consequences of the axioms for ring, field and lattice theory. By a procedure analogous to that which we used for the axioms of group theory, we can define what it means to say that a ring-theoretic, field-theoretic or lattice-theoretic statement A follows from the axioms of ring theory, field theory or lattice theory respectively: This means that A holds true for every ring, field or lattice respectively.

3.7 Consequences of the axioms of Peano. Correspondingly, we can also explain what it means to say that a geometric statement follows from the axioms H1,...,H8 of geometry or that an arithmetical statement follows from the axioms P1,...,P5 of Peano. The only difficulty here, which is more of a technical nature, lies in the fact that mathematics has no generally accepted name for the corresponding analogues to the concept of a group.

We want to consider arithmetic first. As we have already established, Peano's system of axioms contains the three primitive notions natural number, 0 and successor. Thus, an interpretation for the system of axioms consists of a domain of individuals ω, a property $\mathfrak{P}$ over this domain (assigned to the basic concept natural number), an element $\mathfrak{n}$ of ω (assigned to the primitive notion 0) and, lastly, a one-place operation Φ on ω (assigned to the primitive notion successor).

If we compare this situation with that which occurs with the axiom system for group theory treated in 3.5, we see that here, as there, we are in the undesirable position of not being interested in all the elements of the domain of individuals ω: the relevant elements of ω are only those which have the property $\mathfrak{P}$, for only these are interpreted as natural numbers. In the case of group theory, the axiom system G1, G2, G3, which can be thought as being obtained from $G^*0, \ldots, G^*3$ by eliminating the primitive notion group*, is in general preferable. Correspondingly, we shall obtain a more useful axiom system for arithmetic if we avoid the primitive notion natural number. Thus, the first two axioms can be done away with, and we obtain the following new axiom system for arithmetic:

P'1 The successor of x is always different from 0.

P'2 If the successor of x is equal to the successor of y, then x = y.

P'3 (Induction axiom) If 0 has any property P, and if, whenever x has the property P, the successor of x also has this property, then every x has the property P.

An interpretation of the primitive notions which appear in this axiom system consists of a domain of individuals ω, an element $\mathfrak{n}$ of ω (corresponding to 0) and a one-place operation Φ on ω (corresponding to the successor). If we now require that P'1, P'2, and P'3 should hold true for $\mathfrak{n}$ and Φ, we obtain an arithmetical analogue of the concept of a group. For the moment, we want to use the word "ladder" for this concept. Thus, a ladder consists of a domain of individuals ω together with an element $\mathfrak{n}$ of the domain and a one-place operation Φ on the domain such that the arithmetical axioms P'1, P'2, P'3 hold true for $\mathfrak{n}$ and Φ.

In a way analogous to group theory, we shall now say that an arithmetical statement follows from the axioms P'1, P'2, P'3 if and only if it holds true for all ladders.

3.8 Consequences of the axiom system for geometry. Finally, let us consider the axiom system H1,...,H8 for geometry. Just as we spoke of "groups" in group theory and of "ladders" in arithmetic, so we want here to speak of "frames". If we proceed as we did for the axiom systems we have already considered, we shall, in the geometrical case, want the elements of ω to be the points, lines and planes of a frame. As in the case of group theory and arithmetic, we shall expect to obtain an axiom system for geometry which fulfils this requirement if we eliminate the primitive notions point, line, plane. There is, however, one difficulty in the case of geometry. In group theory and arithmetic, we have only one sort of basic object (namely group elements and natural numbers respectively), whereas in geometry we have somehow to distinguish three sorts of basic objects (points, lines and planes) from each other. This could be achieved by using (as is often done in mathematics) basically different sorts of letters for points, lines and planes. If we carried out consistently the ideas above, this in turn would suggest that we considered not just one, but three separate domains of individuals for an interpretation. It is, in fact, possible, to set about the task in this way. However, in logic it has become customary to use only one sort of variable for the basic objects. It is possible to do this for geometry, as for other theories, even if we do without the primitive notions point, line, plane. For these notions can be defined with the help of other primitive notions of geometry in the following way:

Definition 1'. x is a point if and only if there is a y such that x lies on y.

Definition 2'. y is a line if and only if there is an x such that x lies on y.

Definition 3'. y is a plane if and only if there is an x such that x lies in y.

With the aid of these definitions, the axiom system H1,...,H8 for geometry can be transformed into an axiom system which contains only the primitive notions lies on and lies in. It is left to the reader to carry this out (see exercise 3). Thus, a frame consists of a domain of individuals ω together with two relations $\mathfrak{R}_1$ and $\mathfrak{R}_2$, which are interpretations of the basic concepts lies on and lies in, provided that the axioms (transformed as indicated above) of the axiom system for geometry hold true in this interpretation. Thus, we lay down the following definition: A geometrical statement follows from the axiom system for geometry if it holds true for every frame.

3.9 General definition of the notion of consequence. Thus far, we have explained what it means to say that a group-theoretic statement follows from the axioms of group theory, an arithmetical statement from Peano's axioms and a geometrical statement from the axioms of geometry. Now it is easy to find the common factor of these definitions and to say for an arbitrary axiom system $\mathfrak{A}$ and an arbitrary statement A what it means to say that A follows from the axioms of the axiom system $\mathfrak{A}$. In doing this we shall require (as in the examples we have already considered) that the statement A contains as specific notions only those which are primitive notions of the axiom system $\mathfrak{A}$. We shall refer to a statement like this as a specific statement (relative to the axiom system $\mathfrak{A}$ under consideration).

In the case of group theory, we have said that a specific statement A follows from the axioms of group theory if and only if A holds true for every group, i.e. for every interpretation of the primitive notions which appear in the axiom system for which all the axioms of group theory hold true. Similarly, a specific statement A follows from the axioms of arithmetic (geometry) if and only if it holds true for every interpretation of the arithmetical (geometrical) primitive notions for which all the axioms of arithmetic (geometry) hold true. We can generalise this by saying:

A specific statement A follows from an axiom system $\mathfrak{A}$ if and only if it holds true for every interpretation of the primitive notions of $\mathfrak{A}$ in which all the axioms of $\mathfrak{A}$ hold true.

In order to make this formulation easier to deal with, we introduce the concept of a model. Let there be an interpretation in which all the specific notions of a statement A are interpreted. Either the statement holds true in this interpretation or it does not. In the first case, we shall say that the interpretation is a model for A. If an interpretation is a model for every axiom of an axiom system $\mathfrak{A}$, then we shall say that the interpretation is a model for $\mathfrak{A}$. Now, by using the concept of a model, we can explain the notion of consequence by the

Definition. A specific statement A follows from an axiom system $\mathfrak{A}$ if and only if every model for $\mathfrak{A}$ is also a model for A.

Exercises. 1. (Cf. 3.4) Show that:

(a) If ω is a domain of individuals with exactly one element, then there is exactly one function Φ over ω (i.e. one function Φ which is defined just for the elements of ω and such that the values of Φ also lie in ω) such that ω, Φ is a group.

(b) If ω has just two elements, then there are exactly two different functions Φ_1, Φ_2 over ω such that ω, Φ_1 and ω, Φ_2 are groups. These two groups are isomorphic.

(c) If ω has just four elements, then there are two functions Φ_1, Φ_2 over ω such that ω, Φ_1 and ω, Φ_2 are non-isomorphic groups.

2. Show that the following interpretation is a model of the axiom system H1,...,H8 (cf. 2.1). The domain of individuals ω is comprised of the numbers 2,3,5,6,7,10,14, 15,21,30,35,42,70 and 105. (ω is finite, so that this is an example of a "finite geometry" - but, of course, here we have been considering only the connection axioms.) We interpret the primitive notions point, line and plane by the properties of being divisible by exactly one, two or three prime numbers, and lies on and lies in by the relation of divisibility.

3. Using the definitions in 3.8, change the axiom system for geometry H1,...,H8 to make it contain only the primitive notions lies on and lies in.

§4. Remarks on the notion of consequence

4.1 Independence proofs. The considerations of the last paragraph ended with a definition of the notion of consequence, which reads: A statement follows from an axiom system if and only if every model for the axiom system is also a model for the statement. It was only relatively recently that mathematicians realized that the notion of consequence can be characterised in this way. It is interesting that this discovery was made - at least in principle - before the modern axiom systems for group theory, topology etc. were set up, and that it was made in the context of investigations of the euclidean axiom system for geometry. In Hilbert's formulation, the parallel postulate of euclidean geometry reads:

Parallel postulate. Let α be an arbitrary plane, A a point lying in α and c a line lying in α such that A does not lie on c. Then there is exactly one line b which lies in α such that A lies on b (cf. 3.3 as regards the defined notions which appear here).

In comparison with the other axioms of geometry, this axiom is relatively complicated. Thus, even in ancient times, people wondered whether the parallel postulate was perhaps a consequence of the other axioms and was thus dispensable as an axiom.

A statement is said to be independent of an axiom system if it is not a consequence of the axiom system. An axiom system $\mathfrak{A}$ is said to be independent if every statement C of that axiom system is independent of the axiom system $\mathfrak{A}'$ which is obtained from $\mathfrak{A}$ by removing C. The independence postulate is an aesthetic requirement of an axiom system which is often laid down.

The "parallel problem" asks whether the parallel postulate of geometry is independent of the other geometrical axioms or whether it is a consequence of them. Ever since ancient times, people have tried to prove that the parallel postulate follows from the other geometrical axioms. The direct attempts did not lead to the goal. As the direct method was seen not to succeed, Saccheri tried the following indirect method in 1733: Let $\mathfrak{A}$ be the euclidean axiom system without the parallel postulate. Let P' be the negative of the parallel postulate P, i.e. the following statement: There is a line c and a point A which does not lie on c such that there are two different lines b_1 and b_2 which fulfil the following conditions:
(1) A lies on b_1 and b_2. (2) There is a plane in which c and b_1 lie and a plane in which c and b_2 lie. (3) Neither c and b_1 nor c and b_2 meet. If P really were a consequence of $\mathfrak{A}$, then - argued Sacceri - the axiom system $\mathfrak{A}$ together with P' must lead to a "contradiction", i.e. it must be possible to find a geometrical statement A such that both A and the negation of A can be inferred from $\mathfrak{A}$ and P'. Conversely, it could be shown from a "contradiction" like this that P is a consequence of $\mathfrak{A}$.

This sort of argument is, of course, only convincing to a limited extent, as long as it is based on the inexact intuitive notions of consequence and derivation which were all that was available at that time. What is immediately obvious is the existence of a "contradiction" with respect to $\mathfrak{A}$ and P' under the assumption that P can be inferred from $\mathfrak{A}$, for then, clearly P is a statement which, together with its negation (i.e. P') can be inferred from $\mathfrak{A}$ and P' : P can be inferred from $\mathfrak{A}$, therefore a fortiori from $\mathfrak{A}$ and P'. P' can be inferred from P', hence a fortiori from $\mathfrak{A}$ and P'. It is however, probable that Sacceri was not aiming at inferring the particular "contradiction" P, P' from $\mathfrak{A}$ and P'; but rather, he was trying to find another suitable geometrical statement A for which both A and the negation of A could be inferred from $\mathfrak{A}$ and P'. Now let us suppose that such a "contradiction" from $\mathfrak{A}$ and P' has been found, i.e. a statement A which, together with its negation, follows from $\mathfrak{A}$ and P'. We should now have to show that P is a consequence of $\mathfrak{A}$. In order to verify this, we shall take as a basis the notion of consequence which was made precise in §3. In order to prove that P follows from $\mathfrak{A}$, we have to show that every model of $\mathfrak{A}$ is also a model of P. If this assertion were false, then there would be an interpretation of the primitive notions of geometry which is a model for $\mathfrak{A}$ but not for P. Since P does not hold true for this interpretation, clearly the negation P' of P must hold true for it. Thus, the interpretation is a model for P', and therefore also a model for $\mathfrak{A}$ and P', and thus a model of every consequence of $\mathfrak{A}$ and P', and in particular of both A and A'. However, this is not possible, since A' is the negation of A. In this way, Saccheri's method can be justified on the basis of the notion of consequence laid down in §3.

The attempts to derive a contradiction from $\mathfrak{A}$ and P' failed. Instead of this, it turned out that it was possible to construct, on the basis of the axioms $\mathfrak{A}$ and P', a theory which can be developed just as consistently as can euclidean geometry on the basis of the axioms $\mathfrak{A}$ and P. Thus, at the beginning of the nineteenth century, Gauss, Bolyai and Lobatschewskij became convinced that it is not possible to derive a contradiction from $\mathfrak{A}$ and P' and that, consequently, the parallel postulate is independent of the other euclidean axioms. They called the science based on $\mathfrak{A}$ and P' non-euclidean geometry.

These results did not, of course, definitely solve the problem of the independence of the parallel postulate. For it could have happened that a more thorough investigation would, after all, have derived a contradiction from $\mathfrak{A}$ and P'. It was F. Klein, who, in 1871, first gave a definite answer (for three dimensional euclidean geometry, after Beltrami had solved the problem in 1868 for geometry of the plane) by giving a model for the axiom system $\mathfrak{A}$ and P', i.e. an interpretation for which $\mathfrak{A}$ and P' hold true. This shows immediately that P is not a consequence of $\mathfrak{A}$. The given interpretation is a model for $\mathfrak{A}$ and P', and therefore in particular also a model for P'. If P were a consequence of $\mathfrak{A}$, then the interpretation would also be a model for P. However, P cannot hold true in the interpretation, since the negation P' of P holds true in it.

The model for non-euclidean geometry is given with the use of a model of euclidean geometry. Thus, the model exists only if euclidean geometry contains no contradictions. Thus, the result should really be formulated more cautiously thus: If the axiom system of euclidean geometry is free from contradictions, then the parallel postulate is independent of the other axioms.

4.2 Mathematical statements as statement forms. Let us consider the axioms of group theory, given in the previous section, with only one primitive notion times. This word clearly has no fixed meaning. It is interpreted in different ways, according to which group is under consideration. Therefore, it is quite immaterial whether, in an axiom system for group theory, the word times is used or whether it is replaced by some other word which indicates a two-place operation[4]. Since the word times has no fixed meaning, it is meaningless to ask whether, for example, the associative law G1 is true or false. Since Aristotle's time, the attributes true and false have been used to characterise the statements, by saying: A statement in the sense of Aristotle, or an aristotelian statement, is a linguistic structure of which it is meaningful to say that it is true or false (this is the so-called aristotelian two-value principle). Thus, the associative law is not a statement in the aristotelian sense. It could perhaps be called a statement form, in the sense that aristotelian statements can arise from it if the word times is interpreted in different ways and then replaced by a symbol for the operation by which it has been interpreted.

What we have said here for group theory is true mutatis mutandis for all other mathematical theories. Where a mathematical theory is concerned, the mathematician is interested only in what follows from the axioms of this theory. This means that he recognizes as having equal status all interpretations of the primitive notions for which the axioms hold true; no model of the axioms is favoured more than another. Thus, it is clear that the primitive notions of a mathematical theory can have no fixed meaning[5]. Thus, the axioms, theorems and statements of a mathematical theory are not statements in the aristotelian sense, but rather statement forms.

The realization of this fact must come as a shock, if we think of arithmetic and geometry. For, when a mathematician does arithmetic or geometry, he thinks of the natural numbers and of the points, lines and planes of (the) euclidean space; in other words, he conceives of arithmetical and geometrical statements as statements in the aristotelian sense. However, we must keep hold of the fact that for mathematicians, who work axiomatically, euclidean geometry, for example, is determined by its axioms alone, and that it is not permissible to prove that a statement is a theorem of geometry by appealing to "the way we think about euclidean space".

Notice here the contrast with physics. If we interpret the primitive notions of geometry empirically in a certain way, e.g. if we conceive of the lines as rays of light, then, with regard to a fixed interpretation like this, the geometrical statements can be comprehended in the aristotelian sense. But then it is by no means self-evident that in this interpretation, the axioms of geometry go over into true statements (cf. the discussion on relativity theory).

The fact that the primitive notions do not mean anything in particular was (according to Blumenthal) expressed particularly forcefully by Hilbert when he said of the primitive notions of geometry "We must always be able, instead of talking of 'points, lines

---

[4] For example by "composition" or "plus". It is, however, customary to use the word "plus" only when the commutative law is also being taken as an axiom.

[5] An axiom system which has any model at all always has more than one model. Cf. Chap. VI, §2.4 for this.

and planes', to talk even of 'tables, chairs and beer mugs'." What Hilbert meant by this was: Mathematicians who do geometry are not permitted to make use of the fact that the words point, line, plane (and the other primitive notions of geometry) have a meaning in the language of everyday speech. However, if mathematicians ignore the everyday meaning of the primitive notions of geometry, they can just as well replace these primitive notions by others (e.g. point by table), where of course, once again, as long as they are doing geometry, the everyday meaning of the new primitive notions will be of no importance to them.

In order to make it immediately obvious that the everyday meaning of the primitive notions of mathematics is irrelevant, mathematicians often use, in new axiom systems, words of everyday language for which it is quite clear that they are not being used in their traditional meaning. Examples are the word composition for the group operation (which we characterised by times in §2), or the words cup and cap for the basic operations of lattice theory (in view of the symbols $\vee$ and $\wedge$ which represent them).

In formal languages, the primitive notions are not represented by words at all, but variables are used instead, e.g. P for point, Q for line and R for plane. In this way, any misunderstandings which could arise because of everyday meanings are excluded from the start, and the possibility of different sorts of interpretations is indicated.

When mathematical statements are written in such a way that the primitive notions which appear are replaced by variables, it becomes clear that mathematical statements are statement forms, not statements in the aristotelian sense.

As regards the mathematical irrelevence of the everyday meaning of words like point, line, plane, we might ask why we should use these words at all in geometry. However, apart from historical continuity, there are two reasons for this: (1) the everyday meanings of the primitive notions of geometry do point to one model for the axiom system of geometry, which is an invaluable aid to intuitive understanding and, in particular, can lead to new hypotheses. (2) Because of the categoricity of the euclidean axiom system, considered in 4.3, the interpretations of the primitive notions in the different models are isomorphic to the "intuitive model" indicated by the words point, line, plane. If instead of that of geometry another axiom system is considered, then in general there will be no argument analogous to (2), but only one analogous to (1) for the choice of everyday words for the primitive notions, e.g. for the fact that we talk about neighbourhoods in topology.

Russell once formulated the fact that the primitive notions of mathematical theories have no fixed meaning, in the following acute observation: "...mathematics may be defined as the subject in which we never know that we are talking about, nor whether what we are saying is true."

4.3 Categoricity. If, in arithmetical and geometrical investigations, mathematicians think of the numbers and the space, whereas the group does not exist for them, this does, in spite of our previous criticisms, have its roots in reality. For it turns out that (under certain assumptions) any two models of the axioms of arithmetic (and, similarly, of the complete set of axioms of geometry) are "isomorphic" to one another, whereas this is not true for group theory. Because of this property, the axiom system of arithmetic is called categorical or monomorphic. We shall study these situations more closely for arithmetic in Chapter VI.

4.4 The applicability of mathematics. If mathematicians, working axiomatically, confine themselves to drawing consequence from statement forms and refrain from making statements in the aristotelian sense, then this characteristic attitude is at the same time the source of the universal applicability of mathematics. For the proof that a mathematical statement $\underline{A}$ follows from an axiom system $\mathfrak{A}$ means that $\underline{A}$ holds true in every interpretation which is a model for $\mathfrak{A}$. Seen in this light, it will be those axiom systems which have many interesting models which will be especially useful. This is true in particular of the axiom systems of such modern mathematical theories as group theory, topology and lattice theory. The theorems of these theories are indeed applicable in a multiplicity of ways.

4.5 Semantics. In the considerations above, the idea of an interpretation and that of the validity (holding true) of a mathematical statement in an interpretation were fundamental. The ideas of a model and of consequence, as well as others, have been reduced by definitions to these ideas: A model of a mathematical statement is an interpretation in which this statement holds true; a mathematical statement follows from an axiom system if every model of the axiom system (i.e. every model of all the statements which belong to the axiom system) is also a model of the mathematical statement.

The above-named concepts have something in common: They depend on the relation between linguistic structures (the mathematical statements or their components) and entities which are in general of a non-linguistic nature. For the fundamental idea of an interpretation refers to a mapping of the (linguistic) primitive notions of an axiom system onto individuals, properties (relations) or operations. It is customary to call ideas which concern the relationship between linguistic and non-linguistic elements semantic concepts, and to call the corresponding science semantics.

An approach to logic which conceives of logic as the science of the notion of consequence, and introduces the notion of consequence as we have done in these paragraphs, is known as a semantic development of logic. An approach like this assumes that it is permissible to make use of ontological notions like domains of individuals, properties, relations or operations. In comparison with other possible ways of developing logic, it has the advantage of having been carried out particularly extensively and of harmonizing well with the ways of modern mathematical axiomatics.

Semantics can be traced back to Aristotle. The definition of the notion of consequence which we have given here is found, in essence, in Bolzano's "Wissenschaftslehre", which appeared in 1837. However, this work did not become well-known. The notion of consequence was defined again (independently of Bolzano, but this time with the precision which is attainable on the basis of a formal language) in the twentieth century. Here, Tarski's work, "The concept of truth in formalised languages", published in 1935, is particularly worthy of mention.

Exercises. 1. With the aid of the interpretation given in §3, exercise 2, show that the parallel postulate (4.1) is independent of the axioms H1,...,H8 (2.1).

2. Show that the following axiom system with the primitive notion equivalent is independent:

A1. x is always equivalent to itself. (Reflexiveness)

A2. If x is equivalent to y, then y is always equivalent to x. (Symmetry)

A3. If x is equivalent to y and y to z, then x is always equivalent to z. (Transitivity)

(If ω is a domain of individuals and R is a two-place relation such that ω, R is a model of A1, A2, A3, then, as is well-known, R is called an equivalence relation.)

3. Discuss the independence of A1,...,A4 (cf. exercise 2), with

A4. For every x there is a y such that x is equivalent to y or y to x.

4. Prove the independence of the axiom system for group theory given in 2.3.

5. An axiom system for partial orderings (where the primitive notion is represented by $\leqslant$) reads:

01. $x \leqslant x$. (Reflexiveness)

02. If $x \leqslant y$ and $y \leqslant x$, then $x = y$. (Antisymmetry)

03. If $x \leqslant y$ and $y \leqslant z$, then $x \leqslant z$. (Transitivity)

If we add

04. $x \leqslant y$ or $y \leqslant x$. (or in the non-exclusive sense - see 6.2) (Connectedness),

then we obtain an axiom system for total orderings.

(a) Show that 01,...,04 are dependent.

(b) Give an independent axiom system for total orderings.

(c) Give an independent axiom system for total orderings which, by removing one axiom, can be turned into an independent axiom system for partial orderings.

§5. Logic calculi

5.1 Mathematical proofs. The mathematician wants to draw consequences from systems of axioms. However, in a specific case, how does he convince himself that a mathematical statement A follows from a given axiom system $\mathfrak{A}$? It is only in very few cases that this can be seen immediately. Normally he divides the way of seeing that a relationship of consequence holds into a (larger or smaller) number of steps, by inserting suitable mathematical statements. Thus he has before him the typical configuration which we call a mathematical proof. Such a proof consists (if we idealise the situation somewhat) of a finite sequence of mathematical statements written down one after the other, or, better, one beneath the other. The first statements of this sequence are the axioms with which we start, and the last statement of the sequence is the mathematical theorem which is to be proved[6]. Every statement of the sequence (except for the axioms themselves) must be so chosen that it can be seen directly that this statement "follows immediately" from certain of the statements which precede it in the sequence.

If there is a proof of this sort, then the last statement of the proof does indeed follow from the axioms. In order to see this, let us assume that $\mathfrak{J}$ is an arbitrary interpretation which is a model for the axioms. We have to show that $\mathfrak{J}$ is also a model for the last statement of the proof. In fact, we shall show that every statement of the proof is valid in the interpretation $\mathfrak{J}$. If this assertion were false, then there would be a first statement $\underline{A}$ in the proof which was not valid in $\mathfrak{J}$. $\underline{A}$ could not be an axiom, since $\mathfrak{J}$ is a model for all the axioms. Thus $\underline{A}$ must be an immediate consequence of certain statements $\underline{A}_1, \ldots, \underline{A}_r$, which precede the statement $\underline{A}$ in the proof. The statements $\underline{A}_1, \ldots, \underline{A}_r$ are valid in $\mathfrak{J}$ since $\underline{A}$ is the first statement in the proof which is not valid in $\mathfrak{J}$. But $\underline{A}$ is an immediate consequence of $\underline{A}_1, \ldots, \underline{A}_r$. Every model for $\underline{A}_1, \ldots, \underline{A}_r$ is thus a model for $\underline{A}$, and therefore so is $\mathfrak{J}$. But this contradicts the assumption that $\underline{A}$ is not valid in $\mathfrak{J}$. Thus we have shown that every statement of the proof, and therefore, in particular, the last, is a consequence of the axioms.

As a simple example we give the proof for Lemma 1 of 2.2 in the idealised form we have just described[7]:

(1) 0 is a natural number.
(2) P2.
(3) If x is a natural number, then the successor of x is always different from 0.
(4) P4.
(5) P5.
(6) If 0 is a natural number, then the successor of 0 is different from 0.
(7) The successor of 0 is different from 0.
(8) 0 is a natural number and the successor of 0 is different from 0.

This proof consists of eight statements. The first five statements are the axioms. Statement (6) "follows immediately" from (3); (7) "follows immediately" from (1) and (6); (8) "follows immediately" from (1) and (7). The last statement states that 0 has the property P.

---

[6] In general the axioms are not written down at the beginning of the proof every time. However, in a systematic treatment it is convenient to have this requirement. - In this section we want, for the sake of simplicity, always to assume that we are dealing with finite systems of axioms.

[7] In doing this we shall explicitly state only axioms P1 and P3 and shall confine ourselves to abbreviations for P2, P4, P5, since the proof does not depend on these axioms.

5.2 Indefiniteness of the concept of immediate consequence. We spoke above of immediate consequences, where $\underline{A}$ was to be an immediate consequence of $\underline{A}_1, \ldots, \underline{A}_r$ if it could be seen directly that $\underline{A}$ followed from $\underline{A}_1, \ldots, \underline{A}_r$. If we inspect this concept, we have to admit that it is very vague and unsatisfactory, for there is no way of seeing what its objective meaning is. In fact, we have to admit that mathematicians are not at all in agreement as to when the passage to a new mathematical statement in a proof should count as directly obvious. We often speak of more or less "complete proofs," depending on the "degree of obviousness" which has been attained, without actually possessing a criterion for completeness itself. Some considerations which were previously recognized as proofs are today regarded as "defective" and are indeed defective, because the statement which is to be proved does not follow from the axioms from which it is supposed to follow.

Euclidean geometry provides us with an example. It was only in 1882 that Pasch realised that, as well as the original euclidean axioms, further ones, namely the so-called axioms of order, are needed for the proofs of certain geometrical theorems.

With such a state of affairs, the possession of a criterion by which we can judge when a proof can be regarded as "complete" - and, thus, be regarded as a proof at all - must be seen as an urgent necessity. It was the investigations of modern times which first provided a satisfactory solution to this problem. (They were, however, based on ideas to which many generations, going back to ancient times, had contributed.) We shall give a short account of this in 5.3.

5.3 Rules of inference. Let us examine the proof in 5.1 a little more closely. Statement (7) is an "immediate consequence" of the preceding statements (1) and (6); and it is noticeable that (6) contains the statements (1) and (7) as "sub-statements". In order to make this clear, we shall write $\underline{A}_0$ for (1) and $\underline{B}_0$ for (7). Then we can write if $\underline{A}_0$, then $\underline{B}_0$ for (6).

We have just established that $\underline{B}_0$ is an immediate consequence of $\underline{A}_0$ and if $\underline{A}_0$, then $\underline{B}_0$. However, it is obvious that, even for arbitrary statements $\underline{A}$ and $\underline{B}$, the statement $\underline{B}$ is an immediate consequence of $\underline{A}$ and if $\underline{A}$, then $\underline{B}$. We can express this in the form of a rule: In a proof, we may always pass from statements of the form $\underline{A}$ and if $\underline{A}$, then $\underline{B}$ to the statement $\underline{B}$. This rule was known even to Aristotle, and has been called modus ponens since the time of the scholastics.

The order of the two initial statements is not important. Thus, we shall also say that $\underline{B}$ follows from the two statements if $\underline{A}$, then $\underline{B}$ and $\underline{A}$ by modus ponens.

Modus ponens leads from two statements $\underline{A}$ and if $\underline{A}$, then $\underline{B}$ to the statement $\underline{B}$. This statement $\underline{B}$ is a consequence of the two initial statements $\underline{A}$ and if $\underline{A}$, then $\underline{B}$. Because of this, modus ponens is called a sound rule[8], or a rule which preserves validity. Because of the soundness of modus ponens, we can appeal to it in each concrete case. In the example from 5.1 we shall say that line 7 follows from lines 1 and 6 by modus ponens.

A further important property of modus ponens is the following: If we are given any three mathematical statements $\underline{A}$, $\underline{B}$ and $\underline{C}$, we can always decide whether or not $\underline{C}$ follows from $\underline{A}$ and $\underline{B}$ by modus ponens. For it does so if and only if either $\underline{B}$ is identical with if $\underline{A}$, then $\underline{C}$ or $\underline{A}$ is identical with if $\underline{B}$, then $\underline{C}$[9]. We want, in general, to demand that every rule has a similar decidability property.

Modus ponens leads from two initial statements to a new statement, and it is therefore called a rule with two premises. However[10] we do not want to demand that every rule shall have exactly two premises. In general, we want to allow a rule to have some finite number k of premises, where k is characteristic of the rule.

Let a finite system $\mathfrak{R}$ of correct rules be given. Now we can give a precise meaning to the concept of a proof by defining this concept in a way which depends on $\mathfrak{R}$. We shall call a finite sequence of statements a proof relative to $\mathfrak{R}$ with regard to a (finite) system of axioms $\mathfrak{A}$ if the sequence begins with the axioms and if (apart from the axioms themselves) every statement of the sequence can be obtained from previous statements with the aid of one of the rules of $\mathfrak{R}$.

The soundness of the rules of $\mathfrak{R}$ guarantees that all the statements (and thus, in particular, the last statement) of a proof are consequences of the chosen axiom system $\mathfrak{A}$.

Because of the decidability of the rules of $\mathfrak{R}$ we can, given any finite sequence of mathematical statements, decide whether this sequence is a proof relative to $\mathfrak{R}$ from a given finite axiom system $\mathfrak{A}$: First of all, we can decide whether the elements of $\mathfrak{A}$ comprise the beginning of the sequence. If this is the case, then we test the next statement $\underline{A}$. We choose any one of the (finitely many) rules of $\mathfrak{R}$. Let this rule have k premises. Now choose any k statements $\underline{A}_1, \ldots, \underline{A}_k$ which precede $\underline{A}$ (in our case, k axioms)[11]. By our supposition, we can decide whether $\underline{A}$ can be obtained from $\underline{A}_1, \ldots, \underline{A}_k$ by the rule we are considering. If this is not the case, then we test k other preceding statements,

---

[8] Not every rule is sound; consider, for example, the rule which permits us to pass from the two statements $\underline{B}$ and if $\underline{A}$, then $\underline{B}$ to the statement $\underline{A}$.

[9] Here we are assuming that the mathematical statements are written in some sort of standard way, for otherwise there may be syntactically different statements which have the same logical content. This sort of standardisation can be achieved precisely by going over to a formal language.

[10] In contrast to the situation in Aristotle's syllogistics.

[11] Some or all of the $\underline{A}_1, \ldots, \underline{A}_k$ may be identical.

and so on; note that there are only finitely many possibilities. If these are exhausted without its once having been shown that $\underline{A}$ can be obtained from preceding statements with the aid of the rule under consideration, then we pass to another rule in $\mathfrak{R}$ and treat it in the same way. Either this process provides us with a rule in $\mathfrak{R}$ and certain statements which precede $\underline{A}$ in the sequence such that $\underline{A}$ can be obtained from these by the rule, or it does not. In the second case, the sequence we are investigating is not a proof, and in the first case the part of the sequence up to and including $\underline{A}$ is a proof. Now we must examine similarly the statement which follows $\underline{A}$, and so on.

5.4 Complete sets of rules. We have just seen that, given a finite system $\mathfrak{R}$ of sound rules and a particular axiom system $\mathfrak{A}$, we can always decide whether or not a given finite sequence of mathematical statements is a proof from $\mathfrak{A}$ relative to $\mathfrak{R}$. This valuable property makes our precise concept of proof decidedly preferable to the usual, rather vague concept. On the other hand, we are faced with the following problem: Mathematicians are - consciously or unconsciously - convinced that every consequence of an axiom system can be shown to be such by a proof. However, if we restrict the concept of proof by additional demands, as we have just done, this is no longer self-evident.

This suggests the formation of the following concept: A finite system $\mathfrak{R}$ of rules is said to be complete if every consequence of every (finite) axiom system $\mathfrak{A}$ can be obtained from $\mathfrak{A}$ by means of a proof relative to $\mathfrak{R}$. Thus, the problem we have just mentioned is equivalent to the question of the existence of a complete set of sound rules. The possession of such a set of rules - if one exists at all - would undoubtedly be of great value. However, if no such set of rules exists, then there is no finite set of sound rules which is sufficient to obtain all the consequences of an arbitrary system of axioms from it. In this case mathematicians would, as their science developed further, always be "discovering" new sound rules, which would enable them to prove more than was provable before.

For over two thousand years mathematicians have struggled to formulate the problem we have just mentioned and to answer it. In his syllogistics A r i s t o t l e laid down a set of sound rules. However, this set is incomplete. The Greek tradition was carried on by the Arabs, who were concerned in particular with the development of algorithms in mathematics [12].

An algorithm is a schematic computational method of solving problems. The algorithms which are commonly used in mathematics (e.g. the procedure for adding and multiplying numbers which are given in decimal form; the euclidean algorithm for determining the greatest common factor; the well-known algorithm for computing square roots) have each been developed for a very limited range of problems. A complete set of sound rules would also yield an algorithm, and indeed a very universal one. Against the background of Arabic culture, R a y m u n d u s L u l l u s (about 1300) formulated in Spain the idea of an "ars magna", an art of solving schematically not only mathematical

---

[12] The name algorithm comes from the name of the Arabic mathematician
 A l C h w a r i z m i (about 800).

problems but all problems of any sort. This idea of an "ars magna" strongly influenced research throughout many centuries. However, the algorithms which were discovered (e.g. the procedure for solving cubic and quadratic equations) were, again, of limited application.

Leibniz was very deeply involved in the problem of the ars magna. In particular, he thought about the relationship between algorithmic computability and decidability. We can illustrate this relation by a simple example: We know of algorithms for computing, figure by figure, the decimal expansion of π. With the help of such an algorithm, we can produce this expansion up to an arbitrary number of decimal places. However, this certainly does not provide us with the means of deciding whether a given sequence of figures appears in this decimal expansion; If this is the case, then we shall be able to discover this "in a finite length of time" by producing the decimal expansion to sufficiently many decimal places. However, if the given sequence of figures does not appear in the expansion, then naturally we shall not be able to find it however far we develop the decimal expansion of π; we shall never be able to prove by this method that the sequence never appears. There is a completely analogous problem in logic: Let us assume that we have a complete system $\mathfrak{R}$ of sound rules. Then, given any axiom system, we shall be able, by systematic application of the rules, to find all its consequences by means of proofs, in a similar way to that in which we can, with an algorithm, find all the places of the decimal expansion of π[13]. However, this does not provide us with a solution to the decision problem, i.e. the problem of finding a procedure which, given any (finite) axiom system $\mathfrak{A}$ and any given specific mathematical statement $\underline{A}$, enables us to establish in finitely many steps whether or not $\underline{A}$ is a consequence of $\mathfrak{A}$. If $\underline{A}$ does indeed follow from $\mathfrak{A}$, then the systematic application of the rules of the complete set of rules will eventually lead to a proof of $\underline{A}$ from $\mathfrak{A}$. However, if $\underline{A}$ is not a consequence of $\mathfrak{A}$, we shall not be able to prove this by application of the rules of $\mathfrak{R}$ above.

In spite of his efforts, Leibniz was not able to answer the question of the existence of a complete set of rules. His investigations did not influence the following generations; it was only around 1900 that Leibniz's notes on the subject were published.

Today we can see that the use of everyday languages was not the least of the reasons why the problem offered such strong resistance to the efforts of many outstanding research workers. Operating in a natural language with so precise an instrument as an algorithm is like trying to determine with a precision instrument the height of a tree shaken by a gale. So it is no wonder that the first substantial advances were only made when people started using artificial "formalised" languages instead of natural languages.

This change was brought about systematically in the last century. In his The mathematical analysis of logic, G. Boole (1815 - 1864) attempted, from 1847 onwards, a formalisation of language which imitated the formalism of mathematical algebra. Many investigations have been carried out in Boole's formalism; however, it is not now generally used because it differs excessively from the natural languages. (The concepts of cylinder algebras and polyadic algebras are modern developments of the algebraification of logic).

The formalisation created by G. Frege (1848 - 1925) in his Begriffsschrift (1879) was nearer to the natural languages, but did not catch on because of its two-dimensional notation. The (unfortunately not uniform) formalisations which are used today are derived from the formal language created by G. Peano (1858 - 1932). Frege not only built up a formal language, but also set down a system of rules. However, as B. Russell (1872 - 1970) showed, Frege's system of rules is not sound. A revised sound system of rules proved to be sufficient to transform all the important proofs of some basic mathematical theories into proofs relative to this system of rules. This was shown in Whitehead and Russell's monumental work Principia Mathematica (1910/13). Thus it could be suspected that the system of rules used was complete. It was only considerably later, in 1930, that Gödel proved the completeness of a system of rules. This was, however, only for a formal language with limited means of expression, which

---

[13] The only difference is that we obtain the places of the decimal expansion of π in a well-ordered sequence, whereas there is no natural order of succession for the consequences of an axiom system and such an order can only be introduced artificially.

is today known as the language of predicate logic; or, more precisely, as the language of first-order predicate logic. Gödel also showed that, for a richer language, that of second-order predicate logic, there is no complete set of rules [14].

The first of these two results of Gödel brought to a positive conclusion the investigations of an era which had lasted for centuries. However, the second result showed that the first was by no means self-evident. Incidentally, we see again how important the transfer to formal languages is, for the formalisation contributed substantially to the clarification of the difference between first-order and second-order predicate logic.

A complete system of sound rules for the formal language of predicate logic is usually called a p r e d i c a t e c a l c u l u s. The word c a l c u l u s is reminiscent of the c a l c u l i, i.e. the pebbles by means of which addition and other operations used to be carried out on an abacus before the modern methods of calculation, based on the arabic numbers, became customary.

5.5 Assumption calculi. Today, several predicate calculi are known. Some of these cannot immediately be seen to have the simple form which we took as our basis in the general discussion above. But even slightly different calculi such as these do in fact serve to obtain all the consequences from an arbitrary system of axioms.

The predicate calculus which is to be presented in this book is of this more general sort. The reason for the deviation from the form given above is that we want to develop a calculus which, in one characteristic respect, reflects the practices of mathematical proof as truly as possible [15].

The point is this: The axioms of an axiom system $\mathfrak{A}$, which are taken as the basis of a mathematical theory, can be interpreted as assumptions, for these axioms are not justified at all. Moreover, in the course of a proof, a mathematician often introduces further assumptions ad hoc without justifying them in any way. Of course, he then has, later, somehow to "eliminate" these arbitrary further assumptions. We want to give three indications as to the use of these introduced assumptions:

(a) An example is given by the proof in 2.3, which begins by at once introducing the two additional assumptions $xy_1 = z$ and $xy_2 = z$.

(b) If, in a theorem of some mathematical theory, the hypotheses are first enumerated before the actual assertion is made, then this theorem is usually proved by introducing the hypotheses as assumptions at the beginning of the proof.

(c) In fact, every so-called indirect proof consists in introducing as an assumption the negation of the assertion which is to be proved, with the initial aim of deriving a

---

[14] Cf. Chap. VI, §4.5. Apart from the restriction to the first order, Gödel's system of rules is practically identical with that of Principia Mathematica, if we disregard the fact that in Principia not all the rules which are applied are explicitly formulated as such.

[15] Such a true reflection has the advantage that someone who is used to ordinary mathematical proofs can also handle this predicate calculus relatively easily.

"contradiction", in order then to be able to pass immediately to the assertion itself (cf. the proof of Lemma 2 in 2.2).

If it is permitted to introduce additional assumptions in the course of a proof, then we must ensure that, for every assertion which occurs in the proof, we can determine the assumptions on which the assertion "depends". This is most simply and surely done by, in the relevant line of the proof, putting the assumptions belonging to the assertion explicitly at the beginning and the assertion itself at the end. Once we have decided to do this, we can go one step further and treat the axioms just as we do the additional assumptions.

Thus, a proof in an assumption calculus like this consists of a finite sequence of rows, each of which consists of finitely many statements, of which, in each case, the last is regarded as the assertion of the row and the preceding ones as the assumptions. There must be finitely many rules by which we can pass from initial rows (the premises) to another row (the conclusion).

We want also to admit rules without any premises as a special case. We can then write down certain rows immediately by means of such rules. The so-called rule for the introduction of an assumption is an example of such a rule. This rule allows us to write down, at any stage of a proof, a row

(*) $\underline{AA}$

where $\underline{A}$ may be any mathematical statement.

What shall we mean by a sound rule of an assumption calculus? In order to explain this, we introduce the idea of a sound row, or a row which preserves validity: A row is to be called sound if its assertion is a consequence of its assumptions. Now we call a rule sound if, given sound rows as premises, it always leads to a sound row as conclusion.

For a rule without premises, we shall understand the definition of the soundness of a rule as follows: We shall say that such a rule is sound if the rows which it permits us to write down are sound. The above-mentioned rule for the introduction of an assumption is sound: This rule allows us to write down rows of the form (*). But every such row is sound, since the given assumption $\underline{A}$ follows from the assertion, which is also $\underline{A}$.

Clearly, every row of a proof in an assumption calculus which is based on sound rules is sound. This holds in particular for the last row of the proof. Thus, the assertion of this row is a consequence of its assumptions. In this way, an assumption calculus can serve to derive consequences algorithmically. We shall call an assumption calculus complete if every consequence can be obtained as indicated by means of a proof based on the rules of the calculus[16].

---

[16] This refers to the case where there are finitely many axioms from which consequences are to be obtained. For the general case cf. Chap. IV, §1.2.

Now we shall rewrite the proof given in 5.1 in a form which could fit into an assumption calculus. We want to do this using only those axioms of geometry which are really needed for the theorem which is to be proved.

1) 0 is a natural number. 0 is a natural number.

2) If x is a natural number, then the successor of x is always different from 0. If x is a natural number, then the successor of x is always different from 0.

3) If x is a natural number, then the successor of x is always different from 0. If 0 is a natural number, then the successor of 0 is different from 0.

4) 0 is a natural number. If x is a natural number, then the successor of x is always different from 0. The successor of 0 is different from 0.

5) 0 is a natural number. If x is a natural number, then the successor of x is always different from 0. 0 is a natural number and the successor of 0 is different from 0.

Note that this "proof" (like the "proof" in 5.1) is not a proof in the strict sense which we want to give to this concept here, since it is not based on any system of rules. We have to think out a system of rules to go with it.

In the proof we have just given, rows 1) and 2) can be thought of as being obtained by means of the rule of the introduction of an assumption. Rows 3), 4) and 5) correspond to rows (6), (7) and (8) of the proof in 5.1. If we could find a system of sound rules such that the sequence 1),...,5) were a proof relative to these rules, then we should have shown that the assertion of 5), that is, the statement (8), was a consequence of the assumptions of 5), i.e. of the statements (1) and (3), and, thus, a fortiori, a consequence of Peano's axioms.

5.6 Mechanical proof. The frequent repetition of particular statements in the proof 1),...,5) may at first appear pedantic. In fact, in practice abbreviations will be devised. However, at the moment we are trying to work out general principles and therefore need to practise this sort of exactness. When we work with such precision (which can only really be attained by laying down rules and after passing over into a formal language) it becomes clear that carrying out a calculus is really a formal process, which can intuitively be conceived of as moving words and letters. If we think only of the meaning and sense of statements, then slight variations, or even omissions which could easily be filled in, may be unimportant; but if we are trying to manipulate letters like pieces in a puzzle, then every single piece is important.

Given the exactness which we can achieve in this way, we can easily use computers for proofs (in a precise meaning of this word), either to test mechanically proofs which are already given to see whether or not the rules have been applied faultlessly, or to generate proofs according to a given system of rules. It must, however, be said that the limited storage capacity of modern machines provides, at any rate at present, a practical hindrance.

Exercises. 1. Try to find a system, in the sense of 5.3, of sound rules such that the mathematical proof given in 2.3 becomes a proof relative to this system of rules (cf. 5.3).

2. Try (using the rule for the introduction of an assumption) to find a system of sound (in the sense of 5.5) rules such that the sequence 1),...,5) given in 5.5 becomes a proof relative to this system of rules.

3. Find out whether the following rules of inference are sound (5.3):

(a) We may pass from two statements of the form

if A, then B and not A

to the statement not B.

(b) We may pass from two statements of the form

A or (if not A, then B) and not B

to the statement A.

(c) We may pass from seven statements of the form

A_1 or A_2 or A_3 if B_1, then not (B_2 or B_3)

if A_1, then B_1 if B_2, then not (B_3 or B_1)

if A_2, then B_2 if B_3, then not (B_1 or B_2)

if A_3, then B_3

to the statement (if B_1, then A_1) and (if B_2, then A_2) and (if B_3, then A_3).

(d) We may pass from two statements of the form

if (A or B), then C and if (B or C), then A

to the statement if (A and C), then B.

§6. The symbolisation of mathematical statements: Junctors and quantifiers.

6.1 Statement of the problem. In the last few sections we indicated repeatedly that it is necessary to pass from natural languages to a formal language in order to study logic successfully. We could, therefore, start building up such a formal language at once in order to have a basis for our logical investigations. However, the formal languages have grown up against the background of the natural languages. Thus, we shall be able to understand the structure of a formal language better if we know which characteristics of the natural language it reflects and how it does this. We can find this out best by practising "translating" everyday statements into the formal language, or, in other words, by symbolising these statements. The symbolisation of statements is an art which can, to a large extent, be learned by practice. The symbolisation of arbitrary linguistic statements can be very difficult. However, the exercise is simpler if the statements which are to be translated are mathematical statements, since the traditional mathematical statements have a fairly transparent logical structure and since it is precisely this logical structure that we want to bring out in the symbolisation.

Since mathematical statements are usually formulated to some extent in everyday language, we shall only be able to apply the logic we develop here to mathematics if we know how to translate the mathematical statements into the formal language which forms the basis of our logic. The reader is therefore strongly urged to practise symbolising (cf. the exercises at the end of the next section).

In this section and the following one, we want to acquaint the reader with the most essential elements of the language of predicate logic by ascertaining the logical structure of mathematical statements by means of examples. First of all we shall consider universal components of statements (see 3.3); we shall consider the specific components in the following section.

6.2 Junctors. Examples of universal components of statements are words (or word complexes) such as and, or, either-or, not, if-then, if and only if, neither-nor, while, with whose help statements can be connected to form new statements. Words like these are called propositional connectives or (as by Lorenzen) junctors. Apart from the junctor not, the junctors given above are two-place, i.e. we can, by means of them, join together two statements to form a new statement. Thus, for example, from the two statements.

$$3 < 4, \quad 3 = 4$$

we obtain the statement $3 < 4$ or $3 = 4$ (which is abbreviated in arithmetic by $3 \leqslant 4$) through connection by or. Not is a one-place junctor, which, for example, connected with the statement $3 > 4$, gives the statement not $3 > 4$.

Let us assume that we are dealing with an aristotelian statement A (see 4.2). Now, A is either true or false. If A is true, then clearly the statement not A is false, and conversely. We can represent this property of the junctor not by the following so-called truth table (logical matrix):

Truth table for not

A	not A
T	F
F	T

Here, T means truth and F falsehood. T and F are known as the aristotelian truth values.

Because of the above-mentioned property of not, we do not need to know what the statement A is in order to determine the truth value of not A (i.e. in order to discover whether this statement is true or false). It is sufficient to know the truth value of A. For this reason, not is called an extensional junctor.

And is another example of an extensional junctor. If A and B are aristotelian statements[17], then A and B is true if both A and B are true; in every other case, A and B is false. We can represent this by a truth table with two axes:

Truth table for and

A and B		B: T	B: F
A	T	T	F
A	F	F	F

or, more compactly:

and	T	F
T	T	F
F	F	F

If we pass to or, the situation becomes more complicated. It is easy to see from examples that, in everyday speech, this word is used in different ways by different people[18]. If we want to build up a logic, we cannot allow this situation to continue; it is then necessary to define the use of or in some fixed way. We want to do this and to agree to

---

[17] In the following, A and B are always to be aristotelian statements.

[18] This is, in my experience, true even of mathematics students who have been studying differential and integral calculus for some time.

use the word or extensionally, just as we use <u>and</u> and <u>not</u>. <u>A</u> <u>or</u> <u>B</u> is to be false if both A and B are false; in every other case, A or B is to be true. Thus we have the

truth table for <u>or</u>

or	T	F
T	T	T
F	T	F

The <u>or</u> which we have defined precisely in this way, which corresponds to the most usual mathematical usage of this word, is a "non-exclusive or", since <u>A</u> <u>or</u> <u>B</u> is regarded as true even when both <u>A</u> and <u>B</u> are true. This is not the case with the "exclusive or", i.e. with <u>either-or</u>. If we decide to use this connective extensionally too, then we shall decide on the following

truth table for <u>either-or</u>

either-or	T	F
T	F	T
F	T	F

If we connect two statements by <u>and</u>, the resulting statement is true exactly when both its components are true. However, if we connect two statements by (the non-exclusive) <u>or</u>, the resulting statement is false exactly when both its components are false. For this reason, <u>and</u> and <u>or</u> are said to be dual junctors, or junctors which are dual to each other.

Except where we explicitly state otherwise, we shall, in this book, always use the junctors <u>or</u> and <u>either-or</u> in the way shown in the truth tables above. (The same holds for the junctors <u>if-then</u> and <u>if and only if</u>, which we shall discuss shortly.)

By studying the above matrices it is easy to see that the two statements <u>A</u> <u>or</u> <u>B</u> and <u>not</u> (<u>not</u> <u>A</u> <u>and</u> <u>not</u> <u>B</u>) always have the same truth value for arbitrary <u>A</u> and <u>B</u>. It is thus possible to define <u>or</u> from <u>and</u> and <u>not</u>, i.e. to interpret

(1) <u>A</u> <u>or</u> <u>B</u> as an abbreviation for <u>not</u> (<u>not</u> <u>A</u> <u>and</u> <u>not</u> <u>B</u>).

Similarly, we can introduce

(2) <u>either</u> <u>A</u> <u>or</u> <u>B</u> as an abbreviation for (<u>A</u> <u>or</u> <u>B</u>) <u>and</u> <u>not</u> (<u>A</u> <u>and</u> <u>B</u>).

We can show quite generally that every extensional junctor can be defined by means of the junctors <u>and</u> and <u>not</u> in a similar way[19]. This fact allows us to restrict ourselves to the extensional junctors <u>and</u> and <u>not</u> when we are considering logic theoretically. We shall make use of this later.

---

[19] See exercise 5.

We call not the negator and and the conjunctor.

A two-place junctor, which can be represented as not (A and not B) appears relatively often in mathematics. In many cases, mathematicians express it as: if A, then B [20]. If we want to use if-then in this way, we have

(3) if A, then B as an abbreviation for not (A and not B).

This corresponds to the following

truth table for if-then

if-then	T	F
T	T	F
F	T	T

It follows that if A, then B is false exactly when A is true and B is false [21].

It corresponds to the usual way of speaking to interpret

(4) A if and only if B as an abbreviation for (if A, then B) and (if B, then A).

Using the truth table for if-then as a basis, we obtain the

truth table for if and only if

if and only if	T	F
T	T	F
F	F	T

6.3 Symbols, rules for dispensing with brackets. The extensional junctors not, and, or, if-then, if and only if appear so frequently that it is worth introducing symbols to represent them (like + and · in arithmetic). Unfortunately, there is, as yet, no universally accepted symbolism. We shall write

$\neg$ for not (reminiscent of the minus sign).
$\vee$ for or (reminiscent of the latin vel for the non-exclusive or)
$\wedge$ for and (reminds us of the duality of and and or).
$\rightarrow$ for if-then
$\leftrightarrow$ for if and only if [22].

---

[20] However, they also often use if-then in quite a different sense, namely to express a logical consequence.

[21] Thus, of the statements (1) if 1+1=2, then 17 is a prime number, (2) if 1+1=2, then 16 is a prime number, (3) if 1+1=3, then 17 is a prime number, (4) if 1+1=3, then 16 is a prime number, the statement (2) is false; the other statements are true.

[22] Other symbols are, for example, $\sim$ or overlining for not; &, ■ or simply concatenation for and; $\supset$ for if-then, $\equiv$ for if and only if.

It is often necessary to use brackets in order to indicate the construction of a statement clearly. For example, if we write $\underline{A} \wedge \underline{B} \rightarrow \underline{C}$, it is not immediately clear whether we mean $(\underline{A} \wedge \underline{B}) \rightarrow \underline{C}$ or $\underline{A} \wedge (\underline{B} \rightarrow \underline{C})$. However, in order to make our symbolisation easier to take in, it is worth introducing rules for dispensing with brackets. We shall use the following two

Conventions for dispensing with brackets:

(a) $\wedge$ and $\vee$ have precedence over $\rightarrow$ and $\leftrightarrow$ [23].

(b) Iterated $\wedge$-connections and iterated $\vee$-connections are to be bracketed from the left [24].

<u>6.4 Example of a non-extensional junctor.</u> We shall now consider the junctor <u>while</u>, which we have already mentioned in 6.2. First of all, we must observe that the English word <u>while</u> is ambiguous. In fact, this word is used to express (1) a contrast and (2) a coincidence [25].

Thus, we should really speak of two different junctors, namely of <u>while</u> in the contrast-sense and <u>while</u> in the time-sense.

The junctor <u>while</u> in the contrast-sense can usually simply be replaced by <u>and</u>. It is then an extensional junctor.

Strictly speaking, the junctor <u>while</u> in the contrast-sense differs from the junctor <u>and</u> in that <u>while</u> is used particularly to call attention to some difference between the two components connected by <u>while</u>. Thus, the statement <u>$\sqrt{2}$ is algebraic, while π is transcendental</u> means, strictly speaking, the same as the statement <u>$\sqrt{2}$ is algebraic and π is transcendental; note the difference</u>! However, the summons at the end of this statement is mathematically irrelevant.

Now we turn to the junctor <u>while</u> in the time-sense and consider the following true statement:

<u>Milton wrote "Comus" while the great migration of Puritans from England to America was taking place.</u>

If, in this statement, we replace the true substatement:
<u>Milton wrote "Comus"</u> by the statement, likewise true:

---

[23] This is analogous to the convention in algebra that the multiplication sign has precedence over the addition sign, so that ab+c is understood as (ab)+c. By convention (1) $\underline{A} \wedge \underline{B} \rightarrow \underline{C}$ means the same as $(\underline{A} \wedge \underline{B}) \rightarrow$ C. Note that $\wedge$ and $\vee$ (and, similarly, $\rightarrow$ and $\leftrightarrow$) have equal status as regards bracketing, so that, for example, $\underline{A} \wedge \underline{B} \vee \underline{C}$ and $\underline{A} \rightarrow \underline{B} \leftrightarrow \underline{C}$ are ambiguous.

[24] Thus, for example, $\underline{A} \wedge \underline{B} \wedge \underline{C} \wedge \underline{D}$ is an abbreviation for $((\underline{A} \wedge \underline{B}) \wedge \underline{C}) \wedge \underline{D}$.

[25] The two senses of the word may also be expressed by <u>whereas</u> and <u>during the time that</u>. The French language has two distinct junctors <u>tandis que</u> and <u>pendant que</u> corresponding to these two meanings.

Milton wrote "Paradise Lost", then the total statement, which was originally true, becomes false. This shows that the junctor while in the time-sense is not extensional.

Non-extensional junctors are called intensional. In mathematics we can restrict ourselves to extensional junctors.

It is possible to rewrite a statement in which the junctor while in the time-sense appears, to obtain a statement with the same meaning which contains no intensional junctors. Thus, the statement given above in the text has the same meaning as the statement: Every moment in which Milton wrote "Comus" is a moment in which the great migration of Puritans from England to America was taking place.

6.5 The quantifiers in predicate logic. We consider the two statements

(*) All natural numbers are algebraic.

(**) There is an algebraic number which is not rational.

The word all which appears in the first statement is called a generalisor or a universal quantifier and the pair of words there is, which appears in the second statement, a particulisor or an existential quantifier.

The statement (*) can be reformulated in various ways without changing its meaning; e.g. by writing:
Every natural number is algebraic.
A natural number is (always) algebraic.

Thus we shall call the word every and also the word a (an) (at any rate if it is used in the same sense as in the last statement) a universal quantifier. Since, in logic (and mathematics) it is immaterial which universal quantifier is used, we shall only have to consider one universal quantifier in a formal language.
We write

$\bigwedge$ for the universal quantifier.

The symbol " $\bigwedge$ " is a large $\wedge$. In fact, we can always interpret a statement which begins with the universal quantifier as a generalised $\wedge$-connection of statements. Thus, for example, the above statement (*) has the same value as

0 is algebraic and 1 is algebraic and 2 is algebraic and ...

Note, however, that the dots at the end of the line indicate that the and-connection is to be exte~ded infinitely far. Thus, strictly speaking, this is not a statement, since we shall require a statement to have a finite length. It is only when we are dealing with a finite totality that we can really replace the universal quantifier by an and-connection.

We can make comments on the existential quantifier analogous to those on the universal quantifier. For example, the statement (**) can be reformulated in the following ways:

There exists an[26] algebraic number which is not rational.

(At least) one algebraic number is not rational.

Thus, we shall also call there exists and (at least) one existential quantifiers.

In the formal language, we use only one quantifier to express existence, and we write

$\bigvee$ for the existential quantifier.

$\bigvee$ is a large "$\vee$". For we can interpret a statement beginning with the existential quantifier as a generalised or-connection and thus, for example, rewrite the statement there is a natural number which is a prime number as

0 is a prime number or 1 is a prime number or 2 is a prime number or ...[27].

As in the case of the universal quantifier, it is only when we are dealing with a finite totality that it is really possible to replace the existential quantifier by an or-connection.

We can define $\bigvee$ by means of $\bigwedge$ and $\neg$ just as we can define $\vee$ by means of $\wedge$ and $\neg$ (see 6.2). We shall enlarge upon this in the next section.

Exercises. 1. (a) With the help of 6.4, find further examples of non-extensional junctors.

(b) Contrary to mathematical usage, there are examples in everyday language in which and is not used in a commutative way, i.e. in which A and B is true and B and A is false. Such examples stem from the fact that A and B refer to different points in time (without this having been explicitly stated). Find such examples and show that, in such examples, the junctor and is not used extensionally.

2. The following two statements

(1) Paris is the capital of France and London is the capital of England,

(2) Paris is the capital of France or Paris is not the capital of France

are both true. However, there is an important difference between them: In order to recognise the truth of (1), we must have some knowledge about the geographical content of (1). In the case of (2), on the other hand, we can manage without any such knowledge, for, because of our decision about the use of or and not, (2) is - as can be shown easily

---

[26] The words a and an can be used in more than one way; they sometimes act as universal quantifiers and sometimes as existential quantifiers. This is illustrated by the following two statements in the everyday language: An American speaks English. An American has landed on the moon. The ambiguity of the words a and an is another example of the logical incompleteness of the natural languages.

[27] Note that here or cannot be replaced by either-or.

- a true statement independently of whether or not Paris is the capital of France. Because of this, (2) is called a formally true statement. (1), on the other hand, is not formally true, but merely true by virtue of its content. A formally true statement is also, more precisely, called a propositionally formally true statement or a tautology.

If a statement is built up from other statements by means of an extensional junctor, then this statement is said to be (propositionally) decomposable; if not, it is said to be (propositionally) indecomposable. Clearly, every statement can, in a unique way, be built up from indecomposable statements by means of extensional junctors (if we allow the case in which the given statement is itself indecomposable). If we replace the indecomposable components of a statement by different letters A, B, C,... (where, however, components which occur more than once are to be replaced by the same letter each time), then we obtain a structure built up from letters and junctors (and, possibly, brackets to show the way in which the structure is connected together), which we call the (propositional) structure of the statement. The structure of a statement is determined uniquely up to renaming of the letters. Thus, for example, the statement (1) has the structure

$$A \wedge B,$$

whereas the statement (2) has the structure

$$A \vee \neg A.$$

Note that, although the statement

(3) Every town is the capital of France or not the capital of France

contains extensional junctors, it is still (propositionally) indecomposable, because the universal quantifier every blocks the decomposition. (3) is essentially different from the (decomposable) statement

(3') Every town is the capital of France or every town is not the capital of France.

(3) has the structure A and (3') the structure $A \vee B$. Neither of these statements is a tautology.

Let some statement have, for example, the structure $(A \rightarrow B) \vee (C \rightarrow A)$. The test to find out whether the statement is a tautology can be carried out according to the following simple pattern:

A	B	C	$(A \rightarrow B)$	$\vee$	$(C \rightarrow A)$
T	T	T	T	T	T
T	T	F	T	T	T
T	F	T	F	T	T
T	F	F	F	T	T
F	T	T	T	T	F
F	T	F	T	T	T
F	F	T	T	T	F
F	F	F	T	T	T

The column under "$\vee$" shows that the given statement is a tautology.

In many cases, we can replace this systematic method of working out by a shorter, indirect method. In the example given above, this would work as follows: If $(A \rightarrow B) \vee (C \rightarrow A)$ were false in some case, then both $A \rightarrow B$ and $C \rightarrow A$ would have to be false in this case. But $A \rightarrow B$ is false only if A is true (and B is false), and $C \rightarrow A$ is false only if (C is true and) A is false. This contradiction shows that $(A \rightarrow B) \vee (C \rightarrow A)$ can never be false.

Test the following statements to find out whether or not they are tautologies:

(a) If Cardiff is in Wales or Dundee in Scotland, then Cardiff is in Wales.

(b) If it is the case that, if Joe goes to America, he will either stay here or come back a rich man, then Joe will stay in America if he goes there or Joe will come back a rich man if he goes to America.

(c) It is not the case that Caesar is a Gaul if he is a Gaul.

3. We can extend the notion tautology, which, up till now, we have applied only to statements, by also referring to the structures of such statements as tautologies. Test the following structure formulas for this property:

(a) $(C \to A \vee B) \to (C \to A) \vee (C \to B)$

(b) $(A \to (B \to C)) \to (C \to (A \to B))$

(c) $((A \to B) \to A) \to A$

(d) $(A \to B) \wedge (C \to D) \wedge (A \vee C) \wedge \neg (B \wedge D) \to (B \to A) \wedge (D \to C)$

(e) $(A \wedge B \to C) \leftrightarrow (A \to C) \vee (B \to C)$

(f) $(A \vee B \to C) \leftrightarrow (A \to C) \wedge (B \to C)$

4. The connective of alternative denial $|$ is an extensional junctor which is defined by the following matrix:

Truth table for the connective of alternative denial

|	T	F
T	F	T
F	T	T

Show that $\neg$ and $\wedge$ can be defined by means of $|$ (as, for example, $\vee$ can be defined by means of $\neg$ and $\wedge$); see example 2.

5. Show that every n-place extensional junctor can be defined by means of $\neg$ and $\wedge$ (and hence, by exercise 4, also by $|$ alone). (Because of this, the sets of junctors $\{\neg, \wedge\}$ and $\{|\}$ are called functionally complete sets of junctors.)

Sketch of a proof: Let J be an arbitrary n-place extensional junctor. Let its truth table be given in the following form:

$A_1 \ldots A_n$	$J A_1 \ldots A_n$
T ... T	w_1
T ... F	w_2
⋮	⋮
F ... F	w_{2^n}

Each of the truth values w_i appearing in the right-hand column is either T or F. Now we pick out the rows in which $w_i = T$. (The case in which there are no such rows must be considered separately.) Suppose that $w_1' \ldots w_n' \mid T$ is one such row. Then we construct the structure formula

$$(\neg)_1 A_1 \wedge \ldots \wedge (\neg)_n A_n ,$$

where $(\neg)_j$ is to be $\neg$ if $w_j' = F$ and is to be omitted if $w_j' = T$. Now, after enclosing each of the formulas obtained in this way in brackets, we connect them all together by means of $\vee$. The resulting formula can be used as a definition of the junctor J. We have thus solved the problem, since $\vee$ can be defined by means of $\neg$ and $\wedge$.

6. Test the following sets of junctors for functional completeness:

(a) $\{\wedge, \vee, \rightarrow, \leftrightarrow\}$,

(b) $\{\vee, \neg\}$,

(c) $\{\rightarrow, \neg\}$,

(d) $\{\leftrightarrow, \neg\}$,

7. Calculate the number of different n-place extensional junctors for each $n \geqslant 1$.

8. Find all those two-place extensional junctors J such that $\{J\}$ is a functionally complete set of junctors.

§7. The symbolisation of mathematical statements: Individuals, predicates and functions.

7.1 Survey. The most important specific notions which appear in aristotelian statements are names for individuals, names for predicates and names for functions. Notions like these also appear in mathematical statements, but there, as we have seen in 4.2, they are usually to be understood as variables. Thus, correspondingly, we shall have to distinguish between individual variables, predicate variables and function variables. There is, however, a name for a predicate which also appears in mathematical statements and which is not to be understood as a variable; it is the name "=" (known as the equality sign) for the equality predicate.

It appears that it is possible to construct almost all the traditional mathematical statements from the variables we have just mentioned together with the junctors and quantifiers which we considered in the last paragraph and the equality sign. If we replace the names for individuals, predicates and functions which appear in a mathematical statement by letters and the universal notions by the symbols which we introduced in §6, then we obtain a "formula" which is called a symbolisation of the original mathematical statement. In the next chapter, we shall introduce the expressions, as they are called, in a precise way. Every expression can be regarded as a "formula" like this; conversely, all "formulas" (if they satisfy an additional condition, on which we shall enlarge in 7.4) can be taken as expressions.

It is customary to use the letters x, y, z, ..., a, b, ... as individual variables, P, Q, ..., A, B, ... as predicate variables and f, g, ... as function variables. These letters are also often used with indices.

7.2 Names for individuals, predicates and functions. In order to speak about individuals, predicates and functions we have to use names for these entities. In the statement The crown jewels are kept in the Tower of London, The crown jewels and the Tower of London can be understood as names for individuals and are kept in as a name for a predicate. the Tower of London can, however, be analysed further by taking the Tower of as a name for a function and London as a name for an individual[28].

are kept in is a name for a two-place predicate; it can be completed to form an aristotelian statement by the addition of two names for individuals. There are also names for k-place predicates where $k \neq 1$. Thus, for example, is tall is a name for a one-place predicate and lies between - and -[29] a name for a three-place predicate. There are also names for many-place functions. An example of a name for a two-place function is the railway-line between - and -. A name for an n-place function together with n names for individuals yields a name for an individual.

It is sometimes worth admitting no-place predicates and no-place functions as a borderline case. For $n \geq 1$, a name for an n-place predicate together with n names for individuals yields an aristotelian statement. If we want to extend this situation to the case $n = 0$, then a name for a no-place predicate must yield an aristotelian statement without any additional name for an individual. Thus, a name for a no-place predicate is simply an aristotelian statement (i.e. a name for a proposition). For $n \geq 1$, a name for an n-place function together with n names for individuals yields a name for an individual. If we extrapolate this for $n = 0$, we must expect a name for a no-place function to yield a name for an individual without any additional name for an individual. Thus, a name for a no-place function is simply a name for an individual.

The word predicate is also used in grammar, but in not quite the same sense as in logic.

7.3 Examples of aristotelian statements. Now we consider some common types of aristotelian statements and show how to analyse these logically. These analyses can be applied correspondingly to mathematical statements. Example (2) is particularly important.

(1) Sirius is not a planet.

(2) All planets are oblate.

(3) No planet is luminous.

(4) There is at most one winner.

Here, the following analyses suggest themselves:

to (1): Name for an individual: Sirius, name for a predicate: is a planet. Complete statement: Not Sirius is a planet.

to (2): No name for an individual. Names for predicates: is a planet and is oblate. Clearly, statement (2) is trying to express that, for all things, it is not the case that the thing is a planet and is not oblate. Thus we could write provisionally:

---

[28] It may be questioned whether The crown jewels could also be analysed in a similar way.

[29] Consider, for example, the statement: Stratford-on-Avon lies between Oxford and Birmingham. Note that the word and which appears here is not to be understood as a junctor. This is in contrast to the situation in the statement Oxford and Birmingham are towns, which can be regarded as an abbreviation for the statement: Oxford is a town and Birmingham is a town.

For all things: not (the thing is a planet and the thing is not oblate).

By §6, we could use the extensional junctor if-then here and write:

For all things: (if the thing is a planet, then the thing is oblate).

Finally, we ask ourselves what exactly the word pair the thing, which we have introduced here, is. We might at first think that it was a name for an individual we were dealing with here. However, this is not the case, since a name for an individual must be a name for a fixed individual and we cannot say that this is true of the thing here. The part which the expression the thing plays is, rather, that it indicates the places to which the quantifier For all at the beginning relates. We could, instead, use any other symbol, e.g. an asterisk, and write:

For all * (if * is a planet, then * is oblate).

In practice, it has become usual to make the necessary indication by means of a letter. For this purpose, we use the letters which are also used for individual variables, since the components * is a planet and * is oblate can be made into statements by replacing * by a name for an individual. Thus, we finally obtain

(5) For all x (if x is a planet, then x is oblate).

(We could also write "y" or "z" instead of "x".) An individual variable x which is used together with "for all" plays a different part from a variable x as in

(6) x is a planet.

This is superficially obvious from the fact that (5) is an aristotelian statement, whereas (6) is only a statement form (cf. 4.2). We can make a statement from (6) by substituting a suitable name for an individual, e.g. Jupiter, for the individual variable x. However, a replacement like this for x would turn (5) into a meaningless linguistic structure.

In order to make this difference clear, we talk of a free individual variable in the case (6) and of a bound individual variable in case (5). In (5), the variable x is bound by the universal quantifier.

In traditional mathematical formulas, too, the difference between free and bound variables is an essential one. Thus, for example, in $\sin x = y$ the variables x and y are free variables, whereas in formulas such as $\sum_{i=0}^{\infty} \frac{1}{2^i}$ and $\int_0^P \sin x \, dx$ the variables i and x are bound, i by the summation sign and x by the integral sign. In $\sum_{i=0}^{n} \frac{1}{2^i}$ i is a bound variable and n a free variable.

Some authors distinguish the bound variables from the free from the start by using special letters for the bound variables. In order to avoid such a duplication of variables, bound and free variables will not be distinguished typographically here. It is always possible to see whether a variable is bound or not by considering the quantifiers which precede it.

to (3): No name for an individual. Names for predicates: is a planet and is luminous. If we use the variable y as a bound variable in connection with the existential quantifier, then the analysis yields:

not there is a y (y is a planet and y is luminous).

to (4): No name for an individual. Name for a predicate: is a winner. Although the statement (4) begins with there is, it is in fact not an assertion of existence. We can rewrite (4) as: If any x and y are winners, then they are equal. Thus we obtain:

For all x for all y (if x is a winner and y is a winner, then x = y). x and y are bound variables.

7.4 Examples of mathematical statements. We consider Peano's axioms P'1, P'2, P'3 (3.7). Nought appears to be a name for an individual and successor of appears to be a name for a one-place function. In fact, however, it is an individual variable and a function variable that we are dealing with here. We want, therefore, to represent nought by n and successor of by f. 16. If we also use the symbols introduced in the last section for the junctors and the quantifiers, we obtain:

P'1 $\bigwedge x \neg f(x) = n$

P'2 $\bigwedge x \bigwedge y(f(x) = f(y) \rightarrow x = y)$

P'3 $\bigwedge P(Pn \wedge \bigwedge x(Px \rightarrow Pf(x)) \rightarrow \bigwedge xPx)$.

The third axiom needs special clarification: We might represent the fact that 0 has the property P by nP. However, it is usual to put the large letter (the predicate variable) before the small letter (the individual variable). (We proceed correspondingly in the case of many-place predicate variables.) The variable P which occurs in P'3 is clearly a bound (one-place) predicate variable, whereas in previous examples we have dealt only with bound individual variables. It is characteristic of the language of predicate logic, which we are going to build up in the next chapter, that only bound individual variables can appear, and not bound predicate or function variables. The induction axiom as originally given by Peano is not symbolised in the language of predicate logic. However, we can make shift with predicate logic in arithmetic. We shall enlarge on this in Chap. VI, §4.

7.5 Definability of the existential quantifier by means of the universal quantifier. We have already mentioned at the end of 6.5 that the existential quantifier can be defined by means of the universal quantifier. We want to enlarge upon this now. First, let us consider the example:

(*) There is a prime number.

We can interpret this statement as a generalised or-connection:

0 is a prime number or 1 is a prime number or 2 is a prime number or ...

Or can be defined by means of and and not. If we generalise the relation (1) given in 6.2, we obtain:

not (not 0 is prime number and not 1 is a prime number and not 2 is a prime number and ...) .

Now we want to summarise the infinite and-connection by universal quantification:

(**) not for all x not x is a prime number.

Thus, we have defined there is by means of for all and not: The equivalence of (*) and (**) can be seen immediately. In fact, we recognise in general that we can take

(***) there is an x such that ... as an abbreviation for not for all x (not ...).

Thus, for the purpose of theoretical considerations, we can limit ourselves to the universal quantifier and define the existential quantifier, when it is needed, by the relation we have just given.

Exercises. 1. Cauchy's criterion for a function f to be continuous at the point x_0 reads:

$$\bigwedge\varepsilon \bigvee\delta \bigwedge x(|x - x_0| < \delta \rightarrow |f(x) - f(x_0)| < \varepsilon).$$

In writing down this condition, we make use of mathematical symbolism and, in particular, of the convention that x stands for a real number and ε and δ for positive real numbers. If we want to do without mathematical symbolism and the conventions we have mentioned, then we can use d and m for the difference and modulus functions respectively, and L, P and R for the less-than-relation and the properties of being a positive real number and a real number. Then we can reformulate the continuity criterion as:

$$\bigwedge\varepsilon(P\varepsilon \rightarrow \bigvee\delta(P\delta \wedge \bigwedge x(Rx \rightarrow (Lm(d(x, x_0))\delta \rightarrow Lm(d(f(x), f(x_0)))\varepsilon)))).$$

It is characteristic that the predicates P and R which have been added are followed by the junctor → when they occur after a universal quantifier, but by the junctor ∧ when they occur after an existential quantifier. Now symbolise in a corresponding way

(a) the continuity of a function f in an interval I,

(b) the uniform continuity of a function f in an interval I,

(c) the convergence of a sequence of numbers a_n to a number a,

(d) the convergence of a sequence of numbers a_n,

(e) the uniform convergence of a sequence of functions f_n in an interval I.

2. Symbolise Peano's axioms in the form in which they are given in 2.2, using N for the property of being a natural number, n for the zero element and f for the successor function.

3. Symbolise the geometrical axioms H1,...,H8 (2.1), using P, L and Q for the properties of being a point, a line and a plane respectively, and I and 0 for lies in and lies on.

4. Symbolise the axioms for group theory (2.3) and the theorem stated in 2.3, using p for the product function.

5. The reader cannot be reminded often enough how important it is to practise symbolising. The easiest material for practise is provided by mathematical statements. Given any such statement, first of all find the specific notions which occur in it and choose letters to represent them. It is customary to use the letters a,b,c,x,y,z,u,v,w,... (possibly with indices, as also in the following) for individuals, f,g,... for functions and A,...,Z for predicates.

Sometimes, so-called characterisations occur in a statement. Examples are the phrases the point at which the lines a and b meet and the least commun multiple of x and y. A characterisation, such as the x such that Ex is used when we can show (within the framework of a theory) that there is exactly (i.e. at least and at most) one thing which has the property E. In order to represent such characterisations, we can extend our logical symbolism by adding a characterisation operator, which we could, for example, symbolise by ⁊. Thus, the x such that Ex can be expressed by ⁊xEx. If, on the other hand, we want to do without a special characterisation operator, then we shall have to rewrite statements which contain this operator in a way which does not involve characterisation. For the sake of simplicity, we shall confine ourselves here to considering a statement D⁊xEx, which asserts that the x which has the property E (also) has the property D (we proceed analogously if, in place of the one-place predicate D, there is a many-place predicate which is followed by only one characterisation; we do not intend here to go into the difficulties which can arise in the general case). Following a suggestion by Russell, we can interpret this statement to mean that there is exactly one thing which has the property E, and that every (and hence this) thing which has the property E also has the property D. Thus we can rewrite the statement without the use of characterisation as follows:

$$\bigvee xEx \wedge \bigwedge x \bigwedge y(Ex \wedge Ey \rightarrow x = y) \wedge \bigwedge x(Ex \rightarrow Dx)$$

(this, incidentally, can be abbreviated to an equivalent statement; cf. Chap. III, §2, exercise 7).

Symbolise the following statements with and without the characterisation operator:

(a) The point at which the lines a and b meet lies on the line c. (Specific notions: lies on, point, line.)

(b) The greatest common factor of 6 and 8 is a prime number. (Specific notions: 6, 8, is a factor of, prime number.)

(c) There is no greatest natural number. (Specific notions: greater than, natural number.)

6. We have just mentioned the fact that it is possible to do without characterisation. It is also possible to do without functions. [Predicate logic is often built up without functions. However, this often makes quite simple mathematical formulas very difficult to understand. Cf. (a) below.] We shall illustrate this by taking as an example the two-place group-theoretic product operation p. We can introduce a three-place product predicate P such that Pxyz means the same as $p(x,y) = z$. Thus, for example, the commutative law

$$\bigwedge x \bigwedge y\; p(x,y) = p(y,x)$$

can be expressed without the use of the product function as:

$$\bigwedge x \bigwedge y \bigwedge z (Pxyz \to Pyxz) .$$

If we write the group-theoretic axioms with the use of this product predicate instead of a product operation, then we must add to the usual group-theoretic axioms the following two axioms, which express the fact that P has the nature of a function:

$$\bigwedge x \bigwedge y \bigvee z\; Pxyz$$

$$\bigwedge x \bigwedge y \bigwedge z \bigwedge w (Pxyz \wedge Pxyw \to z = w) .$$

Formulate without the use of functions

(a) the group-theoretic axiom system in 2.3 and the theorem given there,

(b) Peano's axiom system in 2.2, and the theorem given there.

II. The Language of Predicate Logic

In § 1, predicate logic is built up as a formal language. We shall define the central semantic concepts for this language in Chap. III, § 2. As a basis for the predicate calculus which we shall give in Chap. IV, § 2, we shall introduce the concept of the free occurrence of an individual variable here in § 4 and that of substitution in § 5. In § 2 and § 4 we shall treat questions of decidability connected with the concepts introduced in this chapter.

§ 1. Terms and expressions

1.1 Survey. We want to build up a formal language. For this, we introduce certain (finite) rows of symbols as expressions (see 1.5). The rows of symbols which appeared in Chap. I, § 7 as symbolisations of mathematical statements can be understood as expressions. Those expressions which contain no junctors and no quantifiers[1], and from which the other expressions can be built up by means of the junctors and quantifiers, are called atomic expressions [i.e. "logically indecomposable" statements (see 1.4)]. As an aid to defining the atomic expressions, we shall introduce (see 1.3) the concept of a term: The terms can be regarded as a generalisation of the terms we know in algebra.

The symbols available to us, from which the expressions are built up, are given in 1.2. Now it is intuitively convenient to assume that there are countably many primitive symbols (also called letters, in generalisation of the usual meaning of this word), from which the rows of symbols are constructed. However, we could (particularly with regard to the possibility of utilising the language mechanically) stipulate that only finitely many primitive symbols should be used. This stipulation can be complied with; we shall return to this in 1.6.

We introduce the terms by setting up a calculus, called the term calculus, and by taking as terms those rows of symbols which are derivable in the term calculus. Correspond-

---

[1] We shall confine ourselves to the junctors $\neg$ and $\wedge$ and the quantifier $\bigwedge$. The other extensional junctors and the quantifier $\bigvee$ can be defined on this basis (cf. Chap. I, § 6.2 and Exercise 5, and also § 7.5). We shall treat the defined notions for the formal language of predicate logic in Chapter VII.

ingly, we set up an expression calculus and take as expressions those rows of symbols which are derivable in it.

In § 2 we shall show that, for every row of symbols, we can decide whether it is a term, or an expression, or neither.

1.2 Primitive symbols and rows of symbols. In the following, we assume that there are countably many primitive symbols, as follows:

(1) For every non-negative number n, countably many n-place function variables. The no-place function variables are also called individual variables[2].

(2) For every non-negative number n, countably many n-place predicate variables. The no-place predicate variables are also called proposition variables[2].

(3) The left and the right bracket.

(4) Two junctors: the negator and the conjunctor.

(5) One quantifier: the universal quantifier (sometimes known as the generalisor).

(6) The equality sign.

We call the junctors and the quantifiers logical symbols.

Rows of symbols are obtained by the concatenation of finitely many primitive symbols (which do not all need to be different from each other). A row of symbols must contain at least one primitive symbol.

For the following, it is not necessary to describe in detail exactly what the primitive symbols are to look like. (Cf., however, a remark at the end of this section and also 1.6.) However, the symbols must be chosen so that they are compatible with the following

Assumptions about the primitive symbols. (a) For every primitive symbol, we can decide whether it is a function variable, a predicate variable, the right or the left bracket, the negator, the conjunctor, the universal quantifier or the equality sign.

(b) Given any function variable or any predicate variable, we can determine the number of places it has.

(c) For each n, there is a 1-1 mapping of the n-place function variables onto the natural numbers and of the n-place predicate variables onto the natural numbers. These

[2] For this, cf. Chap. I, § 7.2.

mappings are effective, i.e. for every n-place function variable or predicate variable the corresponding natural number can be determined, and conversely.

(d) Every row of symbols can be separated effectively and unambiguously into the primitive symbols of which it is composed.

The following example shows that this assumption is by no means always true: Consider two primitive symbols, of which one consists of one stroke and the other of two strokes. Then a row of symbols consisting of three strokes can be built up from the given primitive symbols in three different ways.

These assumptions hold true if we introduce the symbols as in 1.6.

Because of assumption (d), we can determine effectively whether or not a given primitive symbol occurs in a given row of symbols ζ, and how often it occurs in ζ. Let the number of primitive symbols and logical symbols respectively which occur in a row of symbols ζ be called the length $L(\zeta)$ and the rank $R(\zeta)$ respectively of ζ. (Primitive symbols which occur more than once are to be counted a corresponding number of times.)

We use names for certain symbols. Thus, we use:

"(" for the left bracket,
")" for the right bracket,
"$\neg$" for the negator,
"$\wedge$" for the conjunctor,
"$\bigwedge$" for the universal quantifier,
"$=$" for the equality sign.

Further, we shall use variables for arbitrary rows of symbols and also variables for rows of symbols which belong to particular classes of rows of symbols (just as mathematicians use variables for arbitrary real numbers (e.g. "a") and also variables for particular classes of real numbers, e.g. "ε" for positive numbers or "n" for integers). As variables, we use:

"ζ", "ζ_1", "ζ_2", ...	for arbitrary rows of symbols,
"f", "f_1", "f_2", ..., "g", "h", ...	for function variables,
"x", "x_1", "x_2", ..., "y", "z", ...	for individual variables,
"P", "P_1", "P_2", ..., "Q", "R", ...	for predicate variables,
"A", "A_1", "A_2", ..., "B", "C", ...	for proposition variables.

$\zeta_1 \equiv \zeta_2$ is to mean that ζ_1 and ζ_2 are the same row of symbols.

If we write down a row of symbols ζ_2 after the row ζ_1, then we obtain a row of symbols known as the concatenation of ζ_1 and ζ_2. This row of symbols is to be represented in short by $\zeta_1\zeta_2$. Since the associative law $[\zeta_1\zeta_2]\zeta_3 \equiv \zeta_1[\zeta_2\zeta_3]$ holds for concatenation, it is convenient to do away with brackets and write $\zeta_1\zeta_2\zeta_3$.

One more fundamental remark: We must, in principle, distinguish between two languages, namely the language of predicate logic, which we are to build up, and the language in which we talk about the language of predicate logic. The latter language is called a metalanguage relative to the former, the former an object language relative to the latter. We take the language of everyday speech (or, to be more precise, the usual mathematical "purified" everyday language) as our metalanguage. Note that the symbols "(", ")", "$\neg$" etc. which we introduced above are metalinguistic names for the corresponding primitive symbols in the object language. We have said nothing about what these primitive symbols in the object language are to look like, and this is not necessary for our purposes. However, we could settle that the corresponding primitive symbols in the object language should be the same symbols underlined, so that "("is the name for "$\underline{(}$", "$\neg$" the name for "$\underline{\neg}$" etc. This convention would have to be supplemented by one concerning the function and predicate variables; for this cf. 1.6.

Note the difference between "$\equiv$" and "$=$". "$\equiv$" is a metalinguistic predicate, which is not to be confused with the name "$=$" in the metalanguage for the equality sign of the language of predicate logic.

1.3 Terms. We consider a simple calculus, called the term calculus, which is determined by the following two rules:

Rule 1. We are allowed to write down an arbitrary individual variable.

Rule 2. We are allowed to pass from r (not necessarily distinct) rows of symbols $\zeta_1, \ldots, \zeta_r$ $(r \geqslant 1)$ (which have already been obtained) to any row of symbols which is obtained by writing down an r-place function variable followed by the row of symbols $\zeta_1 \ldots \zeta_r$.

Definition. The rows of symbols which can be derived in the term calculus (and only such rows) are called terms.

Examples of terms. In the following, let x, y, z be individual variables, f a one-place and g a two-place function variable. Then, by rule 1, we can derive each of the rows of symbols x, y, z in the term calculus. Putting $r = 1$, we see that we are allowed to pass from x to fx. Since x is derivable in the term calculus, this also holds for fx by rule 2. From fx we can pass to ffx by rule 2. Thus, ffx is also derivable in the term calculus. Putting $r = 2$, we see that we can pass from y and ffx to gyffx. Since y and ffx are derivable, this holds also for gyffx. Thus the rows of symbols x, y, z, ffx and gyffx are terms[3].

---

[3] This way of writing down terms is convenient, but not always easy to understand. If we used brackets, we should write g(y,f(f(x))) instead of gyffx.

In general, it is clearly true that:

Every individual variable is a term. If $\zeta_1, \ldots, \zeta_r$ are terms and f an r-place function variable, then $f\,\zeta_1 \ldots \zeta_r$ is a term.

The terms which are not individual variables are to be called <u>compound terms.</u>

We use "t", "t_1", "t_2",... as <u>variables for terms.</u>

<u>1.4 Atomic expressions.</u> We shall define two sorts of these: Firstly, those rows of symbols which are obtained by writing down an n-place predicate variable P and, after it, n (not necessarily distinct) terms $t_1, \ldots, t_n$, thus obtaining $Pt_1 \ldots t_n$. Taking $n = 0$, we see that every proposition variable (i.e. no-place predicate variable) is also an atomic expression. Secondly, all rows of symbols which are obtained by writing down the equality sign between two terms are to be atomic expressions. Thus, all rows of symbols of the form $t_1 = t_2$ are atomic expressions.

Examples of atomic expressions. If P_0 is a no-place, P a one-place, Q a two-place and R a three-place predicate variable, then, with the aid of the examples of terms given in 1.3, we obtain, for example, the following atomic expressions: P_0, Pffx, Qxy, Rxyz, Rxxz, Rgyffxyfx, x = x, x = y, y = x.

The atomic expressions serve as a jumping-off point for the calculus by means of which we introduce the expressions.

<u>1.5 Expressions.</u> The <u>expression calculus</u> is defined by means of the following four rules:

Rule 1. We are allowed to write down an arbitrary atomic expression.

Rule 2. We are allowed to pass from a row of symbols ζ (which we have already obtained) to the row of symbols $\neg\zeta$.

Rule 3. We are allowed to pass from two (not necessarily distinct) rows of symbols ζ_1, ζ_2 to the row of symbols $(\zeta_1 \wedge \zeta_2)$.

Rule 4. We are allowed to pass from a row of symbols ζ to every row of symbols $\bigwedge x\, \zeta$, where x is an arbitrary individual variable.

<u>Definition.</u> Those rows of symbols which can be derived in the expression calculus (and only such rows) are called <u>expressions.</u>

For optical reasons, and in order to fit in better with the customary notation, we shall write

$$t_1 \neq t_2 \quad \text{for} \quad \neg\, t_1 = t_2$$

and sometimes also

$$\alpha \wedge \beta \quad \text{for} \quad (\alpha \wedge \beta)^4.$$

Examples of expressions (cf. the examples in 1.4).

P_0, Px, Qxy, Qxx, $x = y$, $\neg Px$, $\neg\neg Px$, $x = y$, $(\neg Px \wedge Qxy)$, $\neg(\neg Px \wedge Qxy)$, $(Qxx \wedge \neg(\neg Px \wedge Qxy))$, $(Px \wedge x \neq y)$, $\bigwedge x P_0$, $\bigwedge x Px$, $\bigwedge x Qxy$, $\bigwedge z Qxy$, $\bigwedge x \bigwedge y Qxy$, $\bigwedge x\, x = x$, $\bigwedge x \bigwedge x\, x = x$, $\neg(\bigwedge y Qxy \wedge \bigwedge x(Qxx \wedge \neg Px))$.

In general, it is clearly true that:

Every atomic expression is an expression. If ζ is an expression, then so is $\neg\zeta$. If ζ_1, ζ_2 are expressions, then so is $(\zeta_1 \wedge \zeta_2)$. If ζ is an expression and x an individual variable, then $\bigwedge x\zeta$ is an expression.

We use the Greek letters α, β, γ, δ, ε, on occasion with indices, as variables for expressions.

Moreover, we use the abbreviations (cf. also Chap. I, § 6 and 7):

$(\alpha \to \beta)$ for $\neg(\alpha \wedge \neg\beta)$

$(\alpha \vee \beta)$ for $\neg(\neg\alpha \wedge \neg\beta)$

$(\alpha \leftrightarrow \beta)$ for $((\alpha \to \beta) \wedge (\beta \to \alpha))$

$\bigvee x\alpha$ for $\neg\bigwedge x \neg\alpha$

$(\alpha_1 \wedge \alpha_2 \wedge \alpha_3)$ for $((\alpha_1 \wedge \alpha_2) \wedge \alpha_3)$ (bracketing from the left; correspondingly where there are more conjuncts)

$(\alpha_1 \vee \alpha_2 \vee \alpha_3)$ for $((\alpha_1 \vee \alpha_2) \vee \alpha_3)$

Expressions of the form $Pt_1 \ldots t_r$ are called predicative expressions, expressions of the form $t_1 = t_2$ are called equations. The transition from α to $\neg\alpha$ is called negating α, and $\neg\alpha$ is called the negation of α. The transition from α and β to $(\alpha \wedge \beta)$ is called conjunctive connection of α and β, and $(\alpha \wedge \beta)$ is called the conjunction of α and β. The transition from α to an expression $\bigwedge x\alpha$ is called generalising[5] α (or, sometimes, generalisation of α), and $\bigwedge x\alpha$ is called a generalisation of α.

---

4 We want, accordingly, to omit the outermost brackets of expressions occasionally, except in systematic investigations.

5 An expression α may be generalised in different ways, according to the choice of the individual variable x. Note that we do not require x to occur in α.

An expression is called a negation, a conjunction or a generalisation according to whether it begins with ¬, (or ⋀ respectively[6]. Noting that the atomic expressions begin with a predicate variable or a function variable according to whether they are predicate expressions or equations, we see that every expression is either a predicate expression, or an equation, or a negation, a conjunction or a generalisation.

1.6 Finite alphabet. We want to show that the countably many function and predicate variables can be built up from a finite alphabet. In order to do this, we take four further symbols ○, □, | and *[7] besides (,), ¬, ∧, ⋀. A sequence of strokes (consisting of at least one stroke) is to be called a distinguishing index and a sequence of asterisks (consisting of at least one asterisk) a place index.

A row of symbols is to be called a predicate variable or a function variable respectively if it begins with ○ or □ respectively and is followed by a place index and, finally, a distinguishing index. The number of places of a variable is to be one less than the number of asterisks which appear in it. Thus, for example, ○**|||| is a one-place predicate variable and □*| a no-place function variable (individual variable).

If the primitive symbols are chosen in this way, the assumptions of 1.2 clearly hold[8].

Whether or not we take a finite alphabet as a basis, there are countably many individual variables, countably many terms and countably many expressions.

Exercises. 1. Determine which of the rows of symbols given here are terms or expressions (cf. 1.6):

□|, □***||, □**|*, ○*|, ⊂○*|, □***||□*|□*|, □*|| = □**|,
(○*||| ∧ ○**|□*||), (⋀□*| ○*| ∧ ○*|), ⋀□*|| = ○**|□*.

2. Following on from 1.6, show that the language of predicate logic can be built up from two symbols.

3. In 1.5, brackets are used to build up conjunctions. We can also do without brackets, by writing $\wedge\alpha\beta$ instead of $(\alpha \wedge \beta)$ ("Polish notation"). Determine whether, in this notation, the following rows of symbols are expressions:

∧∧○*| ○*| ○*|, ∧○*| ∧○*| ○*|, ∧○ |*○*| ∧○*|,
¬∧¬¬○*|¬∧○*|¬○*| ○*|.

---

[6] Remember that the outermost brackets of an expression may only be omitted in nonsystematic investigations.

[7] Here, once again, ○, □, | and * are to be understood as names in the metalanguage. The primitive symbols themselves can (as in a remark at the end of 1.2) be taken as the corresponding boldface symbols **○**, **□**, **|** and *****.

[8] Note that the length of a row of symbols, introduced in 1.2, is equal to the number of primitive symbols in it. Thus, every individual variable, and therefore also □*|, has length 1.

§2. Elementary questions of decidability

2.1 Statement of the problem. Let $\mathfrak{S}$ be the set of *all* rows of symbols which can be generated from a given stock of primitive symbols. Let C be a calculus by means of which rows of symbols from $\mathfrak{S}$ can be generated. Let $\mathfrak{D}$ be the set of all rows of symbols which can be derived by the rules of C. Let $\mathfrak{B}$ be the set of the rows of symbols of $\mathfrak{S}$ which do not belong to $\mathfrak{D}$.

There is a natural way of showing that a row ζ of symbols of $\mathfrak{S}$ belongs to $\mathfrak{D}$, namely by giving a derivation of ζ by the rules of C. In fact, we can generate systematically all the possible derivations by the rules of C, and thus obtain a systematic procedure which produces successively all the elements of $\mathfrak{D}$ in an order determined by the procedure. However, as we know today, we cannot, in general, do this for the set $\mathfrak{B}$. In fact, we can do this for $\mathfrak{B}$ if and only if there is a general procedure by means of which, for every ζ in $\mathfrak{S}$, we can decide in finitely many steps whether $\zeta \in D$ or $\zeta \in \mathfrak{B}$. In a case like this, the set $\mathfrak{D}$ and the calculus C which generates it are called *decidable.*

In this section we want, among other things, to show that the term calculus and the expression calculus are decidable.

2.2 Inversion principle. Let C be an *arbitrary* calculus and ζ an arbitrary row of symbols. If ζ is derivable by the rules of C, then, in general, it is derivable in different ways. In every derivation of ζ, in the last step, one of the rules of C leads to ζ. In general, we cannot say which of the rules of C does this. However, it may be that we know a property of ζ which excludes certain rules straight away from being the "last rule leading to ζ". In a particularly favourable case, it may be that there is some property of ζ which means that there is only one rule R of C which can be the last rule in a proof of ζ. In such a case, the conditions for the applicability of the rule R must be fulfilled; thus we have the possibility of ascertaining further properties of ζ. In considerations like this, the rule R is, so to speak, applied "backwards". We therefore speak (following Lorenzen) of an *inversion.*

In the following considerations, such inversions will be used frequently.

2.3 Classification of terms and expressions. A *term* is called *simple* or *compound* according to whether it begins with a no-place or with a more than no-place function variable. A term is simple if and only if it can be obtained by rule 1 of the term calculus (1.3). Thus, a term is simple if and only if it is an individual variable. A term is compound if and only if it can be obtained by rule 2 of the term calculus. A simple term has length 1 and a compound term has length greater than 1. Every term is either simple or compound.

An expression is called (cf. also 1.5)

a predicative expression or predicative if it begins with a predicate variable;

an equation if it begins with a function variable;

a negation if it begins with $\neg$;

a conjunction if it begins with (;

a generalisation if it begins with $\bigwedge$.

Every expression belongs to exactly one of the above-named categories. Because of 1.2 (d), (a), (b) we can decide to which category a given expression belongs.

An expression can be obtained by rule 1 (of the expression calculus; see 1.5) if and only if it is a predicative expression or an equation; by rule 2 if and only if it is a negation; by rule 3 if and only if it is a conjunction; by rule 4 if and only if it is a generalisation.

2.4 Decidability of the property of being a term. We can decide whether or not a row of symbols ζ is a simple term: By 2.3, ζ is a simple term if and only if ζ is an individual variable. By 1.2 (d), (a) and (b), we can decide whether or not this is the case.

We can decide whether or not a row of symbols ζ is a term. We show this by induction on $L(\zeta)$:

If $L(\zeta) = 1$ then, by 2.3, ζ is a term if and only if ζ is a simple term; and, as we have just seen, this is decidable.

If $L(\zeta) > 1$, then ζ is a term if and only if ζ is a compound term. ζ is a compound term if and only if there is an r-place function variable f (with $r \geqslant 1$) and r terms $t_1, \ldots, t_r$ with $\zeta \equiv f t_1 \ldots t_r$. By 1.2 (d), (a) and (b), we can decide whether ζ has the form $f\zeta'$, where f is a more than no-place function variable. If this is the case, find all the possible ways (there are only finitely many!) of representing ζ' as $\zeta' \equiv \zeta_1 \ldots \zeta_r$. ζ is a term if and only if there is a decomposition like this in which $\zeta_1, \ldots, \zeta_r$ are terms. By the induction hypothesis and since $L(\zeta_1), \ldots, L(\zeta_r) < L(\zeta)$, we can decide whether or not this is the case for any given decomposition.

2.5 Decidability of the property of being an expression. We can decide whether or not a row of symbols ζ is an atomic expression: ζ is an atomic expression if and only if ζ is an equation or a predicative expression. ζ is an equation if and only if ζ can be written in the form $\zeta_1\zeta_2\zeta_3$, where ζ_1 and ζ_3 are terms and $\zeta_2 \equiv =$. Clearly, we can decide whether or not such a decomposition exists. ζ is a predicative expression if and only if there is an r-place predicate variable P and r terms $t_1, \ldots, t_r$ such that

$\zeta \equiv Pt_1 \ldots t_r$. We can decide whether or not this is the case (cf. the considerations of 2.4!).

We can decide whether or not a row of symbols ζ is an expression. Proof by induction on $L(\zeta)$:

$L(\zeta) = 1$. ζ can only be an expression if ζ is an atomic expression, and this is decidable.

$L(\zeta) > 1$. ζ is an expression if and only if ζ is an atomic expression, a negation, a conjunction or a generalisation. We can decide whether or not ζ is an atomic expression. ζ is a negation if and only if $\zeta \equiv \neg\zeta'$, where ζ' is an expression. Since $L(\zeta') < L(\zeta)$, we can (by induction hypothesis) decide this. ζ is a conjunction if and only if there is a decomposition $\zeta \equiv (\zeta_1 \wedge \zeta_2)$, where ζ_1 and ζ_2 are expressions. There are only finitely many possible decompositions of the given form. Since $L(\zeta_1)$, $L(\zeta_2) <$ $< L(\zeta)$, we can, in each case, decide whether ζ_1 and ζ_2 are expressions. We prove the assertion similarly for the case where ζ is a generalisation.

2.6 Unique decomposition of sequences of terms. A term cannot be a proper initial segment of another term: We are to show that, for arbitrary t, t', there is no row of symbols ζ with $t\zeta \equiv t'$[9]. Instead we prove (because the proof is technically simpler) the impossibility of

$$t\zeta \equiv t' \quad \text{or} \quad t'\zeta \equiv t$$

by induction on $L(t)$.

If $L(t) = 1$, then t is an individual variable x. Thus t' must also begin with x, and, consequently, we must have $t' \equiv x$. Thus we should, in both cases, have $x\zeta \equiv x$, which is impossible since the lengths of the two rows of symbols are different.

If $L(t) > 1$, then t is a compound term, so that $t \equiv ft_1 \ldots t_r$, where f is an r-place function variable ($r \geqslant 1$). Thus, in both cases, t' also begins with f, so that $t' \equiv ft'_1 \ldots t'_r$. Thus we obtain

$$t_1 \ldots t_r\zeta \equiv t'_1 \ldots t'_r \quad \text{or} \quad t'_1 \ldots t'_r\zeta \equiv t_1 \ldots t_r .$$

In both cases, either $t_1 \equiv t'_1$ or t_1 is a proper initial segment of t'_1 or t'_1 a proper initial segment of t_1. By the induction hypothesis and since $L(t_1) < L(t)$, the last two alternatives are impossible. Thus we have

$$t_2 \ldots t_r\zeta \equiv t'_1 \ldots t'_r \quad \text{or} \quad t'_1 \ldots t'_r\zeta \equiv t_1 \ldots t_r .$$

---

[9] Note that we have agreed that a row of symbols ζ must contain at least one primitive symbol.

If we continue in this way, we finally obtain

$$t_r\zeta \equiv t'_r \quad \text{or} \quad t'_r\zeta \equiv t_r,$$

which is impossible by the induction hypothesis since $L(t_r) < L(t)$.

From the assertion which we have just proved we see immediately that:

(*) If $t_1 \ldots t_r \equiv t'_1 \ldots t'_s$, then $r = s$ and $t_1 \equiv t'_1, \ldots, t_r \equiv t'_r$.

(**) If $t_1\zeta_1 \equiv t_2\zeta_2$, then $t_1 \equiv t_2$ and $\zeta_1 \equiv \zeta_2$.

2.7 Unique decomposition of sequences of expressions. An expression cannot be a proper initial segment of another expression.

We show the impossibility of

$$\alpha\zeta \equiv \alpha' \quad \text{or} \quad \alpha'\zeta \equiv \alpha$$

by induction on $L(\alpha)$.

If $L(\alpha) = 1$, then α is a proposition variable P. Then α' also begins with P, and, consequently, we also have $\alpha' \equiv P$. Thus, in both cases, we should have $P\zeta \equiv P$, which is impossible (compare the lengths!).

Let $L(\alpha) > 1$. α belongs to one of the categories named in 2.3. It can be seen that, in both cases, α' belongs to the same category as α. Thus we have the following cases:

α, α' are predicative expressions: $\alpha \equiv P t_1 \ldots t_r$, $\alpha' \equiv P t'_1 \ldots t'_r$ (clearly with the same P). Because of $L(\alpha) > 1$ we have $r \geqslant 1$. After eradicating P and by repeated application of 2.6 (**), we obtain

$$t_r\zeta \equiv t'_r \quad \text{or} \quad t'_r\zeta \equiv t_r,$$

which contradicts the main result of (2.6).

α, α' are equations: $\alpha \equiv t_1 = t_2$, $\alpha' \equiv t'_1 = t'_2$. We prove the assertion as in the above case.

α, α' are negations: $\alpha \equiv \neg\alpha_1$, $\alpha' \equiv \neg\alpha'_1$. Then we obtain $\alpha_1\zeta \equiv \alpha'_1$ or $\alpha'_1\zeta \equiv \alpha_1$ which contradicts the induction hypothesis since $L(\alpha_1) < L(\alpha)$.

α, α' are conjunctions: $\alpha \equiv (\alpha_1 \wedge \alpha_2)$, $\alpha' \equiv (\alpha'_1 \wedge \alpha'_2)$. Then we obtain

$$\alpha_1 \wedge \alpha_2)\zeta \equiv \alpha'_1 \wedge \alpha'_2) \quad \text{or} \quad \alpha'_1 \wedge \alpha'_2)\zeta \equiv \alpha_1 \wedge \alpha_2).$$

By the induction hypothesis and because $L(\alpha_1) < L(\alpha)$, it follows that $\alpha_1 \equiv \alpha_1'$.

Thus we have

$$\alpha_2)\zeta \equiv \alpha_2') \quad \text{or} \quad \alpha_2')\zeta \equiv \alpha_2) .$$

Because $L(\alpha_2) < L(\alpha)$, $\alpha_2 \equiv \alpha_2'$, so that, in both cases, $)\zeta \equiv)$, which is impossible because the lengths of the two rows of symbols are different.

α, α' are generalisations: $\alpha \equiv \bigwedge x \alpha_1$, $\alpha' \equiv \bigwedge x' \alpha_1'$. We see at once that, in both cases, $x \equiv x'$, so that the assertion can again be proved with the aid of the induction hypothesis.

From the assertion we have just proved we see at once that:

If $\alpha_1 \ldots \alpha_r \equiv \alpha_1' \ldots \alpha_s'$, then $r = s$ and $\alpha_1 \equiv \alpha_1', \ldots, \alpha_r \equiv \alpha_r'$.

2.8 Decomposition of terms and expressions. As a consequence of the above considerations we obtain the

Theorem. (a) If t is a compound term then there is a uniquely determined number $r \geqslant 1$, an r-place function variable f and r terms $t_1 \ldots t_r$ such that $t \equiv ft_1 \ldots t_r$.

(b) If α is a predicative expression, then there is a uniquely determined number $r \geqslant 0$, an r-place predicate variable P and r terms $t_1, \ldots, t_r$ such that $\alpha \equiv Pt_1 \ldots t_r$.

(c) If α is an equation, then there are uniquely determined terms t_1, t_2 such that $\alpha \equiv t_1 = t_2$.

(d) If α is a negation, then there is a uniquely determined expression α_1 with $\alpha \equiv \neg\alpha_1$.

(e) If α is a conjunction, then there are uniquely determined expressions α_1, α_2 with $\alpha \equiv (\alpha_1 \wedge \alpha_2)$.

(f) If α is a generalisation, then there is a uniquely determined individual variable x and an expression α_1 with $\alpha \equiv \bigwedge x \alpha_1$.

In every case, the rows of symbols whose existence is asserted can be found effectively for any given t or α.

Proof. We confine ourselves to the proof of (a) and (e). We need only show that the required decomposition is unique and that we can find it effectively.

to (a): If $t \equiv ft_1 \ldots t_r \equiv f't_1' \ldots t_s'$, then $f \equiv f'$ and $t_1 \ldots t_r \equiv t_1' \ldots t_s'$, from which, by 2.6, the uniqueness follows. f is the first primitive symbol of t. The number of places of f determines r (cf. 1.2b). If we put $t \equiv f\zeta$, then there are only finitely many decompositions of ζ which have the form $\zeta \equiv \zeta_1 \ldots \zeta_r$. Thus we must, in finitely many cases, test whether the given $\zeta_1, \ldots, \zeta_r$ are terms, which is decidable by 2.4.

to (e): If $\alpha \equiv (\alpha_1 \wedge \alpha_2) \equiv (\alpha_1' \wedge \alpha_2')$, then, by 2.7, $\alpha_1 \equiv \alpha_1'$ and, again by 2.7, $\alpha_2 \equiv \alpha_2'$. There are only finitely many decompositions of α of the form $\alpha \equiv (\zeta_1 \wedge \zeta_2)$. Thus we have, in finitely many cases, to test whether the given ζ_1 and ζ_2 are expressions, which is decidable by 2.5.

Exercises. 1. There is a procedure by which, given an arbitrary row of symbols, we can determine in finitely many steps whether or not the row of symbols is a conjunction (of expressions).

2. There is no row of symbols which is both a term and an expression.

3. For any terms t_1, t_2 and any expression α, $t_1 \alpha t_2$ is not an expression.

4. Let a calculus be determined by the following rules:

Rule 1. We may write down one stroke.

Rule 2. We may add a stroke to the end of a row of symbols which has already been obtained.

Rule 3. We may add a circle to the end of a row of symbols which has already been obtained and which ends with a stroke.

Using the inversion principle, prove that $| \bigcirc \bigcirc | \bigcirc |$ cannot be obtained by means of the above rules.

5. Verify that the inversion principle has been used in the solution of exercises 1 and 3 of §1.

§3. Proofs and definitions by induction on the structure of the expressions.

3.1 Analogy to arithmetic. The theory of the natural numbers can be established on the basis of Peano's axiom system (for example, as it is given in Chap. I, §3.7 in the form P'1, P'2, P'3). Now the theory of the expressions of predicate logic can be understood as a sort of generalised arithmetic, in which the atomic expressions correspond to nought. Where, with the natural numbers, we had the successor function, we now have functions with names n, c, g_x, which are defined as follows:

(a) Negating α : $n(\alpha) \equiv \neg\alpha$.

(b) Conjunctive connection of α and β : $c(\alpha,\beta) \equiv (\alpha \wedge \beta)$.

(c) Generalisation of α by means of the individual variable x : $g_x(\alpha) \equiv \bigwedge x\, \alpha$.

In analogy to Peano's axioms, the following axioms are valid:

P*1a $n(\alpha)$ is not an atomic expression.

P*1b $c(\alpha,\beta)$ is not an atomic expression.

P*1c $g_x(\alpha)$ is not an atomic expression.

P*2a If $n(\alpha) \equiv n(\beta)$, then $\alpha \equiv \beta$.

P*2b If $c(\alpha,\beta) \equiv c(\gamma,\delta)$, then $\alpha \equiv \gamma$ and $\beta \equiv \delta$.

P*2c If $g_x(\alpha) \equiv g_x(\beta)$, then $\alpha \equiv \beta$.

P*2d It is never true that: $n(\alpha) \equiv c(\beta,\gamma)$ or $n(\alpha) \equiv g_x(\beta)$ or $c(\alpha,\beta) \equiv g_x(\gamma)$ or $g_x(\alpha) \equiv g_y(\beta)$ (for $x \not\equiv y$).

P*3 (Inductions): Let P be an arbitrary property. Then every expression α has the property P if:

(1) every atomic expression has the property P, and

(2) for arbitrary α, β, x:
if α has the property P, then so does $n(\alpha)$,
if α, β have the property P, then so does $c(\alpha,\beta)$, and
if α has the property P, then so does $g_x(\alpha)$.

If a proof is carried out with the help of P*3, then we speak of an induction on the structure of α.

The axioms P*1a, P*1b, P*1c correspond to Peano's axiom P'1, the axioms P*2a,...,P*2d to the axiom P'2 and the axiom P*3 to the axiom P'3. For the validity of P*1a, P*1b, P*1c and P*2d cf. 2.3. P*2a is valid because of 2.8 (d), P*2b because of 2.8 (e) and P*2c because of 2.8 (f).

In order to prove P*3, we assume that P is a property which fulfils the conditions (1) and (2). We must show that every expression α has the property P. It suffices to prove that, for every number n, all expressions with rank n (see 1.2) have the property P. We can show this by induction on the rank n. In order to do this, we have to show that: If all expressions of rank $< n$ have the property P (induction hypothesis), then every expression of rank n also has the property P[10].

Now let α be a given expression with rank n. We distinguish the following cases:

(a) α is atomic. Then, by (1), α has the property P.

(b) $\alpha \equiv n(\beta)$. Then $R(\beta) < n$ (because α contains one negator more than β). Thus, β has the property P. Then, by (2), α has the property P.

(c) $\alpha \equiv c(\beta,\gamma)$. Then $R(\beta) < n$ and $R(\gamma) < n$. Thus, β and γ have the property P. Then, by (2), α has the property P.

(d) $\alpha \equiv g_x(\beta)$. Then, $R(\beta) < n$. Thus, β has the property P. Then, by (2), α has the property P.

---

[10] This is a variant of induction which is well known in arithmetic. Note that the case $n = 0$ is also covered by the given formulation.

3.2 Inductive definitions. Let us consider, in arithmetic, a system of functional equations for an unknown $(n+1)$-place function f. This is to consist of two equations. The first equation is to define $f(x_1,\ldots,x_n,0)$ for every n-tuple $x_1,\ldots,x_n$ of parameters. The second equation is to give $f(x_1,\ldots,x_n,y')$ in terms of $x_1,\ldots,x_n$, y and $f(x_1,\ldots,x_n,y)$. In arithmetic we can prove, on the basis of Peano's axioms, that there is one and only one function f which satisfies the given system of functional equations. We say that f is inductively defined by the two equations.

By analogy to arithmetic we can, in the generalised arithmetic given above, consider a system of functional equations for an unknown $(n+1)$-place function f which is defined for expressions as arguments. The system of equations is to consist of four equations.

The first equation is to determine $f(\alpha_1,\ldots,\alpha_n,\beta)$ for arbitrary expressions $\alpha_1,\ldots,\alpha_n$ and for every atomic expression β.

The second equation is to give $f(\alpha_1,\ldots,\alpha_n, n(\beta))$ in terms of $\alpha_1,\ldots,\alpha_n$, β and $f(\alpha_1,\ldots,\alpha_n,\beta)$.

The third equation is to give $f(\alpha_1,\ldots,\alpha_n, c(\alpha,\beta))$ in terms of $\alpha_1,\ldots,\alpha_n$, α, β, $f(\alpha_1,\ldots,\alpha_n,\alpha)$ and $f(\alpha_1,\ldots,\alpha_n,\beta)$.

The fourth equation is to give $f(\alpha_1,\ldots,\alpha_n, g_x(\beta))$ in terms of $\alpha_1,\ldots,\alpha_n$, x, β and $f(\alpha_1,\ldots,\alpha_n,\beta)$.

We say that such a function f is defined by induction on the structure of the expressions.

The proof of the existence and uniqueness of such functions f can be conducted using methods similar to those used in arithmetic. We shall not carry out the proof here.

Exercises. 1. To each symbol (cf. 1.2) we assign a whole number: 1 if the symbol is the left bracket, -1 if it is the right bracket and 0 otherwise. If we run through the symbols of a row of symbols ζ from left to right then we obtain, by this method, a sequence of the numbers 0, 1, -1. From this sequence we form a new sequence $\Phi(\zeta)$, by replacing each element of the original sequence by the sum of the elements which occur up to that point. $\Phi(\zeta)$ has $L(\zeta)$ members (for $L(\zeta)$ see 1.2). Show that: If ζ is an expression, then $\Phi(\zeta)$ has no negative elements and the last element of $\Phi(\zeta)$ is 0.

2. As in exercise 1, we assign a whole number to each symbol, but this time 2 if the symbol is the left bracket, -1 if it is the right bracket or the conjunctor and 0 otherwise. Thus, as before, we can assign a sequence $\Psi(\zeta)$ of whole numbers to every row of symbols ζ. Show that: If ζ is an expression, then $\Psi(\zeta)$ contains no negative elements and the last element of $\Psi(\zeta)$ is 0.

3. By induction on the structure of the expressions, we can assign to every expression α the set of the subexpressions of α, as follows: If α is atomic, then α is the only subexpression of α. If $\alpha \equiv \neg\beta$ or $\alpha \equiv \bigwedge x\beta$, then α and the subexpressions of β are the subexpressions of α. Finally, if $\alpha \equiv (\beta \wedge \gamma)$, then α and the subexpressions of β and of γ are the subexpressions of α. Let c, n and g be the number of conjunctors, nega-

tors and universal quantifiers respectively which occur in α. Show that α has at most $2c + n + g + 1$ subexpressions.

4. By induction on the structure of the expressions, we can assign to every expression α the set of the subexpressions of α localised with respect to α. The localised subexpressions are rows of symbols which are built up from the symbols given in 1.2 together with the symbols 0, 1, 2, which are used for localising. If α is atomic, then α is the only subexpression of α localised with respect to α. If $\alpha \equiv \neg\beta$ or $\alpha \equiv \bigwedge x\beta$, then the subexpressions of α localised with respect to α are the following: (1) α, (2) 0λ, where λ is a subexpression of β localised with respect to β. If $\alpha \equiv (\beta \wedge \gamma)$, then the subexpressions of α localised with respect to α are the following: (1) α, (2) 1λ, where λ is a subexpression of β localised with respect to β, (3) 2μ, where μ is a subexpression of γ localised with respect to γ. Show (cf. exercise 3) that α has exactly $2c + n + g + 1$ subexpressions localised with respect to α.

§4. Free and bound variables

4.1 Introduction. Let us take as an example the expression

$$(\bigwedge x(Qx \wedge \neg Rxy) \wedge \neg \bigwedge y Syz),$$

where we assume that x, y, z are pairwise distinct individual variables. Let us first direct our attention to the variable x. This appears after the universal quantifier as well as after Q and after R. We shall say that the variable x in the latter two positions is in the scope of the universal quantifier and is bound by this universal quantifier. In the same way, the variable y which appears after S is, in this position, bound by the universal quantifier which precedes y. By contrast, the variable y which appears after x is not in the scope of a universal quantifier; it is not bound, and is thus free. The same holds for the variable z. In the expression we are considering, the variable x occurs only bound, the variable y occurs both bound and free and the variable z occurs only free. Every individual variable which is different from x, y and z occurs neither free nor bound in the expression.

By far the most important of the concepts which we have just introduced in an intuitive way is that of the free occurrence of an individual variable x in an expression α. We shall define this concept here.

In order to do this, we lay down the rules of a calculus which we shall call the calculus of free occurrence. Certain rows of symbols of the form $x\alpha$ are derivable in this calculus. If $x\alpha$ is derivable in the calculus of free occurrence, we shall say that the individual variable x occurs free in the expression α.

We shall, simultaneously, lay down the rules of another calculus in which a row of symbols of the form αx is derivable if and only if x does not occur free in α. We shall call this calculus the calculus of the negation of free occurrence.

4.2 The rules of the calculi

Calculus of free occurrence

Rule 1. If α is atomic and x occurs in α, then we may write down $x\alpha$.

Rule 2. We may pass from $x\alpha$ to $x\neg\alpha$.

Rule 3a. We may pass from $x\alpha$ to $x(\alpha \wedge \beta)$.

Rule 3b. We may pass from $x\beta$ to $x(\alpha \wedge \beta)$.

Rule 4. We may pass from $x\alpha$ to $x \bigwedge y\alpha$ if $x \not\equiv y$.

Calculus of the negation of free occurrence

Rule 1. If α is atomic and x does not occur in α, then we may write down αx.

Rule 2. We may pass from αx to $\neg\alpha x$.

Rule 3. We may pass from αx and βx to $(\alpha \wedge \beta)x$.

Rule 4a. We may pass from αx to $\bigwedge y\alpha x$.

Rule 4b. We may write down $\bigwedge x\alpha x$.

Examples. Let P, Q and f be one-place and let $x \not\equiv y$. In the calculus of free occurrence, we can derive:

$$xPx,\ xPfx,\ yQy,\ x\neg Px,\ x(Px \wedge Qy),\ y(Px \wedge Qy),\ y\bigwedge x(Pfx \wedge Qy)\ ;$$

in the calculus of the negation of free occurrence, we can derive:

$$Pxy,\ Pyx,\ Pfxy,\ Qyx,\ (Py \wedge Qy)x,\ \bigwedge xPfxx,\ \bigwedge yQyx.$$

4.3 Decidability. In accordance with the intention with which we set up the above calculi, we prove the

Theorem. For every x and every α, either $x\alpha$ is derivable in the calculus K of free occurrence or αx is derivable in the calculus K' of the negation of free occurrence. We can decide effectively which of these two cases holds.

Proof. We prove the assertion for x and α; we may assume that the assertion has already been proved for x and for all expressions γ with $L(\gamma) < L(\alpha)$. First of all we ascertain whether α is atomic, a negation, a conjunction or a generalisation.

(a) If α is atomic, then either x occurs in α or it does not. By 1.2, assumption (d), we can ascertain effectively which of these cases holds. If the first case holds, $x\alpha$ is derivable in K; if the second case holds, αx is derivable in K'. It cannot happen that both $x\alpha$ is derivable in K and αx is derivable in K'. For, if $x\alpha$ is derivable in K, then this must be by means of rule 1 (inversion!) since α is atomic. But the condition for rule 1 to be applied is that x occurs in α. In the same way, we see that x does not occur in α if αx is derivable in K'. Thus, the assertion is proved.

(b) If α is a negation, then there is a β with $\alpha \equiv \neg\beta$. Since $L(\beta) < L(\alpha)$, by hypothesis either $x\beta$ is derivable in K or βx is derivable in K', and we can decide effectively which of these two cases holds. If $x\beta$ is derivable in K, then, by rule 2, $x\alpha$ is also derivable in K. If βx is derivable in K', then, by rule 2, αx is also derivable in K'. It is not possible for both $x\alpha$ to be derivable in K and αx to be derivable in K'. For, if $x\alpha$ were derivable in K and αx in K', then, in both cases, this must be by means of rule 2 (inversion!), so that $x\beta$ would be derivable in K and βx in K', which contradicts the hypothesis. We prove the assertion similarly in the remaining cases.

4.4 Free occurrence of function and predicate variables. We have defined above what we mean by saying that an individual variable occurs free in an expression. In ordinary predicate logic, in contrast to the situation in second-order predicate logic, the individual variables are the only variables which can be bound by a quantifier. Thus we shall say that a function variable with more than 0 places or a predicate variable occurs free in an expression α if the variable occurs in the expression at all. Incidentally, note that any individual variable x which occurs free in an expression α also occurs in α (see exercise 4).

Exercises. 1. Complete the proof of the theorem in 4.3.

2. Set up the rules for a calculus in which a row of symbols ζ is derivable if and only if ζ has the form $x\alpha$, where x occurs in α (calculus of occurrence).

3. In the same way, set up a calculus of bound occurrence.

4. Using the calculi of occurrence, of free occurrence and of bound occurrence, show that every variable which occurs free or bound in an expression α also occurs in α, and that every variable which occurs in α also occurs either free or bound in α.

5. If ζ is derivable in one of the calculi mentioned in the previous exercise, then there is exactly one x and one α such that $\zeta \equiv x\alpha$.

6. For each of the given calculi, show that it is decidable whether a row of symbols is derivable in this calculus.

§ 5. Substitution

5.1 Introduction. Substitution plays an important part in the development of logic by means of calculi. We shall, under certain conditions, assign an expression β to an expression α by substitution of a term t for an individual variable x. If $x \equiv t$ we shall have $\beta \equiv \alpha$, whereas in general β will be different from α. Not every α, x, t will possess a β such that α is transformed into β by substitution of t for x. However, if there is such a β it will be uniquely determined.

First, we shall consider a few examples; we shall use the abbreviation Subst $\alpha\, x\, t\, \beta$ to indicate that α is transformed into β by substitution of t for x. In most cases, we obtain β by starting with α and writing down the term t instead of x everywhere

where x occurs in α. If, for example, in the following, x and y are variables which are different from each other, then

$$\text{Subst } Pxyy \; x \; t \; Ptyy,$$
$$\text{Subst } Pxyy \; y \; t \; Pxtt.$$

If the variable for which we want to substitute does not occur free in α, then the substitution can always be carried out and we have $\beta \equiv \alpha$. For example:

$$\text{Subst } Px \; y \; t \; Px,$$
$$\text{Subst } \bigwedge xPx \; x \; t \; \bigwedge xPx.$$

In a rather more complicated case, we have

$$\text{Subst } (Pxy \wedge \bigwedge xPxz) \; x \; t \; (Pty \wedge \bigwedge xPxz).$$

Finally, let us consider the expression $\bigwedge xPxy$. Let x, y, z be pairwise distinct. Then we have, for example,

$$\text{Subst } \bigwedge xPxy \; y \; z \; \bigwedge xPxz,$$
$$\text{Subst } \bigwedge xPxy \; y \; y \; \bigwedge xPxy.$$

However, we shall not want to have Subst $\bigwedge xPxy \; y \; x \; \bigwedge xPxx$. One reason for this is the following[11]: In all the examples we have considered up till now, we can see that: If the variable for which we are to substitute occurs free (for this cf. the introduction to §4) in some position in the expression α with which we start, then all the variables of the term t which is to be substituted also occur free in the corresponding positions in the resulting expression β. This observation would no longer hold if Subst $\bigwedge xPxy \; y \; x \; \bigwedge xPxx$: The variable y for which we are to substitute occurs free in the expression $\bigwedge xPxy$ with which we start. x is to be substituted for y. The second variable x which appears in Pxx is, however (like the first), bound by the universal quantifier at the beginning of $\bigwedge xPxx$. In order to escape from this dilemma, we shall stipulate that, in the example we have just considered and in similar cases, the substitution cannot be carried out, i.e. that, in such a case, α, x, t do not possess any β such that Subst α x t β. Compensation for the fact that substitution cannot be carried out in such cases will be provided by a theorem which we shall prove in Chap. VII, §6.3.

5.2 The operator Δ_x^t. In order to define substitution it is convenient to introduce the operator Δ_x^t as an auxiliary concept. This operator can be applied to an arbitrary term t_0, and $\Delta_x^t t_0$ is, likewise, always a term. We define $\Delta_x^t t_0$ inductively on the structure of t_0 as following (let f be an r-place function variable):

(*)
$$\Delta_x^t z \equiv \begin{cases} t & \text{if} \quad x \equiv z \\ z & \text{if} \quad x \not\equiv z \end{cases}$$

(**)
$$\Delta_x^t f t_1 \ldots t_r \equiv f \Delta_x^t t_1 \ldots \Delta_x^t t_r .$$

Thus, $\Delta_x^t t_0$ is the term which is obtained from t_0 by substituting the term t for the variable x everywhere.

---

[11] For a further reason cf. Chap. VII, §6.1.

The following assertions hold for the operator Δ_x^t:

1) If x does not occur in t_0, then $\Delta_x^t t_0 \equiv t_0$.

2) If z occurs in t_0 and $x \not\equiv z$, then z also occurs in $\Delta_x^t t_0$.

3) If x occurs in t_0 and z in t, then z also occurs in $\Delta_x^t t_0$.

4) If y does not occur in t_0, then $\Delta_y^x \Delta_x^y t_0 \equiv t_0$.

5) If z occurs in $\Delta_x^t t_0$, then one of the following two cases holds:

   a) $x \not\equiv z$ and z occurs in t_0, or
   b) x occurs in t_0 and z in t.

The proof can be carried out by induction corresponding to the definition of Δ_x^t. Here, we confine ourselves to the proof of the assertion 5); for the proof, we assume that z occurs in $\Delta_x^t t_0$.

If t_0 is an individual variable, i.e. $t_0 \equiv u$, then $\Delta_x^t t_0 \equiv t$ or u, according to whether $x \equiv u$ or $x \not\equiv u$. In the first case, x occurs in t_0 and z in t. In the second case, $u \equiv z$ (since z occurs in u), hence $x \not\equiv z$ and, clearly, z occurs in t_0.

If $t_0 \equiv f t_1 \ldots t_r$, then we have $\Delta_x^t t_0 \equiv f \Delta_x^t t_1 \ldots \Delta_x^t t_r$. Since z occurs in $\Delta_x^t t_0$, there must be a j such that z occurs in $\Delta_x^t t_j$. But then, by induction hypothesis, we have: ($x \not\equiv z$ and z occurs in t_j) or (x occurs in t_j and z in t). This yields: ($x \not\equiv z$ and z occurs in t_0) or (x occurs in t_0 and z in t).

Note that, here, we have defined $\Delta_x^t t_0$ by induction on the structure of t_0. However, as is the case with all such definitions, we could equally well have defined Δ_x^t by means of a calculus (see exercise 1).

5.3 Substitution calculus. In order to define substitution, we proceed in a way analogous to the definition of free occurrence in the last section. We set up a calculus in which certain rows of symbols of the form $\alpha x t \beta$ are derivable. If $\alpha x t \beta$ is derivable in this calculus, then we shall say that α is transformed into β by substitution of t for x. This will be the case if and only if the relationship which we described in 5.1 holds between α, β, x and t (note in particular the distinction of cases in rule 4a and rule 4b). The rules 4b and 4a of the substitution calculus make use of the concept of free occurrence and of the negation of free occurrence respectively of an individual variable in an expression. Thus, the substitution calculus is built up on the basis of the calculus of free occurrence and the calculus of the negation of free occurrence. We can express this by saying that the substitution calculus is a layered calculus.

The substitution calculus has the following rules:

Rule 1a. If P is an n-place predicate variable, then we may write down all rows of symbols of the form

$$P t_1 \ldots t_n \, x \, t \, P \Delta_x^t t_1 \ldots \Delta_x^t t_n .$$

In particular, for n = 0, every row of symbols of the form P x t P, if P is a no-place predicate variable (a proposition variable). Note that, for example, $\Delta_x^t t_1$ is not a row of symbols which begins with "Δ", but, rather, that Δ_x^t is an operation symbol of the metalanguage. The fact that the row of symbols given above may be written down means, intuitively, that the expression $P t_1 \ldots t_n$ can be transformed into the expression $P \Delta_x^t t_1 \ldots \Delta_x^t t_n$ by substituting the term t for the individual variable x. In order to make it easier to see which parts of this row of symbols belong together, it is convenient to write $P t_1 \ldots t_n \;\; x \; t \;\; P \Delta_x^t t_1 \ldots \Delta_x^t t_n$, and correspondingly in more general cases. Note, however, that, strictly speaking, the gaps are not there.

Rule 1b. We may write down all rows of symbols of the form

$$t_1 = t_2 \;\; x \; t \;\; \Delta_x^t t_1 = \Delta_x^t t_2 .$$

Rule 2. We may pass from a row of symbols of the form

$$\alpha \, x \, t \, \beta$$

to the row of symbols

$$\neg\alpha \, x \, t \, \neg\beta .$$

Rule 3. We may pass from two rows of symbols of the forms

$$\alpha_1 \, x \, t \, \beta_1$$

$$\alpha_2 \, x \, t \, \beta_2$$

to the row of symbols

$$(\alpha_1 \wedge \alpha_2) \, x \, t \, (\beta_1 \wedge \beta_2) .$$

Rule 4a. If x does not occur free in $\bigwedge z\alpha$, then we may write down the row of symbols

$$\bigwedge z\alpha \;\; x \, t \, \bigwedge z\alpha .$$

Rule 4b. If x occurs free in $\bigwedge z\alpha$ and if z does not occur in t, then we may pass from the row of symbols

$$\alpha \, x \, t \, \beta$$

to the row of symbols

$$\bigwedge z \alpha \, x t \bigwedge z \beta \, .$$

We shall say that α is transformed into β by substitution of t for x (abbreviated as Subst $\alpha\, x t \beta$) if the row of symbols $\alpha\, x t \beta$ is derivable in the substitution calculus.

Clearly, the following <u>inversion theorem</u> holds: The row of symbols $\alpha\, x t \beta$ can be derived in the substitution calculus only by means of

rule 1a if α is a predicative expression,

rule 1b if α is an equation,

rule 2 if α is a negation,

rule 3 if α is a conjunction,

rule 4a if α is a generalisation and x does not occur free in α,

rule 4b if α is a generalisation and x occurs free in α.

Hence, in particular: if Subst $\alpha\, x t \beta$ and, in addition,

(i) $\alpha \equiv \neg \alpha_1$, then there is a β_1 with Subst $\alpha_1 x t \beta_1$ and $\beta \equiv \neg \beta_1$.

(ii) $\alpha \equiv (\alpha_1 \wedge \alpha_2)$, then there is a β_1 and a β_2 with Subst $\alpha_1 x t \beta_1$ and Subst $\alpha_2 x t \beta_2$ and $\beta \equiv (\beta_1 \wedge \beta_2)$.

(iii) $\alpha \equiv \bigwedge z \alpha_1$ and x occurs free in α, then there is a β_1 with Subst $\alpha_1 x t \beta_1$ and $\beta \equiv \bigwedge z \beta_1$.

<u>5.4 Theorems about substitution.</u> We note the following theorems here for use later; we shall prove them in 5.5.

<u>Theorem 1.</u> For every α, x, t there is at most one β with Subst $\alpha\, x t \beta$.

<u>Theorem 2.</u> If α is atomic, then, for every x and t, there is exactly one β with Subst $\alpha\, x t \beta$.

<u>Theorem 3.</u> Subst $\alpha\, x x \alpha$ always holds.

<u>Theorem 4.</u> If x does not occur free in α, then Subst $\alpha\, x t \alpha$ always holds.

<u>Theorem 5.</u> If Subst $\alpha\, x t \beta$ and if x occurs free in α, then t is a component of β and every individual variable which occurs in t occurs free in β.

Theorem 6. If y does not occur in α, then there is a (and hence, by theorem 1, exactly one) β with Subst $\alpha\, x\, y\, \beta$ and Subst $\beta\, y\, x\, \alpha$.

Theorem 7. Let Subst $\alpha\, x\, t\, \beta$ and let z be an arbitrary individual variable. Then z occurs free in β if and only if at least one of the following two cases holds:

(i) z occurs free in α and $z \not\equiv x$.
(ii) x occurs free in α and z occurs in t.

Theorem 8. If Subst $\alpha\, x\, t\, \beta$ and x does not occur in t, then x does not occur free in β.

Theorem 9. If Subst $\alpha\, x\, y\, \beta$, then y occurs free in β if and only if x occurs free in α or y occurs free in α.

5.5 Proof of the above theorems

Theorem 1 can be proved by induction on the structure of α.

(1) If α is atomic, then $\alpha\, x\, t\, \beta$ can only have been derived by rule 1a or rule 1b. In both cases, β is uniquely determined by α, x, t.

(2) If $\alpha \equiv \neg\alpha_1$, then $\alpha\, x\, t\, \beta$ must have been derived by rule 2. Then there is a β_1 such that Subst $\alpha_1\, x\, t\, \beta_1$ and $\beta \equiv \neg\beta_1$. By induction hypothesis, β_1 is determined uniquely; hence, so is β.

(3) If $\alpha \equiv (\alpha_1 \wedge \alpha_2)$, then the proof follows analogously to (2).

(4) If $\alpha \equiv \bigwedge z\, \alpha_1$, then we distinguish two cases: (a) If x does not occur free in α, then $\alpha\, x\, t\, \beta$ must have been obtained by rule 4a. Thus, $\beta \equiv \alpha$, and β is therefore uniquely determined. (b) If x occurs free in α, then $\alpha\, x\, t\, \beta$ must have been derived by rule 4b. In this case, the uniqueness of β follows in a way analogous to (2).

Theorem 2. The existence follows from rule 1a or 1b. The uniqueness follows from theorem 1.

Theorem 3 and Theorem 4 can be proved easily by induction on the structure of α.

Theorem 5 can be proved by induction on the structure of α as follows:

(1) If $\alpha \equiv P\, t_1 \ldots t_r$ (the proof for the case $\alpha \equiv t_1 = t_2$ is similar), then

$$\beta \equiv P\, \Delta_x^t t_1 \ldots \Delta_x^t t_r .$$

If x occurs free in α, then there is a j such that x occurs in t_j. Then, by 5.2 3), every individual variable which occurs in t also occurs in $\Delta_x^t t_j$ and thus occurs free in β. Moreover, t is a component of β.

(2) If $\alpha \equiv \neg\alpha_1$, then 5.3 (i) holds. If x occurs free in α, then it also occurs free in α_1. Thus, by induction hypothesis, every individual variable which occurs in t occurs free in β_1 and therefore also in β. Moreover, t is a component of β_1, and thus also of β.

(3) If $\alpha \equiv (\alpha_1 \wedge \alpha_2)$, then 5.3 (ii) holds. If x occurs free in α, then it also occurs free in α_1 or α_2. Thus, by induction hypothesis, every individual variable which occurs in t occurs free in β_1 or in β_2, and, thus, in β. Moreover, t is a component of β_1 or of β_2, and, hence, of β.

(4) If $\alpha \equiv \bigwedge z \alpha_1$ and if x occurs free in α, then $\alpha x t \beta$ must have been derived by rule 4b and 5.3 (iii) holds. Then, clearly, x also occurs free in α_1. Thus, by induction hypothesis, every individual variable y which occurs in t occurs free in the expression β_1 which exists by 5.3 (iii). If y did not occur free in β, then we should have $y \not\equiv z$. But then z would occur free in t, which cannot be the case since rule 4b has been applied. Moreover, t is a component of β_1, and therefore also of β.

Theorem 6 can be proved by induction on the structure of α as follows:

(1) Let $\alpha \equiv P t_1 \ldots t_r$ (analogously for the case $\alpha \equiv t_1 = t_2$). We have

$$\text{Subst } P t_1 \ldots t_r \; x \; y \; P \Delta_x^y t_1 \ldots \Delta_x^y t_r$$

$$\text{and} \qquad \text{Subst } P \Delta_x^y t_1 \ldots \Delta_x^y t_r \; y \; x \; P \Delta_y^x \Delta_x^y t_1 \ldots \Delta_y^x \Delta_x^y t_r .$$

y does not occur free in $P t_1 \ldots t_r$. Thus, y does not occur in any of the t_j. Thus, for each j we have, by 5.2 4), $\Delta_y^x \Delta_x^y t_j \equiv t_j$, from which the assertion follows.

(2) Let $\alpha \equiv \neg\alpha_1$. y does not occur free in α, and therefore also not in α_1. By induction hypothesis, there is a β_1 with Subst $\alpha_1 x y \beta_1$ and Subst $\beta_1 y x \alpha_1$. We obtain the assertion by applying rule 2.

(3) Let $\alpha \equiv (\alpha_1 \wedge \alpha_2)$. Since y does not occur free in α, it does not occur free in α_1 or α_2 either. By induction hypothesis, there are expressions β_1 and β_2 with Subst $\alpha_1 x y \beta_1$, Subst $\beta_1 y x \alpha_1$, Subst $\alpha_2 x y \beta_2$, Subst $\beta_2 y x \alpha_2$. The assertion follows by application of rule 3.

(4) Let $\alpha \equiv \bigwedge z \alpha_1$. Since y does not occur in α, it does not occur in α_1 either and therefore does not occur free in α_1. We distinguish two cases:

(4a) x does not occur free in α. Then, by rule 4a, we have Subst $\bigwedge z \alpha_1$ x y $\bigwedge z \alpha_1$ and Subst $\bigwedge z \alpha_1$ y x $\bigwedge z \alpha_1$.

(4b) x occurs free in α. By induction hypothesis, there is (since y does not occur in α_1) a β_1 with Subst $\alpha_1 x y \beta_1$ and Subst $\beta_1 y x \alpha_1$. Since $y \not\equiv z$, z does not occur in the term y. By applying rule 4b, we obtain Subst $\bigwedge z \alpha_1 x y \bigwedge z \beta_1$. Hence, by theorem 5 and since x occurs free in $\bigwedge z \alpha_1$, the variable y occurs free in $\bigwedge z \beta_1$. Since x occurs free in α whereas z clearly does not, $z \neq x$ and z does not occur in the term x. By applying rule 4b, we obtain Subst $\bigwedge z\beta_1 yx \bigwedge z\alpha_1$. Thus, we have proved the assertion.

T h e o r e m 7 is a generalisation of Theorem 5. Under the assumption Subst $\alpha x t \beta$, we have to show three things:

(I) If z occurs free in α and $z \neq x$, then z occurs free in β.
(II) If x occurs free in α and z occurs in t, then z occurs free in β.
(III) If z occurs free in β, then z occurs free in α and $z \neq x$, or x occurs in α and z occurs in t.

Assertion (II) follows from Theorem 5. We prove (I) and (III) by induction on the structure of α.

to (I)

(1) Let $\alpha \equiv P t_1 \ldots t_r$. There is a j such that z occurs in t_j. Now $\beta \equiv P \Delta^t_x t_1 \ldots \Delta^t_x t_r$. By 5.2 2), z occurs in $\Delta^t_x t_j$ and therefore occurs free in β. (The proof is similar when α is an equation.)

(2) If $\alpha \equiv \neg\alpha_1$, then there is a β_1 with Subst $\alpha_1 x t \beta_1$, and $\beta \equiv \neg\beta_1$. z occurs free in α_1. Hence, by induction hypothesis, z occurs free in β_1. Hence, z also occurs free in β.

(3) If $\alpha \equiv (\alpha_1 \wedge \alpha_2)$, then there are expressions β_1 and β_2 with Subst $\alpha_1 x t \beta_1$ and Subst $\alpha_2 x t \beta_2$, and $\beta \equiv (\beta_1 \wedge \beta_2)$. z occurs free in α, and therefore it occurs free in α_1 or in α_2. Let us assume that z occurs free in α_1 (we prove the assertion analogously in the other case). Then, by induction hypothesis, z occurs free in β_1. Since $\beta \equiv (\beta_1 \wedge \beta_2)$, z also occurs free in β.

(4) Finally, let $\alpha \equiv \bigwedge u \alpha_1$. We distinguish two cases:

C a s e 1. x does not occur free in $\bigwedge u \alpha_1$. Then Subst $\alpha x t \alpha$, and hence $\beta \equiv \alpha$. Together with the assumption that z occurs free in α, this yields the assertion.

C a s e 2. x occurs free in $\bigwedge u \alpha_1$. Then u does not occur in t, and there is (by the inversion theorem in 5.3) a β_1 such that Subst $\alpha_1 x t \beta_1$ and $\beta \equiv \bigwedge u \beta_1$. Since z occurs free in $\bigwedge u \alpha_1$, z also occurs free in α_1. Thus, by induction hypothesis, z occurs free in β_1. Now $z \neq u$, since z occurs free in $\bigwedge u \alpha_1$. Hence, z occurs free in $\bigwedge u \beta_1$, i.e. in β.

to (III)

(1) Let $\alpha \equiv P t_1 \ldots t_r$. Then $\beta \equiv P \Delta^t_x t_1 \ldots \Delta^t_x t_r$. Since z occurs free in β, z occurs free in one of the $\Delta^t_x t_j$. Then, by 5.2 5), we have: ($x \neq z$ and z occurs in t_j) or (x occurs in t_j and z occurs in t). This yields: ($x \neq z$ and z occurs free in α) or (x occurs free in α and z occurs in t), as we claimed. (We prove the assertion similarly if α is an equation.)

(2) If $\alpha \equiv \neg\alpha_1$, then there is a β_1 with Subst $\alpha_1 x t \beta_1$ and $\beta \equiv \neg\beta_1$. z occurs free in β, and therefore also in β_1. By induction hypothesis, either z occurs free in α_1 and $z \not\equiv x$ or x occurs free in α_1 and z occurs in t. The assertion for α follows immediately.

(3) If $\alpha \equiv (\alpha_1 \wedge \alpha_2)$, then there are expressions β_1 and β_2 with Subst $\alpha_1 x t \beta_1$, Subst $\alpha_2 x t \beta_2$ and $\beta \equiv (\beta_1 \wedge \beta_2)$. z occurs free in β, and therefore also in β_1 (or in β_2, in which case we can prove the assertion similarly). By induction hypothesis, either z occurs free in α_1 and $z \not\equiv x$ or x occurs free in α_1 and z occurs in t. The assertion for α follows immediately.

(4) If $\alpha \equiv \bigwedge u \alpha_1$, then we distinguish two cases:

Case 1: x does not occur free in $\bigwedge u \alpha_1$. Then $\beta \equiv \alpha$. Thus z occurs free in α and, since x does not occur free in α, $z \equiv x$. Thus we have proved the assertion, since we have proved its first disjunct.

Case 2: x occurs free in $\bigwedge u \alpha_1$. Then there is a β_1 with Subst $\alpha_1 x t \beta_1$ and $\beta \equiv \bigwedge u \beta_1$. Since z occurs free in β, z also occurs free in β_1. Hence, by induction hypothesis, we have: z occurs free in α_1 and $z \not\equiv x$, or x occurs free in α_1 and z occurs in t. In the first case, z also occurs free in α (note that, since z occurs free in β, $z \not\equiv u$), and in the second case x occurs free in α (note that, since x occurs free in $\bigwedge u \alpha_1$, $x \not\equiv u$). In both cases, the assertion for case 2 follows from the induction hypothesis.

Theorem 8 can be obtained as a special case of Theorem 7 by putting $z \equiv x$ there. If z occurred free in β, then (i) or (ii) would have to hold. However, (i) is false because $z \equiv x$ and (ii) is false because x does not occur free in t.

Theorem 9 can also be obtained from Theorem 7: For, putting $t \equiv z \equiv y$, theorem 7 states that y occurs free in β if and only if

(*) (i) y occurs free in α and $y \not\equiv x$ or (ii) x occurs free in α.

(Here, in (ii), we have left out the trivial condition that y occurs in y, which is always true.) Now, if y occurs free in β, then, by (*), either x or y occurs free in α. Conversely, let us assume that either x or y occurs free in α. If x occurs free in α, then, by the above assertion of equivalence, y occurs free in β. If y occurs free in α, then, for $y \equiv x$, the theorem follows as above; for $y \not\equiv x$, we can use the above equivalence to prove that y occurs free in β.

5.6 Further theorems about substitution. The following two theorems can now easily be proved by induction on the length of α (cf. 1.2 for the concepts of length and of rank).

Theorem 1. If Subst $\alpha x t \beta$, then $R(\alpha) = R(\beta)$.

Theorem 2. For arbitrary α, x, t, β we can always decide whether Subst α x t β or not.

Exercises. 1. Set up a calculus in which the rows of symbols which are derivable are precisely those which have the form t_0 x t t_1, where $\Delta_x^t t_0 \equiv t_1$ (cf. 5.2).

2. Carry out the proofs of the two theorems in 5.6.

3. Let x, y, z, u, v be pairwise distinct individual variables, P, Q predicate variables and f, g function variables. Test whether or not the following assertions about substitution hold:

(a) Subst Pxy y x Pyx

(b) Subst $\bigwedge$yPxfy x gzx $\bigwedge$yPgzxfy

(c) Subst $\bigwedge$x(Pxy $\wedge$ $\bigwedge$uQzxu) z v $\bigwedge$x(Pxy $\wedge$ $\bigwedge$uQvxu)

(d) Subst $\bigwedge$yPxy x y $\bigwedge$yPyy

(e) Subst $\bigwedge$x(Pxy $\wedge$ $\bigwedge$u $\neg$ Qzxu) z x $\bigwedge$x(Pxy $\wedge$ $\bigwedge$u $\neg$ Qxxu)

(f) Subst ($\bigwedge$xPxy $\wedge$ $\bigwedge$yPxy) x fz ($\bigwedge$xPxy $\wedge$ $\bigwedge$yPfzy)

4. If Subst α x t β and if x occurs free in α, then x is free in β if and only if x is in t.

5. Suppose that Subst α x y β and Subst β y t γ, and that y does not occur free in α. Then we have Subst α x t γ.

6. (Generalisation of 5.) Suppose that Subst α x s β and Subst β y t γ, and that y does not occur free in α. Then we have Subst α x Δ_y^ts γ.

7. Show that, in exercise 5, the condition that y does not occur free in α may not be omitted.

8. Suppose that Subst α x_1 t_1 β and Subst β x_2 t_2 γ, and, moreover, that $x_1 \not\equiv x_2$ and that x_1 does not occur free in t_2 nor x_2 in t_1. Then there is an expression δ such that Subst α x_2 t_2 δ and Subst δ x_1 t_1 γ (juxtaposition of substitutions). Show also that the last three conditions of the hypothesis cannot be dispensed with.

III. The Semantics of Predicate Logic

The central idea in logic is that of consequence. In Chap. I, § 3.9 we saw how the notion of consequence for mathematical statements can be introduced on the basis of semantic ideas. The expressions of predicate logic correspond to the mathematical statements. In this chapter, we shall introduce the notion of consequence and the other semantic concepts which are necessary for its definition with the precision which is now possible.

§ 1. Introduction to the semantics of the language of predicate logic

This section contains preparatory considerations which are to help us to understand better the definition of the important concept "model" which we shall give in § 2.

1.1 Expressions as statement forms. In general, the expression of predicate logic contain free variables (e.g. Px, $\bigwedge x Px$, $fx = y$). Thus, such expressions are not statements. However, they can be understood as statement forms, as can "mathematical statements" (cf. Chap. I, § 4.2). Thus, it is meaningful to develop the important semantic concepts, e.g. that of interpretation and that of a model, for predicate logic too (cf. Chap. I, § 4.5). Since, by contrast with the everyday mathematical language, the language of predicate logic is one which has been built up with systematic exactness, the semantic concepts can be defined precisely for it. Although "most" of the expressions of predicate logic (e.g. all the atomic expressions) contain free variables, there are also expressions without free variables. Such an expression α can contain neither predicate variables nor more-than-no-place function variables. Thus, the atomic components of α must be equations in which the terms on either side of the equality sign are individual variables. All these individual variables must be bound by a universal quantifier. Examples of such expressions are: $\bigwedge x\, x = x$, $\bigwedge x \bigwedge y\, x \neq y$, $\bigwedge x(\bigwedge y\, x = y \wedge x = x)$.

We could take the expressions without free variables as aristotelian statements; however, we do not in fact do this, for the following reason: Every such expression α contains at least one universal quantifier. Hence, the meaning of α only becomes clear when we determine the totality to which the universal quantifier relates. We could get round this difficulty by assuming that there is a "biggest" totality which is laid down once and for all, and that every universal quantifier relates to this totality. However, thinking of the universal quantifier in this way would have the disadvantage that, in every theory, we should - albeit via the universal quantifier - be talking about all things, and not only about those things with which the theory is particularly concerned. Moreover, in the light of the antimonies of set theory, we tend to be somewhat wary of taking the idea of a "biggest" totality as a basis for our thinking.

If, for the reasons given above, we do not assume the existence of a biggest totality, then we must always say to which totality the universal quantifier which occurs in a given expression is to relate. These totalities, which are different from case to case, are none other than the domains of individuals of which we have already spoken in Chap. I, § 3.5. Thus, an expression in which universal quantifiers appear only obtains a meaning when a domain of individuals, to which the universal quantifiers relate, is given.

We have already noted above that universal quantifiers occur in all the expressions of predicate logic which contain no free variables. Because the domain of individuals has to be determined each time, these expressions cannot be taken as statements in the aristotelian sense any more than can those which contain free variables. We can only talk about the truth or falsehood of such expressions if, in addition, we are given a domain of individuals. In this sense we can regard these expressions, too, as statement forms.

As an example, we consider the expressions $\bigwedge x \bigwedge y\ x = y$. We shall regard this statement as true if we take a domain of individuals with only one element as our basis, and as false if our domain of individuals has at least two elements.

1.2 Interpretations. In order to give a meaning to an expression of predicate logic, we have to take a domain of individuals as a basis and interpret the variables which occur free (cf. Chap. I, § 3, nos. 4 and 9). An interpretation $\mathfrak{J}$ gives a meaning to the free variables by assigning suitable entities to them (Chap. I, § 4.5). Thus, an interpretation is a mapping. What are the variables mapped onto? If, for example, we have an atomic expression of the form $Pfxy$, where x and y are distinct individual variables, then an interpretation $\mathfrak{J}$ must be defined for the arguments P, f, x and y. It is natural to require that $\mathfrak{J}(x)$ and $\mathfrak{J}(y)$ should be elements of the domain of individuals ω which we have taken as a basis, that $\mathfrak{J}(f)$ should be a function over ω and, finally, that $\mathfrak{J}(P)$ should be a two-place predicate (relation) between elements of ω.

We spoke above of functions and predicates "over ω". By this, we mean the following:

(1) We require of an n-place predicate $\mathfrak{P}$ over ω that, for any n elements $\mathfrak{x}_1, \ldots, \mathfrak{x}_n$ of ω (in that order), either $\mathfrak{P}$ fits $\mathfrak{x}_1, \ldots, \mathfrak{x}_n$ ($\mathfrak{x}_1, \ldots, \mathfrak{x}_n$ are in the relation $\mathfrak{P}$) or not. Thus, for example, the less-than relation is a two-place predicate over the domain of the natural numbers, since, for every pair $\mathfrak{x}, \mathfrak{y}$ of natural numbers either $\mathfrak{x} < \mathfrak{y}$ or not.

A one-place predicate $\mathfrak{P}$ over ω is a property over ω; we require that, for every $\mathfrak{r}$ in ω; either $\mathfrak{P}$ fits $\mathfrak{r}$ ($\mathfrak{r}$ has the property $\mathfrak{P}$) or not. Since we have no-place predicate variables, we shall also need no-place predicates. We require of a no-place predicate $\mathfrak{P}$ over ω (by extrapolating the general definition) that either $\mathfrak{P}$ fits (without reference to an element of ω) or it does not.

(2) We require of an n-place <u>function $\mathfrak{f}$ over ω</u> that, for any n given elements $\mathfrak{r}_1, \ldots, \mathfrak{r}_n$ of ω is that order, there exist a unique element of ω which is the <u>value</u> of the function. We shall, as is usual, denote this value by $\mathfrak{f}(\mathfrak{r}_1, \ldots, \mathfrak{r}_n)$. Note that, for example, the quotient function over the domain of the real numbers only becomes a (two-place) function over this domain in our sense of the word if values of the function are also laid down for the case when the divisor is 0. We require of a no-place function over ω (by extrapolating the general definition) that there exist a unique element of ω which is the value of the function. Thus, the no-place functions can be identified with the individuals (we have already anticipated this by regarding the individual variables as no-place function variables).

Thus, an interpretation is always tied to a domain of individuals ω. This domain of individuals plays a dual role: Firstly, it determines the totality to which the universal quantifiers relate; and, secondly, for every P and every f, $\mathfrak{J}(P)$ must be a predicate over ω and $\mathfrak{J}(f)$ a function over ω.

We required above that, given an n-place predicate $\mathfrak{P}$, either it fits n given individuals or it does not, and that, given an n-place function $\mathfrak{f}$ and any n arguments, a unique value of the function <u>exists</u>. We do <u>not</u> mean by this that we can always <u>decide effectively</u> whether or not $\mathfrak{P}$ fits $\mathfrak{r}_1, \ldots, \mathfrak{r}_n$, nor that $\mathfrak{f}(\mathfrak{r}_1, \ldots, \mathfrak{r}_n)$ can always be <u>computed</u>. For example, we shall certainly admit the one-place predicate $\mathfrak{P}$ over the domain of the diophantine equations which is defined as follows: $\mathfrak{P}$ fits any given diophantine equation if and only if the equation has a solution. (We know now that there is no procedure by means of which we can decide, for an arbitrary diophantine equation, whether or not it is soluble.)

With regard to a given expression α it is really only necessary to interpret the variables which occur free in α; with regard to α, an interpretation $\mathfrak{J}$ would not need to be defined for the other variables. However, for technical reasons, it is easier to require of an interpretation $\mathfrak{J}$ that it should be defined for all the variables which occur in the language of predicate logic; for, otherwise, we should always have to say for which arguments the mapping $\mathfrak{J}$ is defined. Of course, in order to answer the question whether or not an interpretation $\mathfrak{J}$ is a model of an expression α, we only need to know how the variables which occur free in α are interpreted in $\mathfrak{J}$. For this, cf. the <u>coincidence theorem</u> (3.1).

1.3 Models. The basic concept of semantics is that of a model, which we have already discussed in Chap. I, § 3.9 on the basis of the everyday mathematical language. Now we want to carry this over to the language of predicate logic. Thus, we have to explain what we mean by saying that an interpretation $\mathfrak{J}$ is a model of an expression α, or - which is to mean the same thing - that an expression α is valid in an interpretation $\mathfrak{J}$.

The most obvious way of doing this is by induction on the structure of the expression α. Let us consider as an example an interpretation $\mathfrak{J}$ over the domain of individuals ω of the natural numbers. Let P be two-place and let x be different from y. Let $\mathfrak{J}(P)$ be the less-than relation and let $\mathfrak{J}(x) = 2$, $\mathfrak{J}(y) = 7$. (For what follows, it is irrelevant how $\mathfrak{J}$ is defined for the other variables.) Then we shall say that this interpretation $\mathfrak{J}$ is a model of the expression Pxy, because the predicate $\mathfrak{J}(P)$ which has been assigned to the predicate variable P by the interpretation $\mathfrak{J}$ fits the natural numbers $\mathfrak{J}(x)$ and $\mathfrak{J}(y)$ (in that order) which have been assigned to the individual variables x and y by $\mathfrak{J}$ - i.e. because the less-than relation fits 2 and 7 (in that order). On the other hand, we shall say that $\mathfrak{J}$ is not a model of the expressions Pyx, Pxx and Pyy. Moreover, we shall say that $\mathfrak{J}$ is a model of the expressions $x = x$ and $y = y$, because $\mathfrak{J}$ assigns the same element of ω to the individual variables on either side of the equality sign. We shall also say that $\mathfrak{J}$ is not a model of $x = y$, because $\mathfrak{J}$ assigns different numbers to the variables x and y.

The property of being a model can be defined in the way we have indicated for all atomic expressions. It is, however, important to note that the individual variables may also be replaced by arbitrary terms. But an element $\mathfrak{J}(t)$ of the domain of individuals ω may be assigned to each term t in a natural way, so that the definition of a model which we have sketched above can also be carried over for the case where the individual variables are replaced by arbitrary terms. Let us consider, as an example, the term $t \equiv fx$. We assume that $\mathfrak{J}(f)$ is the squaring function, where $\mathfrak{J}$ is the interpretation considered above. Since $\mathfrak{J}$ maps the individual variable x onto the number 2 and the one-place function f onto the squaring function, it is natural to take $\mathfrak{J}(fx)$ as the square of 2, in other words as $\mathfrak{J}(f)(\mathfrak{J}(x))$, i.e. as the number 4. $\mathfrak{J}(t)$ can be defined inductively for every term t in a similar way.

We shall say that $\mathfrak{J}$ is a model of $\neg\alpha$ if $\mathfrak{J}$ is not a model of α, and that $\mathfrak{J}$ is a model of $(\alpha \wedge \beta)$ if $\mathfrak{J}$ is both a model of α and a model of β.

Thus, in the example we considered above, $\mathfrak{J}$ is a model of $\neg Pyx$ and of $(Pxy \wedge Pfxy)$, but it is not a model of $\neg Pxy$.

It only remains to define

1.4 Models of generalisations. As a simple example, we consider the expression $\bigwedge xQx$, where Q is a one-place predicate variable. The variable x does not occur free in this expression. If $\mathfrak{J}$ is an arbitrary interpretation then, by the remark at the end of 1.2, whether or not $\mathfrak{J}$ is a model of the expression $\bigwedge x\, Q\, x$ does not depend on how the variable x is interpreted by $\mathfrak{J}$. We shall say that $\mathfrak{J}$ is a model of $\bigwedge x\, Q\, x$ if and only if the predicate $\mathfrak{J}(Q)$ fits all the things in the domain of individuals. In the light of the fact that we intend to define the property of being a model inductively, we want to reformulate this so that, in the definition, we talk about the property of being a model of the expression $Q\, x$ which is obtained from $\bigwedge x\, Q\, x$ by leaving out $\bigwedge x$. It is clearly not enough to require merely that $\mathfrak{J}$ should be a model of $Q\, x$, for this would only ensure that the predicate $\mathfrak{J}(Q)$ fitted the individual $\mathfrak{J}(x)$. The fact that $\mathfrak{J}(Q)$ is to fit every individual can clearly be formulated by demanding that all interpretations $\mathfrak{J}'$ which coincide with $\mathfrak{J}$ for all arguments other than x (but not necessarily for x) should be models of $Q\, x$. For then, for different choices of $\mathfrak{J}'$, $\mathfrak{J}'(x)$ runs through all the elements of the domain of individuals. In order to be able to express this requirement conveniently, we shall introduce the notation $\mathfrak{J}_x^{\mathfrak{r}}$. By $\mathfrak{J}_x^{\mathfrak{r}}$, we mean the interpretation which is defined over the same domain of individuals as $\mathfrak{J}$ and which coincides with $\mathfrak{J}$ for all arguments except, possibly, x, and which maps x onto the individual $\mathfrak{r}$. Thus, we have the required definition: The interpretation $\mathfrak{J}$ is a model of $\bigwedge x\, Q\, x$ if and only if, for every individual $\mathfrak{r}$, the interpretation $\mathfrak{J}_x^{\mathfrak{r}}$ is a model of $Q\, x$. If, more generally, we replace $Q\, x$ by α, then we shall, correspondingly, want $\mathfrak{J}$ to be a model of $\bigwedge x \alpha$ if and only if, for every $\mathfrak{r}$, the interpretation $\mathfrak{J}_x^{\mathfrak{r}}$ is a model of α. If we assume that we have already defined what we mean by saying that an arbitrary interpretation $\mathfrak{J}$ is a model of α, then the conditions we have laid down above make it possible to define what we mean by saying that $\mathfrak{J}$ is a model of $\bigwedge x \alpha$.

§ 2. Definition of the most important semantic concepts

In this section we shall give the definition of the property of being a model of a given expression. We have already explained the concept of an interpretation and the most important steps in the definition of a model in the last section, so that, here, we can concentrate on the technical implementation of the program we sketched there. We begin with a few observations about the foundations of semantics. Then, on this basis, we define the concept of an interpretation.

2.1 The foundations of semantics. We take the concept of a domain of individuals as our starting-point. Every set except the empty set can be taken as a domain of individuals. Thus, a domain of individuals is never empty.

Let us consider any domain of individuals ω. The totality of all predicates over ω and the totality of all functions over ω are completely determined by ω.

If $\mathfrak{P}$ is an n-place predicate over ω $(n \geqslant 0)$ and if $\mathfrak{r}_1, \ldots, \mathfrak{r}_n$ are (not necessarily distinct) elements of ω, then either $\mathfrak{P}$ fits $\mathfrak{r}_1, \ldots, \mathfrak{r}_n$ (in the given order) or it does not.

If we want, we can characterize the predicate $\mathfrak{P}$ by the set P of those n-tuples of elements of ω which $\mathfrak{P}$ fits. For every predicate $\mathfrak{P}$, there is such a set P which is assigned to $\mathfrak{P}$. Conversely, every set P of n-tuples of elements of ω is assigned in this way to one (and only one) n-place predicate $\mathfrak{P}$, namely the predicate which fits all the n-tuples of elements of ω which belong to P, and none which do not belong to P.

For $n = 0$ there are, over any ω, two predicates, namely the predicate which fits and the predicate which does not fit. These predicates are often called truth and falsehood respectively.

If $\mathfrak{f}$ is an n-place function over ω $(n \geqslant 1)$ and $\mathfrak{x}_1, \ldots, \mathfrak{x}_n$ are (not necessarily distinct) elements of ω, then there is exactly one element $\mathfrak{x}$ of ω such that $\mathfrak{x} = \mathfrak{f}(\mathfrak{x}_1, \ldots, \mathfrak{x}_n)$. If we want, we can characterize the function $\mathfrak{f}$ by the set F of those $(n+1)$-tuples $\mathfrak{x}_1, \ldots, \mathfrak{x}_{n+1}$ of elements of ω for which $\mathfrak{x}_{n+1} = \mathfrak{f}(\mathfrak{x}_1, \ldots, \mathfrak{x}_n)$. For every function $\mathfrak{f}$, there is such a set F which is assigned to $\mathfrak{f}$. Conversely, every set F of n-tuples of elements of ω which has the property that, for every n-tuple $\mathfrak{x}_1, \ldots, \mathfrak{x}_n$ of elements of ω, there is exactly one element $\mathfrak{x}_{n+1}$ of ω such that the $(n+1)$-tuple $\mathfrak{x}_1, \ldots, \mathfrak{x}_{n+1}$ belongs to F is assigned in this way to one (and only one) n-place function $\mathfrak{f}$, namely the function $\mathfrak{f}$ for which $\mathfrak{f}(\mathfrak{x}_1, \ldots, \mathfrak{x}_n) = \mathfrak{x}$ if and only if the n-tuple $\mathfrak{x}_1, \ldots, \mathfrak{x}_n, \mathfrak{x}$ lies in F.

It is convenient to regard the elements of ω as no-place functions over ω.

For every $n \geqslant 0$ and for every ω, there is at least one n-place predicate and at least one n-place function over ω.

R e m a r k 1. Above, we took the extensionalist point of view, by assuming that an n-place predicate is uniquely determined by the set of the n-tuples which it fits, and a function by the set of all the $(n+1)$-tuples for which $\mathfrak{x}_{n+1} = \mathfrak{f}(\mathfrak{x}_1, \ldots, \mathfrak{x}_n)$. According to the intensionalist point of view, it is possible to have different n-place predicates which fit exactly the same n-tuples and different functions which always have the same value given the same argument. The difference between two predicates which fit exactly the same n-tuples can be seen, for example, in their differing definitions (the same holds for functions). Thus, for example, under the intensionalist way of thinking, we could differentiate between the one-place predicate which fits only the smallest even number and the one-place predicate which fits only the smallest prime number. Both these predicates fit just the number two, and are therefore extensionally identical.

For the sake of simplicity we shall, in this book, take the extensionalist point of view. It should, however, be mentioned explicitly that, for normal predicate logic (but not for second-order predicate logic), the intensionalist point of view can just as well be taken as a basis.

R e m a r k 2. Above, we took the attitude that, given a domain of individuals ω, all the predicates and functions over ω are determined. Thus, ω determines the domain ε_ω of predicates and functions which belongs to ω. Recently, however, people have also considered subdomains ε of ε_ω, and have limited themselves to those predicates and functions over ω which belong to ε. For such non-standard views, cf. also Chap. V, §5, exercise 1.

2.2 Definition of the concept of an interpretation. By an interpretation over a domain of individuals ω, we mean a mapping $\mathfrak{J}$ which is defined for all function variables and for all predicate variables, and such that:

(a) For every $n \geqslant 0$ and for every n-place function variable f, $\mathfrak{J}(f)$ is an n-place function over ω.

(b) For every $n \geqslant 0$ and for every n-place predicate variable P, $\mathfrak{J}(P)$ is an n-place predicate over ω.

Because of our semantic presuppositions (cf. 2.1), there is, for every domain of individuals ω, at least one interpretation over ω.

If $\mathfrak{J}$ is an interpretation over a domain of individuals ω, x an arbitrary individual variable and $\mathfrak{x}$ an arbitrary element of ω, then let $\mathfrak{J}_x^{\mathfrak{x}}$ be the mapping which coincides with $\mathfrak{J}$ for all arguments other than x and for which $\mathfrak{J}_x^{\mathfrak{x}}(x) = \mathfrak{x}$. $\mathfrak{J}_x^{\mathfrak{x}}$ is, like $\mathfrak{J}$, an interpretation over ω. For every f which is at least one-place, we have $\mathfrak{J}_x^{\mathfrak{x}}(f) = \mathfrak{J}(f)$, and for every P we have $\mathfrak{J}_x^{\mathfrak{x}}(P) = \mathfrak{J}(P)$.

We write $\mathfrak{J}_{x\,y}^{\mathfrak{x}\,\mathfrak{y}}$ as an abbreviation for $(\mathfrak{J}_x^{\mathfrak{x}})_y^{\mathfrak{y}}$. We note the following relationships, which are easy to verify, for use later:

$$(*) \qquad \mathfrak{J}_x^{\mathfrak{J}(x)} = \mathfrak{J}, \quad \mathfrak{J}_{x\,x}^{\mathfrak{x}\,\mathfrak{y}} = \mathfrak{J}_x^{\mathfrak{y}}, \quad \mathfrak{J}_{x\,y}^{\mathfrak{x}\,\mathfrak{y}} = \mathfrak{J}_{y\,x}^{\mathfrak{y}\,\mathfrak{x}} \quad \text{for} \quad x \not\equiv y .$$

Up till now we have defined $\mathfrak{J}(t)$, not for any arbitrary term t, but only for the case in which t is an individual variable. However, taking this as a basis, we can define $\mathfrak{J}(t)$ inductively for every term t in a natural way by setting, for an r-place function variable $f (r \geqslant 1)$:

$$\mathfrak{J}(ft_1 \ldots t_r) = \mathfrak{J}(f)(\mathfrak{J}(t_1), \ldots, \mathfrak{J}(t_r)) .$$

We can see at once that $\mathfrak{J}(t)$ is always an element of ω whenever t is a term. It is also true that (for Δ cf. Chap. II, §5.2)

$$(**) \qquad \mathfrak{J}_x^{\mathfrak{J}(t)}(t_0) = \mathfrak{J}(\Delta_x^t t_0) .$$

This can be proved by induction on the length of t_0. If $t_0 \equiv z$, then (**) can be obtained immediately from the definitions of $\mathfrak{J}_x^{\mathfrak{x}}$ and Δ_x^t, by distinguishing the two cases $z \equiv x$ and $z \not\equiv x$. If, on the other hand, $t_0 \equiv ft_1 \ldots t_r$, then, by Chap. II, §5.2 (**) and by making use of the induction hypothesis, we have

$$\mathfrak{I}_x^{\mathfrak{I}(t)}(ft_1\dots t_r) = \mathfrak{I}_x^{\mathfrak{I}(t)}(f)(\mathfrak{I}_x^{\mathfrak{I}(t)}(t_1),\dots,\mathfrak{I}_x^{\mathfrak{I}(t)}(t_r))$$
$$= \mathfrak{I}(f)(\mathfrak{I}(\Delta_x^t t_1),\dots,\mathfrak{I}(\Delta_x^t t_r))$$
$$= \mathfrak{I}(f\Delta_x^t t_1\dots\Delta_x^t t_r)$$
$$= \mathfrak{I}(\Delta_x^t ft_1\dots t_r)\,.$$

2.3 Definition of a model. We define by induction on the structure of α what we mean by saying that α is valid in an interpretation $\mathfrak{I}$ over ω. We also say, synonymously, that $\mathfrak{I}$ is a model of α over ω, abbreviated as: $\mathrm{Mod}_\omega \mathfrak{I}\alpha$. For the sake of convenience, we do not always mention ω explicitly[1].

(0a) $\mathrm{Mod}_\omega \mathfrak{I} P t_1\dots t_r$ if and only if $\mathfrak{I}(P)$ fits $\mathfrak{I}(t_1),\dots,\mathfrak{I}(t_r)$.

(0b) $\mathrm{Mod}_\omega \mathfrak{I} t_1 = t_2$ if and only if $\mathfrak{I}(t_1)$ coincides with $\mathfrak{I}(t_2)$.

(1) $\mathrm{Mod}_\omega \mathfrak{I}\neg\alpha$ if and only if not $\mathrm{Mod}_\omega \mathfrak{I}\alpha$.

(2) $\mathrm{Mod}_\omega \mathfrak{I}(\alpha\wedge\beta)$ if and only if $\mathrm{Mod}_\omega \mathfrak{I}\alpha$ and $\mathrm{Mod}_\omega \mathfrak{I}\beta$.

(3) $\mathrm{Mod}_\omega \mathfrak{I}\bigwedge x\alpha$ if and only if $\mathrm{Mod}_\omega \mathfrak{I}_x^{\mathfrak{r}}\alpha$ for every element $\mathfrak{r}$ of ω.

This definition defines $\mathrm{Mod}_\omega \mathfrak{I}\alpha$ for every expression α and every interpretation $\mathfrak{I}$ over ω. We want to extend this definition by defining, for every set $\mathfrak{M}$ of expressions, what it means to say that $\mathfrak{M}$ is valid in $\mathfrak{I}$ over ω. This is to mean that every element of $\mathfrak{M}$ is valid in $\mathfrak{I}$ over ω:

$\mathrm{Mod}_\omega \mathfrak{I}\mathfrak{M}$ if and only if $\mathrm{Mod}_\omega \mathfrak{I}\alpha$ for every α in $\mathfrak{M}$.

An interpretation over ω which is a model of $\mathfrak{M}$ is to be referred to, briefly, as an ω-model of $\mathfrak{M}$. Correspondingly, an interpretation which is a model of $\mathfrak{M}$ is referred to as a model of $\mathfrak{M}$.

2.4 Definition of consequence and of universal validity. We want to say that an expression α follows from a set $\mathfrak{M}$ of expressions over ω if, for every interpretation $\mathfrak{I}$ over ω: If $\mathrm{Mod}_\omega \mathfrak{I}\mathfrak{M}$, then $\mathrm{Mod}_\omega \mathfrak{I}\alpha$; in other words, if every ω-model of $\mathfrak{M}$ is also an ω-model of α. We say that α follows from $\mathfrak{M}$ (without reference to a domain of individuals) if, for every domain of individuals ω, the expression α follows from $\mathfrak{M}$ over ω; i.e. if, for every domain of individuals ω and for every interpretation $\mathfrak{I}$ over ω: If $\mathrm{Mod}_\omega \mathfrak{I}\mathfrak{M}$, then $\mathrm{Mod}_\omega \mathfrak{I}\alpha$, i.e. if every model of $\mathfrak{M}$ is also a model of α. We use α is a consequence of $\mathfrak{M}$ as a synonym for α follows from $\mathfrak{M}$.

---

[1] The inductive definition is so arranged that we define for arbitrary interpretations $\mathfrak{I}$ over ω what $\mathrm{Mod}_\omega \mathfrak{I}\alpha$ means.

In order to express concisely the fact that α follows from $\mathfrak{M}$ over ω or that α follows from $\mathfrak{M}$, we write: $\mathfrak{M} \vDash_{\omega} \alpha$, $\mathfrak{M} \vDash \alpha$ respectively[2].

If $\mathfrak{M} = \{\alpha_1, \ldots, \alpha_r\}$, $r \geqslant 1$, then we abbreviate $\{\alpha_1, \ldots, \alpha_r\} \vDash_{\omega} \alpha$ and $\{\alpha_1, \ldots, \alpha_r\} \vDash \alpha$ by $\alpha_1, \ldots, \alpha_r \vDash_{\omega} \alpha$, $\alpha_1, \ldots, \alpha_r \vDash \alpha$ respectively. If $\mathfrak{M}$ is empty, then we abbreviate $\mathfrak{M} \vDash_{\omega} \alpha$, $M \vDash \alpha$ by $\vDash_{\omega} \alpha$, $\vDash \alpha$ respectively.

We say that an expression α is <u>universally valid over ω</u> or that it is <u>universally valid</u> if $\vDash_{\omega} \alpha$ or $\vDash \alpha$ respectively.

<u>2.5 Definition of satisfiability.</u> If an expression α or a set $\mathfrak{M}$ of expressions possesses a model over ω, then α or $\mathfrak{M}$ is said to be <u>satisfiable over ω</u>. α or $\mathfrak{M}$ is said simply to be <u>satisfiable</u> if there is a domain of individuals ω such that α or $\mathfrak{M}$ respectively is satisfiable over ω.

There is an important connection between consequence and satisfiability:

<u>Theorem 1.</u> $\mathfrak{M} \vDash_{\omega} \alpha$ if and only if $\mathfrak{M} \cup \{\neg\alpha\}$ is not satisfiable over ω.

P r o o f. $\mathfrak{M} =_{\omega} \alpha$ iff every interpretation $\mathfrak{J}$ over ω which is a model of $\mathfrak{M}$ is also a model of α,
iff there is no interpretation $\mathfrak{J}$ over ω which is a model of $\mathfrak{M}$ but is not a model of α,
iff there is no interpretation $\mathfrak{J}$ over ω which is a model of $\mathfrak{M}$ and a model of $\neg\alpha$,
iff $\mathfrak{M} \cup \{\neg\alpha\}$ is not satisfiable over ω.

With the aid of Theorem 1, we can show that the same connection holds if we leave out ω:

<u>Theorem 1'.</u> $\mathfrak{M} \vDash \alpha$ if and only if $\mathfrak{M} \cup \{\neg\alpha\}$ is not satisfiable.

P r o o f. $\mathfrak{M} \vDash \alpha$ iff, for every $\omega : \mathfrak{M} \vDash_{\omega} \alpha$,
iff, for every $\omega : \mathfrak{M} \cup \{\neg\alpha\}$ is not satisfiable over ω,
iff $\mathfrak{M} \cup \{\neg\alpha\}$ is not satisfiable over any ω,
iff $\mathfrak{M} \cup \{\neg\alpha\}$ is not satisfiable.

The connection provided by the last two theorems shows that it is possible to reduce the two-place relation of consequence to the one-place property of satisfiability[3]. We shall make use of this reduction in the proof of G ö d e l ' s completeness theorem.

---

[2] The symbol "$\vDash$" is derived from the symbol "$\vdash$". For this, cf. Chap. IV, § 1.1.

[3] There is an analogous situation in group theory, where it is possible to reduce every two-place congruence relation to the one-place property of being an element of a certain normal subgroup.

We note also the special cases which are obtained from theorems 1 and 1' for an empty set of expressions:

Theorem 1a. α is universally valid over ω if and only if $\neg\alpha$ is not satisfiable over ω.

Theorem 1'a. α is universally valid if and only if $\neg\alpha$ is not satisfiable.

Theorem 1 shows how the relation of consequence could be defined with the aid of satisfiability. Conversely, the relation of satisfiability can be defined from the consequence relation. This is shown by

Theorem 2. Let α be an expression and $\mathfrak{M}$ a set of expressions. Then $\mathfrak{M}$ is not satisfiable over ω if and only if $\mathfrak{M} \models_{\omega} (\alpha \wedge \neg\alpha)$.

Proof. $\mathfrak{M} \models_{\omega} (\alpha \wedge \neg\alpha)$ iff every ω-model of $\mathfrak{M}$ is also an ω-model of $(\alpha \wedge \neg\alpha)$,
iff every ω-model of $\mathfrak{M}$ is also an ω-model of α and of $\neg\alpha$,
iff every ω-model of $\mathfrak{M}$ is also both an ω-model of α and not an ω-model of α,
iff there is no ω-model of $\mathfrak{M}$,
iff $\mathfrak{M}$ is not satisfiable over ω.

Theorem 2'. Let α be an expression and $\mathfrak{M}$ a set of expressions. Then $\mathfrak{M}$ is not satisfiable if and only if $M \models (\alpha \wedge \neg\alpha)$.

Exercises. In the following, $(\alpha \vee \beta)$, $(\alpha \rightarrow \beta)$, $(\alpha \leftrightarrow \beta)$ and $\bigvee x\alpha$ are to be understood as abbreviations - cf. Chap. II, § 1.5. x, y, z, x_1, x_2, x_3 are assumed to be pairwise distinct.

1. Let ω be the domain of individuals consisting of all people. Let P be a two-place predicate variable. Let $\mathfrak{J}_1$, $\mathfrak{J}_2$, $\mathfrak{J}_3$ be interpretations such that: $\mathfrak{J}_1(P)$, $\mathfrak{J}_2(P)$, $\mathfrak{J}_3(P)$ respectively) is the two-place relation between people which fits the people $\mathfrak{x}$, $\mathfrak{y}$ if and only if $\mathfrak{x}$ is the father (sister, ancestor respectively) of $\mathfrak{y}$, while the other variables may be interpreted by $\mathfrak{J}_1$, $\mathfrak{J}_2$, $\mathfrak{J}_3$ in any (fixed) way. Determine which of the following expressions hold true in $\mathfrak{J}_1$ ($\mathfrak{J}_2$, $\mathfrak{J}_3$ respectively):

$$\bigwedge x \bigwedge y\, Pxy, \quad \bigwedge x\, Pxx, \quad \bigwedge x \bigwedge y \bigwedge z(Pxy \wedge Pyz \rightarrow Pxz), \quad \bigwedge x \bigwedge y(Pxy \rightarrow Pyx) .$$

2. Find an interpretation $\mathfrak{J}$ over the domain ω of the natural numbers such that $\mathrm{Mod}_{\omega}\,\mathfrak{J}(\bigvee x \bigwedge y\, fxy = x \wedge \bigvee x \bigwedge y\, fxy = y)$.

3. Show that $\models_{\omega}(\bigwedge x \bigvee y\, Pxy \leftrightarrow \bigvee y \bigwedge x\, Pxy)$ if and only if ω has exactly one element.

4. Let ω be a domain with exactly two elements. Find interpretations $\mathfrak{J}_1$, $\mathfrak{J}_2$ over ω in which $(\bigwedge x \bigvee y\, Pxy \leftrightarrow \bigvee y \bigwedge x\, Pxy)$ holds, such that $\mathfrak{J}_1(P) \neq \mathfrak{J}_2(P)$.

5. (a) Let $\alpha_1 \equiv \bigwedge x_1 \bigvee y\, x_1 \neq y$. Show that α_1 is satisfiable over a domain of individuals ω if and only if ω has at least two elements.

(b) Let $\alpha_2 \equiv \bigwedge x_1 \bigwedge x_2 \bigvee y(x_1 \neq y \wedge x_2 \neq y)$. Show that α_2 is satisfiable over ω if and only if ω has at least three elements.

(c) For each $n \geqslant 1$, find an expression α_n which is satisfiable over a domain of individuals ω if and only if ω has more than n elements.

6. (a) Show that $\bigwedge x_1 \bigwedge x_2\, x_1 = x_2$ is satisfiable over ω if and only if ω has at most one element.

(b) Show that $\bigwedge x_1 \bigwedge x_2 \bigwedge x_3 ((x_1 = x_2 \vee x_1 = x_3) \vee x_2 = x_3)$ is satisfiable over ω if and only if ω has at most two elements.

(c) For each $n \geqslant 1$, find an expression α_n which is satisfiable over a domain of individuals ω if and only if ω has at most n elements.

7. For every $n \geqslant 1$, find an expression α_n which is satisfiable over a domain of individuals ω if and only if ω has precisely n elements.

8. For every $n \geqslant 2$, find an expression α_n which is satisfiable over a finite domain of individuals ω if and only if the number of elements in ω is a multiple of n.

9. For every $n \geqslant 2$ and for every m such that $0 \leqslant m < n$, find an expression α_{nm} which is satisfiable over a finite domain of individuals ω if and only if, when the number of elements of ω is divided by n, the remainder is m.

10. Show that

$$((\bigwedge x \bigvee y\, Pxy \wedge \bigwedge x \bigwedge y (Pxy \to \neg Pyx)) \wedge \bigwedge x \bigwedge y \bigwedge z ((Pxy \wedge Pyz) \to Pxz))$$

is not satisfiable over ω if ω is finite. Show also that the expression is satisfiable over the domain of individuals of the natural numbers.

11. Show that

(a) If $\alpha \models \beta$ and $\beta \models \gamma$, then $\alpha \models \gamma$.

(b) If $\mathfrak{M} \models \beta$ for all $\beta \in \mathfrak{N}$ and $\mathfrak{N} \models \gamma$, then $\mathfrak{M} \models \gamma$.

Refute:

(c) If $\mathfrak{M} \models \alpha$ and $\mathfrak{N} \models \alpha$, then $\mathfrak{M} \cap \mathfrak{N} \models \alpha$.

(d) If $\mathfrak{M} \underset{\omega_1}{\models} \alpha$ and $\mathfrak{M} \underset{\omega_2}{\models} \alpha$, then $\mathfrak{M} \underset{\omega_1 \cup \omega_2}{\models} \alpha$.

§ 3. Theorems about models

In the following, we shall derive some important theorems about models.

The coincidence theorem states that the existence or otherwise of the relation $\mathrm{Mod}_\omega \mathfrak{I}\alpha$ only depends on how $\mathfrak{I}$ maps those variables which occur free in α (cf. 1.2 for this).

If an expression α is transformed into an expression β by substitution of t for x, then, by the substitution theorem, it is possible, for every interpretation $\mathfrak{I}$ in which β is valid, to find a uniquely determined interpretation $\mathfrak{I}'$, dependent on $\mathfrak{I}$ and over the same domain of individuals, in which α is valid.

Since, in the following, we shall always be talking about interpretations over the same domain of individuals ω, we shall leave out the index ω in $\mathrm{Mod}_\omega \mathfrak{I}\alpha$ for the sake of simplicity.

3.1 The coincidence theorem. We shall use $\mathfrak{V}(\alpha)$ to mean the set of those variables (of any sort) which occur free in the expression α. $\mathfrak{V}(\alpha)$ is also called the vocabulary of α. We shall say that the two interpretations $\mathfrak{J}$ and $\mathfrak{K}$ coincide with respect to the set of variables $\mathfrak{V}$, in symbols $\mathfrak{J} \underset{\mathfrak{V}}{=} \mathfrak{K}$, if $\mathfrak{J}$ and $\mathfrak{K}$ are interpretations over the same domain of individuals and have the same values for the variables in $\mathfrak{V}$. Using this notation, we have the

Coincidence theorem: If $\mathfrak{J} \underset{\mathfrak{V}(\alpha)}{=} \mathfrak{K}$, then (Mod $\mathfrak{J}\alpha$ iff Mod $\mathfrak{K}\alpha$).

Proof. We prove the assertion: For all $\mathfrak{J}, \mathfrak{K}$: If $\mathfrak{J} \underset{\mathfrak{V}(\alpha)}{=} \mathfrak{K}$, then Mod $\mathfrak{J}\alpha$ iff Mod $\mathfrak{K}\alpha$ by induction on the structure of α.

(0a) Mod $\mathfrak{J} P t_1 \ldots t_n$ iff $\mathfrak{J}(P)$ fits $\mathfrak{J}(t_1), \ldots, \mathfrak{J}(t_n)$

iff $\mathfrak{K}(P)$ fits $\mathfrak{J}(t_1), \ldots, \mathfrak{J}(t_n)$

iff Mod $\mathfrak{K} P t_1 \ldots t_n$.

(The passage from the first line to the second is justified by the fact that, since $\mathfrak{J} \underset{\mathfrak{V}(P t_1 \ldots t_n)}{=} \mathfrak{K}$, in particular $\mathfrak{J}(P) = \mathfrak{K}(P)$ and $\mathfrak{J}(t_j) = \mathfrak{K}(t_j)$ for $j = 1, \ldots, n$.)

(0b) Mod $\mathfrak{J}\, t_1 = t_2$ iff $\mathfrak{J}(t_1) = \mathfrak{J}(t_2)$

iff $\mathfrak{K}(t_1) = \mathfrak{K}(t_2)$

iff Mod $\mathfrak{K}\, t_1 = t_2$.

(1) If $\mathfrak{J} \underset{\mathfrak{V}(\neg\alpha)}{=} \mathfrak{K}$, then also $\mathfrak{J} \underset{\mathfrak{V}(\alpha)}{=} \mathfrak{K}$. Hence

Mod $\mathfrak{J} \neg \alpha$ iff not Mod $\mathfrak{J}\alpha$

iff not Mod $\mathfrak{K}\alpha$ (induction hypothesis)

iff Mod $\mathfrak{K} \neg \alpha$.

(2) If $\mathfrak{J} \underset{\mathfrak{V}(\alpha \wedge \beta)}{=} \mathfrak{K}$, then $\mathfrak{J} \underset{\mathfrak{V}(\alpha)}{=} \mathfrak{K}$ and $\mathfrak{J} \underset{\mathfrak{V}(\beta)}{=} \mathfrak{K}$. Hence

Mod $\mathfrak{J}(\alpha \wedge \beta)$ iff Mod $\mathfrak{J}\alpha$ and Mod $\mathfrak{J}\beta$

iff Mod $\mathfrak{K}\alpha$ and Mod $\mathfrak{K}\beta$ (induction hypothesis)

iff Mod $\mathfrak{K}(\alpha \wedge \beta)$.

(3) If $\mathfrak{J} \underset{\mathfrak{V}(\bigwedge x \alpha)}{=} \mathfrak{K}$, then we claim, first of all, that $\mathfrak{J}_x^{\mathfrak{r}} = \mathfrak{J}_x^{\mathfrak{r}}$ for arbitrary $\mathfrak{r}$. For, if a variable occurs free in α, then either it occurs free in $\bigwedge x\alpha$, from which, since $\mathfrak{J} \underset{\mathfrak{V}(\bigwedge x \alpha)}{=} \mathfrak{K}$, the assertion follows for this variable; or it is identical with x, and then trivially $\mathfrak{J}_x^{\mathfrak{r}}(x) = \mathfrak{r} = \mathfrak{K}_x^{\mathfrak{r}}(x)$. Hence:

$\text{Mod}\,\mathfrak{J}\bigwedge x\alpha$ iff for every $\mathfrak{r}$: $\text{Mod}\,\mathfrak{J}_x^{\mathfrak{r}}\alpha$

iff for every $\mathfrak{r}$: $\text{Mod}\,\mathfrak{K}_x^{\mathfrak{r}}\alpha$ (induction hypothesis, applied to $\mathfrak{J}_x^{\mathfrak{r}}$ and $\mathfrak{K}_x^{\mathfrak{r}}$)

iff $\text{Mod}\,\mathfrak{J}\bigwedge x\alpha$.

Corollary 1. If x does not occur free in α, then

$$\text{Mod}\,\mathfrak{J}\alpha \quad \text{iff} \quad \text{Mod}\,\mathfrak{J}_x^{\mathfrak{r}}\alpha.$$

Corollary 2. $\text{Mod}\,\mathfrak{J}\bigwedge x\alpha$ iff $\text{Mod}\,\mathfrak{J}_x^{\mathfrak{r}}\bigwedge x\alpha$.

3.2 The substitution theorem reads:

If $\text{Subst}\,\alpha x t\beta$, then $\text{Mod}\,\mathfrak{J}_x^{\mathfrak{J}(t)}\alpha$ iff $\text{Mod}\,\mathfrak{J}\beta$.

Remark. The substitution theorem enables us to transform an arbitrary model of β into a model of α, namely $\mathfrak{J}_x^{\mathfrak{J}(t)}$.

Proof. We prove the assertion: If $\text{Subst}\,\alpha x t\beta$, then for all $\mathfrak{J}$: $\text{Mod}\,\mathfrak{J}_x^{\mathfrak{J}(t)}\alpha$ iff $\text{Mod}\,\mathfrak{J}\beta$ by induction on the structure of α.

(0a) $\text{Subst}\,Pt_1\ldots t_n\; x\; t\; P\Delta_x^t t_1\ldots\Delta_x^t t_n$. We have

$$\text{Mod}\,\mathfrak{J}_x^{\mathfrak{J}(t)}\,Pt_1\ldots t_n \quad \text{iff} \quad \mathfrak{J}_x^{\mathfrak{J}(t)}(P) \text{ fits } \mathfrak{J}_x^{\mathfrak{J}(t)}(t_1),\ldots,\mathfrak{J}_x^{\mathfrak{J}(t)}(t_n)$$

$$\text{iff} \quad \mathfrak{J}(P) \text{ fits } \mathfrak{J}(\Delta_x^t t_1),\ldots,\mathfrak{J}(\Delta_x^t t_n)$$

(because $\mathfrak{J}_x^{\mathfrak{J}(t)}(P) = \mathfrak{J}(P)$ and using 2.2(**))

$$\text{iff} \quad \text{Mod}\,\mathfrak{J}P\Delta_x^t t_1\ldots\Delta_x^t t_n.$$

(0b) We prove the theorem analogously for the case $\alpha \equiv t_1 = t_2$.

(1) If $\text{Subst}\,\neg\alpha x t\gamma$, then there is a β such that $\gamma\equiv\neg\beta$ and $\text{Subst}\,\alpha x t\beta$. Hence, by induction hypothesis,

$$\text{Mod}\,\mathfrak{J}_x^{\mathfrak{J}(t)}\alpha \quad \text{iff} \quad \text{Mod}\,\mathfrak{J}\beta.$$

From this we obtain

$$\text{Mod}\,\mathfrak{J}_x^{\mathfrak{J}(t)}\neg\alpha \quad \text{iff} \quad \text{Mod}\,\mathfrak{J}\neg\beta.$$

(2) If Subst $(\alpha_1 \wedge \alpha_2)\,x\,t\,\gamma$, then there are expressions β_1 and β_2 such that $\gamma \equiv (\beta_1 \wedge \beta_2)$ and Subst $\alpha_1\,x\,t\,\beta_1$ and Subst $\alpha_2\,x\,t\,\beta_2$. By induction hypothesis,

$$\text{Mod } \mathfrak{J}_x^{\mathfrak{J}(t)}\alpha_1 \text{ iff Mod } \mathfrak{J}\beta_1 \quad \text{and} \quad \text{Mod } \mathfrak{J}_x^{\mathfrak{J}(t)}\alpha_2 \text{ iff Mod } \mathfrak{J}\beta_2.$$

This gives us

$$\text{Mod } \mathfrak{J}_x^{\mathfrak{J}(t)}(\alpha_1 \wedge \alpha_2) \text{ iff Mod } \mathfrak{J}(\beta_1 \wedge \beta_2).$$

(3) If α is a generalisation, we distinguish two cases:

Case 1. x does not occur free in $\bigwedge z\alpha$. Then Subst $\bigwedge z\alpha\,x\,t\,\bigwedge z\alpha$. The assertion Mod $\mathfrak{J}_x^{\mathfrak{J}(t)} \bigwedge z\alpha$ iff Mod $\mathfrak{J} \bigwedge z\alpha$ can then be obtained from Corollary 1 to the coincidence theorem, since, by hypothesis, x does not occur free in $\bigwedge z\alpha$.

Case 2. x occurs free in $\bigwedge z\alpha$. Then, if Subst $\bigwedge z\alpha\,x\,t\,\gamma$, z does not occur in t and there is a β with $\gamma \equiv \bigwedge z\beta$ and Subst $\alpha\,x\,t\,\beta$. Moreover, $x \not\equiv z$, since x occurs free in $\bigwedge z\alpha$ but z does not. Now we have

Mod $\mathfrak{J}_x^{\mathfrak{J}(t)} \bigwedge z\alpha$ iff for all z: Mod $\mathfrak{J}_x^{\mathfrak{J}(t)}{}_z^{\mathfrak{z}}\alpha$

iff for all $\mathfrak{z}$: Mod $\mathfrak{J}_z^{\mathfrak{z}}{}_x^{\mathfrak{J}(t)}\alpha$, by 2.2, since $x \not\equiv z$

iff for all $\mathfrak{z}$: Mod $\mathfrak{J}_z^{\mathfrak{z}}{}_x^{\mathfrak{J}_z^{\mathfrak{z}}(t)}\alpha$, since $\mathfrak{J}(t) = \mathfrak{J}_z^{\mathfrak{z}}(t)$, because z does not occur in t

iff for all $\mathfrak{z}$: Mod $\mathfrak{J}_z^{\mathfrak{z}}\beta$, by induction hypothesis, applied to the interpretation $\mathfrak{J}_z^{\mathfrak{z}}$ instead of $\mathfrak{J}$

iff Mod $\mathfrak{J} \bigwedge z\beta$, q.e.d.

Exercises. 1. Let P and P' be predicate variables with the same number of places. Let α' be obtained from α by replacing P by P' everywhere in α. We write: Subst $\alpha\,P\,P'\alpha'$.

(a) Define Subst $\alpha\,P\,P'\alpha'$ by induction on the structure of α.

(b) Prove the following analogue to the substitution theorem:
If Subst $\alpha\,P\,P'\alpha$, then (Mod $\mathfrak{J}_P^{\mathfrak{J}(P')}\alpha$ iff Mod $\mathfrak{J}\alpha'$).

(c) Show that: If P' occurs neither in α nor in β and if Subst $\alpha\,P\,P'\alpha'$ and Subst $\beta\,P\,P'\beta'$, then (if $\alpha' \models \beta'$, then $\alpha \models \beta$).

2. In exercise 1, instead of considering predicate variables P and P' with the same number of places, consider function variables f and f' with the same number (greater than 0) of places.

IV. A Predicate Calculus

§ 1. Preliminary remarks about the rules of the predicate calculus. The concept of derivability

1.1 The relation $\alpha_1, \ldots, \alpha_n \vdash \alpha$, or $\vdash \alpha_1 \ldots \alpha_n \alpha$. As we have already explained in Chap. I, § 5, we want to lay the rules of down a calculus with the help of which we can obtain algorithmically all the consequences of arbitrary sets of expressions. Various different calculi of this sort are known today; every such calculus is called a predicate calculus or, more precisely, a first-order predicate calculus. (For predicate caluculus of higher order cf. Chap. VI, § 1.) In the next section, we shall give a particularly simple calculus of this sort. For the sake of simplicity, we shall call this calculus the predicate calculus (instead of a predicate calculus).

The predicate calculus we shall introduce here takes the form of an assumption calculus (for this, cf. the remarks in Chap. I, § 5.5). The rows of symbols which we can derive consist of linear sequences of finitely many expressions, where every such sequence must consist of at least one expression. Thus, every row of symbols which is derivable in the predicate calculus has the form

$$\alpha_1 \ldots \alpha_r$$

with $r \geqslant 1$. The connection with the notion of consequence, which was our reason for creating the calculus, can be described by two assertions:

(1) If $\alpha_1 \ldots \alpha_r$ is derivable in the predicate calculus, then α_r follows from $\alpha_1, \ldots, \alpha_{r-1}$ (in the particular case that the sequence consists of a single element α_1, this is to mean that α_1 follows from the empty set, i.e. that α_1 is universally valid).

(2) If, conversely, an expression α follows from a set $\mathfrak{M}$, then there are finitely many elements $\alpha_1, \ldots, \alpha_s$ of $\mathfrak{M}$ such that $\alpha_1 \ldots \alpha_s \alpha$ is derivable in the predicate calculus.

From this, we see that the last expression of a derivable sequence of expressions has a different status from the preceding ones. We want to call attention to this by using

(**) α_r is derivable from $\alpha_1, \ldots, \alpha_{r-1}$ in the predicate calculus as a synonym for

(*) $\alpha_1 \ldots \alpha_r$ is derivable in the predicate calculus.

(If the sequence $\alpha_1, \ldots, \alpha_r$ consists of only one element, then the two sentences are the same.) In analogy to the symbol "$\models$" with which we represent the relation of consequence, we shall represent the relation of derivability by "$\vdash$".

"$\vdash$" was introduced by Frege, the most important logician of the last century (1848-1925), in his "Begriffsschrift" (1879) to denote the relation of inference.

Thus, we abbreviate (*) and (**) by:

(***) $\vdash \alpha_1 \ldots \alpha_r$ and $\alpha_1, \ldots, \alpha_{r-1} \vdash \alpha_r$ respectively
(for $r = 1$, we have $\vdash \alpha_1$ in both cases).

In order to make the special status of the last element of a sequence of expressions obvious, we shall sometimes write $\alpha_1 \ldots \alpha_{r-1}\ \alpha_r$, with a small gap between the last two expressions, instead of $\alpha_1 \ldots \alpha_{r-1} \alpha_r$. We shall also often write $\alpha_1 \ldots \alpha_{r-1} : \alpha_r$. (This colon is, strictly speaking, not a component of the row of symbols; it is there merely for the convenience of the reader.)

1.2 The relation $\mathfrak{M} \vdash \alpha$. In order to make the analogy between the relation of consequence and that of derivability even clearer, we shall introduce the syntactic relation $\mathfrak{M} \vdash \alpha$ (in words: α is derivable from the set of expressions $\mathfrak{M}$) in analogy to the semantic relation $\mathfrak{M} \models \alpha$ by means of the

Definition. $\mathfrak{M} \vdash \alpha$ if and only if there are finitely many elements $\alpha_1, \ldots, \alpha_s$ $(s \geqslant 0)$ such that $\alpha_1, \ldots, \alpha_s \vdash \alpha$.

The main result of theoretical predicate logic can now be formulated as follows:

(3) $\mathfrak{M} \vdash \alpha$ if and only if $\mathfrak{M} \models \alpha$.

It states that the semantic relation of consequence and the syntactic relation of derivability coincide with each other.

We can split the assertion (3) into:

(1′) <u>If</u> $\mathfrak{M} \vdash \alpha$, <u>then</u> $\mathfrak{M} \models \alpha$ <u>Soundness of the predicate calculus</u>.

(2′) <u>If</u> $\mathfrak{M} \models \alpha$, <u>then</u> $\mathfrak{M} \vdash \alpha$ <u>Completeness of the predicate calculus</u>.

The names for these two statements can be explained as follows:

(1′) states that we cannot derive anything which is not also a consequence. If we could derive from a set $\mathfrak{M}$ an expression α which was not a consequence of $\mathfrak{M}$, then we would describe the calculus as unsound.

(2′) asserts that <u>every</u> consequence of a set $\mathfrak{M}$ can also be <u>derived</u> from $\mathfrak{M}$ (in the sense of the definition for $\mathfrak{M} \vdash \alpha$ given above). If there were a set $\mathfrak{M}$ and an expression α such that $\mathfrak{M} \models \alpha$ but not $\mathfrak{M} \vdash \alpha$, then we would describe the calculus as incomplete in this respect.

Of the two assertions of soundness and completeness, that of soundness is easier to prove (see § 3). It was only in 1930 that the completeness of a predicate calculus was first proved by Gödel. The assertion (2′) is therefore also known as <u>Gödel's completeness theorem</u>. We shall give a proof of the completeness of the predicate calculus we have introduced here in Chap. V.

Incidentally, we note that (1) is equivalent to (1′) and (2) to (2′) (see 1.1), and hence (1) and (2) combined to (3):

<u>If</u> (1), <u>then</u> (1′). Let $\mathfrak{M} \vdash \alpha$. Then there are elements $\alpha_1, \ldots, \alpha_s$ of $\mathfrak{M}$ such that $\alpha_1 \ldots \alpha_s \alpha$ is derivable. Then, by (1), α follows from $\alpha_1, \ldots, \alpha_s$. This means that every model of $\alpha_1, \ldots, \alpha_s$ is also a model of α. But then certainly every model of $\mathfrak{M}$ is a model of α. Thus, $\mathfrak{M} \models \alpha$.

<u>If</u> (1′), <u>then</u> (1). Let $\alpha_1 \ldots \alpha_r$ be derivable in the predicate calculus. Then

$$\alpha_1, \ldots, \alpha_{r-1} \vdash \alpha_r.$$

Then, by (1′), $\alpha_1, \ldots, \alpha_{r-1} \models \alpha_r$.

<u>If</u> (2), <u>then</u> (2′). Let $\mathfrak{M} \models \alpha$. By (2), there are finitely many elements $\alpha_1, \ldots, \alpha_s$ of $\mathfrak{M}$ such that $\alpha_1 \ldots \alpha_s \alpha$ is derivable. Thus, $\alpha_1, \ldots, \alpha_s \vdash \alpha$. This yields $\mathfrak{M} \vdash \alpha$.

<u>If</u> (2′), <u>then</u> (2). Let α follow from $\mathfrak{M}$. Then, by (2′), $\mathfrak{M} \vdash \alpha$. This means that there are finitely many elements $\alpha_1, \ldots, \alpha_s$ of $\mathfrak{M}$ such that $\alpha_1 \ldots \alpha_s \alpha$ is derivable in the predicate calculus.

§ 2. The rules of the predicate calculus

2.1 Introduction. The system of rules which we shall lay down in the following consists of eleven rules. By means of these rules we can derive rows of symbols which - as we have already explained in the last section - consist of a concatenation of finitely many expressions. In the following, we shall call such rows of symbols sequents. The expressions of which a sequent is composed are called the members of the sequent. Every sequent has at least one member. We use the letter σ, possibly with indices, to indicate sequents. The last expression of a sequent σ is called the end (succedent) of σ, in symbols ε_σ. Those which precede ε_σ are called the beginning (antecedent) of σ. The beginning of σ is either empty or another sequent. The (possibly empty) set of those expressions of which the beginning is composed is called the initial set of σ, in symbols $\mathfrak{A}_\sigma$.

In the following system of rules, there are two rules which allow us to write down sequents immediately. The other rules can only be applied if we have already derived one or two sequents (starting sequents); these rules allow us to pass from the starting sequents - which, in this context, are called premises - to a further sequent, which is called the conclusion.

All the rules - except for the assumption rule and the substitution rule - are associated with logical connectives or with the equality symbol, as follows: Three rules with the conjunctor, two with the negator, two with the universal quantifier and two with the equality symbol.

In what follows, the reader should, above all, try to understand what the application of the eleven rules means in a purely formal sense. (For this, see also the schematic representation of the rules in 2.3.) The fact that the rules are also unobjectionable as regards their content will be shown by the theorem of the soundness of the predicate calculus, which we shall prove in § 3.

2.2 Formulation of the rules. 1) The assumption rule (A) is a rule without premises. It allows us to write down the sequent $\alpha\alpha$ for an arbitrary expression α.

2) The rule (C) for introducing the conjunctor has two premises. It allows us to pass from any two sequents σ_1, σ_2 to every sequent σ which has the properties:

(a) $$\varepsilon_\sigma \equiv (\varepsilon_{\sigma_1} \wedge \varepsilon_{\sigma_2})$$ ("introduction of the conjunctor")

(b) $$\mathfrak{A}_\sigma = \mathfrak{A}_{\sigma_1} \cup \mathfrak{A}_{\sigma_2}.$$

Remark: Note that the condition (b) determines only the initial set of σ, not the beginning of σ itself. This means, in particular, that the order of the members in the beginning of σ can be chosen arbitrarily. Also, an element of $\mathfrak{A}_\sigma$ may appear more than once in the beginning of σ. In general, however, we shall choose the beginning of σ to be as short as possible, i.e. without any repetition of members. - This remark also holds for all the subsequent rules.

3) The first rule (C′) for removing the conjunctor has one premise. It allows us to pass from any sequent σ_1 whose end ε_{σ_1} is a conjunction $(\alpha \wedge \beta)$ to every sequent σ which has the properties:

(a) $\varepsilon_\sigma \equiv \alpha$ ("removal of the conjunctor"),

(b) $\mathfrak{A}_\sigma = \mathfrak{A}_{\sigma_1}$.

4) The second rule (C″) for removing the conjunctor has one premise. It is formulated like (C′), but with $\varepsilon_\sigma \equiv \beta$ instead of $\varepsilon_\sigma \equiv \alpha$.

5) The removal rule (R) is a rule with two premises. The starting sequents σ_1, σ_2 are assumed to be such that $\varepsilon_{\sigma_1} \equiv \varepsilon_{\sigma_2}$ and that there is an expression α with $\alpha \in \mathfrak{A}_{\sigma_1}$ and $\neg\alpha \in \mathfrak{A}_{\sigma_2}$. The rule allows us to pass from σ_1 and σ_2 to every sequent σ which has the properties:

(a) $\varepsilon_\sigma \equiv \varepsilon_{\sigma_1}$ $(\equiv \varepsilon_{\sigma_2})$,

(b) $\mathfrak{A}_\sigma = (\mathfrak{A}_{\sigma_1} - \{\alpha\}) \cup (\mathfrak{A}_{\sigma_2} - \{\neg\alpha\})$.

(The "removal" consists in taking α out of $\mathfrak{A}_{\sigma_1}$ and $\neg\alpha$ out of $\mathfrak{A}_{\sigma_2}$.)

6) The contradiction rule (X) has two premises. The starting sequents σ_1, σ_2 are assumed to be such that $\varepsilon_{\sigma_2} \equiv \neg\varepsilon_{\sigma_1}$. (i.e. the end expressions of the starting sequents "contradict" each other.) The rule allows us to pass from σ_1 and σ_2 to every sequent σ which has the properties:

(a) ε_σ is an arbitrary expression,

(b) $\mathfrak{A}_\sigma = \mathfrak{A}_{\sigma_1} \cup \mathfrak{A}_{\sigma_2}$.

7) The uncritical rule (G) for removing the generalisor is a rule with one premise. It allows us to pass from a sequent σ_1 whose end is a generalisation $\bigwedge x\alpha$ to every sequent σ which has the properties:

(a) $\varepsilon_\sigma \equiv \alpha$ ("removal of the generalisor"),

(b) $\mathfrak{A}_\sigma = \mathfrak{A}_{\sigma_1}$.

The description "uncritical" refers to the fact that, by contrast to the "critical" rule (G_x) which follows, no additional conditions are laid down for the application of the rule (G).

8) The <u>critical rule</u> (G_x) <u>for introducing the generalisor</u> is a rule with one premise. For a given x, it can be applied to a sequent σ_1 only if the variable x does not occur free in any member of the beginning of σ_1 ("critical condition"). If this condition is fulfilled, then the rule (G_x) allows us to pass from the sequent σ_1 to every sequent σ which has the properties:

(a) $\varepsilon_\sigma \equiv \bigwedge x \varepsilon_{\sigma_1}$ ("introduction of the generalisor"),

(b) $\mathfrak{A}_\sigma = \mathfrak{A}_{\sigma_1}$.

Strictly speaking, (G_x) consists of infinitely many rules, one for each choice of x. The same is true of the rules (S_x^t) and (E_x^t) which follow. In fact, it is true also of (A) (where α may take infinitely many values), (X) (where the same holds) and (E) (where t may be any term).

9) The <u>substitution rule</u> (S_x^t) depends on the individual variable x and the term t. It is a rule with one premise. The rule (S_x^t) is applicable to a sequent σ_1 if σ_1 satisfies the following condition: For every member γ of σ_1 there is an expression δ such that Subst $\gamma x t \delta$. Now let Subst $\varepsilon_{\sigma_1} x t \varepsilon$ and let A be the set of expressions which is obtained from the set $\mathfrak{A}_{\sigma_1}$ of expressions by the above-mentioned substitution of t for x. The rule (S_x^t) allows us to pass from σ_1 to every sequent σ which has the properties:

(a) $\varepsilon_\sigma \equiv \varepsilon$,

(b) $\mathfrak{A}_\sigma = \mathfrak{A}$.

10) <u>The first equality rule</u> (E) is a rule without premises. It allows us to write down the expressions $t = t$ for an arbitrary term t.

11) <u>The second equality rule</u> (E_x^t) depends on the individual variable x and the term t. It is a rule with one premise. The rule (E_x^t) is applicable to a sequent σ_1 if there is an expression ε such that Subst $\varepsilon_{\sigma_1} x t \varepsilon$. The rule allows us to pass from σ_1 to every sequent σ which has the properties:

(a) $\varepsilon_\sigma \equiv \varepsilon$,

(b) $\mathfrak{A}_\sigma = \mathfrak{A}_{\sigma_1} \cup \{x = t\}$.

2.3 Schematic representation of the rules. In the following table, the rules which we formulated precisely above are noted clearly and compactly (but less precisely).

The rules of the predicate calculus

$$\text{(A)} \quad \alpha : \alpha$$

$$\text{(C)} \quad \frac{\begin{array}{l}\Sigma_1 : \alpha \\ \Sigma_2 : \beta\end{array}}{\Sigma_{12} : (\alpha \wedge \beta)} \qquad \text{(C}'\text{)} \quad \frac{\Sigma : (\alpha \wedge \beta)}{\Sigma : \alpha} \qquad \text{(C}''\text{)} \quad \frac{\Sigma : (\alpha \wedge \beta)}{\Sigma : \beta}$$

$$\text{(R)} \quad \frac{\begin{array}{l}\Sigma_1 \alpha : \beta \\ \Sigma_2 \neg \alpha : \beta\end{array}}{\Sigma_{12} : \beta} \qquad \text{(X)} \quad \frac{\begin{array}{l}\Sigma_1 : \alpha \\ \Sigma_2 : \neg \alpha\end{array}}{\Sigma_{12} : \beta}$$

$$\text{(G)} \quad \frac{\Sigma : \bigwedge x \alpha}{\Sigma : \alpha} \qquad (G_x) \quad \frac{\Sigma : \alpha}{\Sigma : \bigwedge x \alpha}\text{, if } x \text{ does not occur free in } \Sigma$$

$$(S_x^t) \quad \frac{\Sigma : \alpha}{\Sigma' : \alpha'}\text{, if Subst } \Sigma \alpha x t \Sigma' \alpha'$$

$$\text{(E)} \quad : t = t \qquad (E_x^t) \quad \frac{\Sigma : \alpha}{\Sigma x{=}t : \beta}\text{, if Subst } \alpha x t \beta$$

Remarks on the notation in the representation:

(a) The (possibly empty) beginnings of the sequents (or parts of these beginnings) are represented symbolically by Σ, Σ_1, Σ_2 or Σ_{12} respectively.

(b) In the conclusions of the rules, the beginnings may be permuted arbitrarily. Moreover, expressions which occur in these beginnings may be written down arbitrarily often (in particular, they may also be written down once only). The notation "Σ_{12} is to mean that any given expression is a member of Σ_{12} if and only it is a member of Σ_1 or of Σ_2.

(c) x occurs free in Σ is to mean that x occurs free in at least one member of Σ.

(d) Subst $\Sigma \alpha x t \Sigma' \alpha'$ is to mean that Subst $\alpha x t \alpha'$ and that every member of Σ is transformed into the corresponding member of Σ' by substitution of t for x.

(e) For ":" cf. the last remark in § 1.1.

2.4 Derivations. By a derivation we mean a finite sequence of sequence, written down one beneath the other, each of which can either be written down by means of the assumption rule (A) or the first equality rule (E) or can be obtained by means of one of the other rules from one or two of the sequents which precede it in the sequence.

In the case of rules with two premises, the order in which the starting sequents appear in the sequence of sequents is to be immaterial.

A sequent is said to be derivable if there is a derivation whose last member is this sequent.

We can see at once from the definition of a derivation that, if we write down two or more derivations one beneath the other, the result is another derivation.

By the length of a derivation we mean the number of sequents of which the derivation consists.

An important principle of proof is the following: In order to show that every derivable sequent σ has a property $\mathfrak{P}$, it clearly suffices to show:

(1) The sequents which can be produced by (A) and (E) (i.e. the sequents $\alpha\alpha$ and $t = t$) have the property $\mathfrak{P}$.

(2) The sequents which can be produced by any of the other rules have the property $\mathfrak{P}$ whenever the relevant premises have the property $\mathfrak{P}$.

2.5 Examples of derivations. In order to practise applying the rules, we want to give a few simple examples of derivations. At the end of each line we shall make a note of the rule by means of which the sequent in that line was obtained and the premises to which this rule was applied. In order to facilitate the making of these notes we shall number the lines of the derivation, starting from the top. In the following, we assume that x, y, and z are pairwise distinct individual variables.

Example 1.

1.	$Px : Px$	(A)
2.	$Qy : Qy$	(A)
3.	$PxQy : (Px \wedge Qy)$	(C) 1,2

Example 2.

1.	$(Px \wedge Qx) : (Px \wedge Qx)$	(A)
2.	$(Px \wedge Qx) : Px$	(C') 1
3.	$(Px \wedge Qx) : Qx$	(C'') 1
4.	$(Px \wedge Qx) : (Qx \wedge Px)$	(C) 2,3 [1]

Example 3.

1.	$(Px \wedge \neg Px) : (Px \wedge \neg Px)$	(A)
2.	$(Px \wedge \neg Px) : Px$	(C') 1
3.	$(Px \wedge \neg Px) : \neg Px$	(C'') 1
4.	$(Px \wedge \neg Px) : \neg(Px \wedge \neg Px)$	(X) 2,3
5.	$\neg(Px \wedge \neg Px) : \neg(Px \wedge \neg Px)$	(A)
6.	$: \neg(Px \wedge \neg Px)$	(R) 4,5

Example 4.

1.	$\bigwedge x(Px \wedge Qx) : \bigwedge x(Px \wedge Qx)$	(A)
2.	$\bigwedge x(Px \wedge Qx) : (Px \wedge Qx)$	(G) 1
3.	$\bigwedge x(Px \wedge Qx) : Px$	(C') 2
4.	$\bigwedge x(Px \wedge Qx) : \bigwedge x Px$	(G_x) 3 [2]

---

[1] Note that the order in which the starting sequents used in the application of (C) occur in the derivation is immaterial. Here, in the notation of 2.2, $\sigma_1 \equiv (Px \wedge Qx) : Qx$ and $\sigma_2 \equiv (Px \wedge Qx) : Px$.

[2] Note that the condition for the application of this critical rule (namely that x should not occur free in $\bigwedge x(Px \wedge Qx)$) is fulfilled.

Example 5.

1.	$\bigwedge x \bigwedge y P x y : \bigwedge x \bigwedge y P x y$	(A)
2.	$\bigwedge x \bigwedge y P x y : \bigwedge y P x y$	(G) 1
3.	$\bigwedge x \bigwedge y P x y : P x y$	(G) 2
4.	$\bigwedge x \bigwedge y P x y : P x x$	(S^x_y) 3[3]
5.	$\bigwedge x \bigwedge y P x y : \bigwedge x P x x$	(G_x) 4[4]

Example 6.

1.	$x = z : x = z$	(A)
2.	$x = z \quad x = y : y = z$	(E^y_x) 1
3.	$x = x \quad x = y : y = x$	(S^x_z) 2
4.	$: x = x$	(E)
5.	$\neg\ x = x : \neg x = x$	(A)
6.	$\neg\ x = x : y = x$	(X) 4,5
7.	$x = y : y = x$	(R) 3,6

<u>2.6 Three general remarks.</u> (I) <u>If</u> $\alpha_1 \ldots \alpha_r \alpha$ <u>is derivable and if</u> $\beta_1 \ldots \beta_r$ <u>is any permutation of</u> $\alpha_1 \ldots \alpha_r$, <u>then</u> $\beta_1 \ldots \beta_r \alpha$ <u>is derivable</u>.

Proof: Let there be a derivation of $\alpha_1 \ldots \alpha_r \alpha$. Extend this by one line by applying the rule (S^x_x), where x is an arbitrary individual variable. The assertion now follows immediately, since, by Chap. II, § 5.5, Theorem 3, every expression is transformed into itself by substitution of x for x and since, in the application of the rule (S^x_x), the beginning may be permuted.

(II) <u>If</u> $\alpha_1 \ldots \alpha_r \alpha$ <u>is derivable</u>, then $\alpha_1 \ldots \alpha_r \beta \alpha$ <u>is also derivable for arbitrary</u> β.

Proof. Start from a derivation of $\alpha_1 \ldots \alpha_r \alpha$. Extend this as follows:

...	.... : ...	
n.	$\alpha_1 \ldots \alpha_r : \alpha$	
n + 1.	$\beta : \beta$	(A)
n + 2.	$\alpha_1 \ldots \alpha_r \beta : (\alpha \wedge \beta)$	(C) n, n + 1
n + 3	$\alpha_1 \ldots \alpha_r \beta : \alpha$	(C') n + 2

This is a derivation of $\alpha_1 \ldots \alpha_r \beta \alpha$.

(III) <u>If</u> $\mathfrak{M} \cup \{\alpha\} \vdash \beta$, <u>then there are expressions</u> $\alpha_1, \ldots, \alpha_r$ in $\mathfrak{M}$ ($r \geqslant 0$) <u>such that the sequent</u> $\alpha_1 \ldots \alpha_r \alpha \beta$ <u>is derivable</u>.

Proof. By the definition of the relation $\vdash$ in § 1, there are expressions $\alpha_1, \ldots, \alpha_r$ in $\mathfrak{M} \cup \{\alpha\}$ such that $\alpha_1 \ldots \alpha_r \beta$ is derivable. If one of these $\alpha_j \equiv \alpha$, then the assertion follows from (I). If, however, none of the α_j is identical with α, then the assertion can be obtained from (II).

---

[3] Note that Subst $\bigwedge x \bigwedge y P x y\ y x \bigwedge x \bigwedge y P x y$.

[4] x does not occur free in $\bigwedge x \bigwedge y P x y$.

2.7 Decidability of the property of being a derivation. Given a finite sequence of sequents, we can decide whether or not this sequence is a derivation.

In order to show that this is true, it clearly suffices to prove the following three assertions:

(a) Given an arbitrary sequent σ, we can decide whether or not it can be written down by means of (A) or (E).

(b) We can decide whether a sequent σ can be derived from a sequent σ_1 or from two sequents σ_1, σ_2 (whichever the case may be) by means of one of the rules (C), (C'), (C''), (R), (X), (G).

(c) We can decide whether there is an x (an x and a t) such that σ is derivable from a sequent σ_1 by means of the rule (G_x) (by means of one of the rules (S_x^t), (E_x^t)).

In each case, the sequent σ must first be separated into its members, which is effectively possible by Chap. II, § 2.7. Thus, $\mathfrak{A}_\sigma$ and ε_σ can be determined effectively.

to (a): σ can be derived by (A) if the antecedent of σ contains exactly one element and if this element is identical with ε_σ. This is decidable. σ can be derived by (E) if the antecedent of σ is empty and the succedent is an equation of the form $t = t$. This can also be decided.

to (b): As a typical example, we consider the question whether, for given sequents σ_1, σ_2, σ, σ can be derived from σ_1 and σ_2 by means of (C). First of all, we determine $\mathfrak{A}_{\sigma_1}$, $\mathfrak{A}_{\sigma_2}$, $\mathfrak{A}_\sigma$, ε_{σ_1}, ε_{σ_2}, ε_σ. σ can be obtained from σ_1 and σ_2 by means of (C) if and only if $\varepsilon_\sigma \equiv (\varepsilon_{\sigma_1} \wedge \varepsilon_{\sigma_2})$ or $\varepsilon_\sigma \equiv (\varepsilon_{\sigma_2} \wedge \varepsilon_{\sigma_1})$, and if $\mathfrak{A}_\sigma = \mathfrak{A}_{\sigma_1} \cup \mathfrak{A}_{\sigma_2}$. Clearly, we can decide whether these conditions are fulfilled.

to (c) · (c_1): σ can be obtained from σ_1 by means of the rule (G_x) if and only if the following three conditions are fulfilled:

1) $\varepsilon_\sigma \equiv \bigwedge x \varepsilon_{\sigma_1}$,

2) x does not occur free in the expressions of $\mathfrak{A}_{\sigma_1}$,

3) $\mathfrak{A}_\sigma = \mathfrak{A}_{\sigma_1}$.

Condition 1) shows that, for a given σ, there is only one individual variable x for which the conditions can possibly be fulfilled. But then we can test effectively whether or not the above conditions are fulfilled [for 2) cf. Chap. II, § 4.3].

(c_2): σ can be obtained from σ_1 by means of the rule (S_x^t) if and only if the following conditions are fulfilled:

1) Subst ε_{σ_1} x t ε_σ

2) For every $\gamma \in \mathfrak{A}_{\sigma_1}$ there is a δ with $\delta \in \mathfrak{A}_\sigma$ and Subst γ x t δ.

3) Every δ in $\mathfrak{A}_\sigma$ can be obtained as in 2).

It is easy to test whether these conditions are fulfilled for a given x, t (for this, cf. particularly Chap. II, §4.3). However, we must note that, in general, x and t are not uniquely determined by the fact that σ_1 is transformed into σ by substituting t for x. All the same, we can limit ourselves to considering only finitely many pairs x, t, as is shown by the following: If x does not occur free in any of the members of σ_1, then Subst γ x t γ for every member γ of σ_1, so that the above three conditions read: $\varepsilon_\sigma \equiv \varepsilon_{\sigma_1}$ and $\mathfrak{A}_\sigma = \mathfrak{A}_{\sigma_1}$, which can be effectively tested for its truth or otherwise. Thus, we need only consider those individual variables x which occur free in at least one of the members of σ_1. There are only finitely many such individual variables, and they can be effectively found by Chap. II, §4.3. Let x be such an individual variable and let γ be a member of σ_1 in which x occurs free. It suffices to show that there are only finitely many terms t for which the conditions could be fulfilled and that these terms can be determined effectively. Let Subst γ x t δ. Then δ must be one of the members of σ. By Chap. II, §5.4, Theorem 5, t is a component of δ. Thus, we can determine from the start all the terms t for which the conditions could possibly be fulfilled.

(c_3): The rule (E_x^t) can be treated analogously.

Exercises 1. Show, by giving derivations, that the following sequents are derivable:

(a) $\bigwedge x \bigwedge y Pxy : \bigwedge y \bigwedge x Pxy$
(b) $(P \wedge \bigwedge x Qx) : (\bigwedge x P \wedge Qx)$
(c) $\bigwedge x (P \wedge Qx) : (P \wedge \bigwedge x Qx)$
(d) $\neg x = y : \neg y = x$
(e) $\bigwedge x \bigwedge y (x \neq y \wedge Px) : \bigwedge x \neg Px$

2. (a) Show that, instead of the first equality rule (E), we could manage with the weaker rule

(E') We may write down $x_0 = x_0$,

where x_0 is a fixed individual variable.

(b) Show analogously that, instead of the second equality rule (E_x^t), we could manage with the weaker rule (E_x^y) (where x, y are arbitrary individual variables).

3. Let $\mathfrak{C}$ be the class of those expressions which can be built up from equations of the type t = t by means of conjunctors and universal quantifiers. Examples:

$$t = t, \qquad \bigwedge y (t_1 = t_1 \wedge \bigwedge x\, t_2 = t_2).$$

Let α be a sequent, consisting of a single expression, which is derivable without the use of the removal rule (R). Show that:

(a) $\alpha \in \mathfrak{C}$,

(b) every proof of α can, by eradicating certain rows, be transformed into a proof of α in which neither the rule (A), nor the rule (X), nor the rule (E_x^t) is used.

(c) there is at least one proof of α in which none of the rules (C'), (C''), (G), (S_x^t) are applied.

4. In Chap. V we show that the predicate calculus is complete. Show that the predicate calculus without the rule (A) is incomplete.

5. Justify the "principle of proof" given at the end of 2.4. (cf. Chap. II, § 3.)

§ 3. The soundness of the predicate calculus

The reader is remined of the introductory considerations in Chap. IV, § 1. Here, we prove the

3.1 Theorem of the soundness of the predicate calculus. If $\mathfrak{M} \vdash \alpha$, then $\mathfrak{M} \models \alpha$.

The proof proceeds as follows: We call a sequent σ sound if $\mathfrak{A}_\sigma \models \varepsilon_\sigma$. We call a rule sound if, applied to sound starting sequents, it always yields a sound sequent (for a rule without premises, this is to mean that it only allows us to write down sound sequents). In 3.2 we show that each of the eleven rules which constitute the predicate calculus (see Chap. IV, § 2) is a sound rule. It then follows immediately that every derivable sequent is sound.

Now the theorem of the soundness of the predicate calculus can be proved as follows: Let $\mathfrak{M} \vdash \alpha$. By definition, this means that there are finitely many elements $\alpha_1, \ldots, \alpha_r$ of $\mathfrak{M}$ such that the sequent $\alpha_1 \ldots \alpha_r \alpha$ is derivable. By the above considerations, this sequent is sound. This means that $\alpha_1, \ldots, \alpha_r \models \alpha$. A fortiori, $\mathfrak{M} \models \alpha$, q.e.d.

3.2 The soundness of the rules of predicate logic. First of all, we note the following: If the rule under consideration leads from some starting sequents to a sequent σ, then we have to prove that σ is sound if the starting sequents are sound. In order to prove the soundness of σ, we start from some domain of individuals ω and from an interpretation $\mathfrak{J}$ over ω for which $\mathrm{Mod}_\omega \mathfrak{J} \mathfrak{A}_\sigma$. We have (using the soundness of the starting sequents) to show that $\mathrm{Mod}_\omega \mathfrak{J} \varepsilon_\sigma$. In all the proofs which follow, we do not need to leave the domain of individuals ω which we have chosen. Thus we shall not need to mention ω explicitly.

We shall continue to use the notation of 2.2.

The soundness of (A): Here, $\mathfrak{A}_\sigma = \{\varepsilon_\sigma\} = \{\alpha\}$, so that, trivially: If $\mathrm{Mod}\, \mathfrak{J} \mathfrak{A}_\sigma$, then $\mathrm{Mod}\, \mathfrak{J} \varepsilon_\sigma$.

The soundness of (C): Let $\mathrm{Mod}\,\mathfrak{J}\mathfrak{A}_\sigma$. Then both $\mathrm{Mod}\,\mathfrak{J}\mathfrak{A}_{\sigma_1}$ and $\mathrm{Mod}\,\mathfrak{J}\mathfrak{A}_{\sigma_2}$. By the soundness of σ_1 and σ_2, this yields $\mathrm{Mod}\,\mathfrak{J}\varepsilon_{\sigma_1}$ and $\mathrm{Mod}\,\mathfrak{J}\varepsilon_{\sigma_2}$. It follows from this that $\mathrm{Mod}\,\mathfrak{J}(\varepsilon_{\sigma_1} \wedge \varepsilon_{\sigma_2})$, i.e. $\mathrm{Mod}\,\mathfrak{J}\varepsilon_\sigma$.

The soundness of (C'): Let $\mathrm{Mod}\,\mathfrak{J}\mathfrak{A}_\sigma$. Then $\mathrm{Mod}\,\mathfrak{J}\mathfrak{A}_{\sigma_1}$. By the soundness of σ_1, this yields $\mathrm{Mod}\,\mathfrak{J}\varepsilon_{\sigma_1}$, i.e. $\mathrm{Mod}\,\mathfrak{J}(\alpha \wedge \beta)$. It follows from this that $\mathrm{Mod}\,\mathfrak{J}\alpha$, i.e. $\mathrm{Mod}\,\mathfrak{J}\varepsilon_\sigma$.

The soundness of (C'') can be shown analogously.

The soundness of (R): Let $\mathrm{Mod}\,\mathfrak{J}\mathfrak{A}_\sigma$. Then

$$\mathrm{Mod}\,\mathfrak{J}\mathfrak{A}_{\sigma_1} - \{\alpha\} \text{ and } \mathrm{Mod}\,\mathfrak{J}\mathfrak{A}_{\sigma_2} - \{\neg\,\alpha\}.$$

Now either $\mathrm{Mod}\,\mathfrak{J}\alpha$ or not $\mathrm{Mod}\,\mathfrak{J}\alpha$.
If $\mathrm{Mod}\,\mathfrak{J}\alpha$, then, by $\mathrm{Mod}\,\mathfrak{J}\mathfrak{A}_{\sigma_1} - \{\alpha\}$, $\mathrm{Mod}\,\mathfrak{J}\mathfrak{A}_{\sigma_1}$. Then, because of the soundness of σ_1, $\mathrm{Mod}\,\mathfrak{J}\varepsilon_{\sigma_1}$, i.e. $\mathrm{Mod}\,\mathfrak{J}\varepsilon_\sigma$.
If not $\mathrm{Mod}\,\mathfrak{J}\neg\,\alpha$, then $\mathrm{Mod}\,\mathfrak{J}\neg\,\alpha$ and hence, because of $\mathrm{Mod}\,\mathfrak{J}\mathfrak{A}_{\sigma_2} - \{\neg\,\alpha\}$, $\mathrm{Mod}\,\mathfrak{J}\mathfrak{A}_{\sigma_2}$. Then, by the soundness of σ_2, $\mathrm{Mod}\,\mathfrak{J}\varepsilon_{\sigma_2}$, i.e. $\mathrm{Mod}\,\mathfrak{J}\varepsilon_\sigma$.

Thus, the assertion has been proved for both the two cases.

The soundness of (X): Let $\mathrm{Mod}\,\mathfrak{J}\mathfrak{A}_\sigma$. Then $\mathrm{Mod}\,\mathfrak{J}\mathfrak{A}_{\sigma_1}$ and $\mathrm{Mod}\,\mathfrak{J}\mathfrak{A}_{\sigma_2}$. Because of the soundness of σ_1 and σ_2, this yields $\mathrm{Mod}\,\mathfrak{J}\varepsilon_{\sigma_1}$ and $\mathrm{Mod}\,\mathfrak{J}\varepsilon_{\sigma_2}$. But, since $\varepsilon_{\sigma_2} \equiv \neg\,\varepsilon_{\sigma_1}$, $\mathrm{Mod}\,\mathfrak{J}\neg\,\varepsilon_{\sigma_1}$ and thus not $\mathrm{Mod}\,\mathfrak{J}\varepsilon_{\sigma_1}$. This is a contradiction to $\mathrm{Mod}\,\mathfrak{J}\varepsilon_{\sigma_1}$. Thus there is no interpretation $\mathfrak{J}$ with $\mathrm{Mod}\,\mathfrak{J}\mathfrak{A}_\sigma$. Hence, trivially, $\mathrm{Mod}\,\mathfrak{J}\varepsilon_\sigma$ for all interpretations $\mathfrak{J}$ which are Models of $\mathfrak{A}_\sigma$.

The soundness of (G): Let $\mathrm{Mod}\,\mathfrak{J}\mathfrak{A}_\sigma$. Then $\mathrm{Mod}\,\mathfrak{J}\mathfrak{A}_{\sigma_1}$. By the soundness of σ_1, this yields $\mathrm{Mod}\,\mathfrak{J}\varepsilon_{\sigma_1}$, i.e. $\mathrm{Mod}\,\mathfrak{J}\bigwedge x\alpha$. This means that, for all $\mathfrak{x}$, $\mathrm{Mod}\,\mathfrak{J}_x^{\mathfrak{x}}\alpha$. If we put $\mathfrak{x} = \mathfrak{J}(x)$ and note that $\mathfrak{J}_x^{\mathfrak{J}(x)} = \mathfrak{J}$ (see Chap. III, § 2.2), then we have $\mathrm{Mod}\,\mathfrak{J}\alpha$, i.e. $\mathrm{Mod}\,\mathfrak{J}\varepsilon_\sigma$.

The soundness of (G_x): Let $\mathrm{Mod}\,\mathfrak{J}\mathfrak{A}_\sigma$. Then $\mathrm{Mod}\,\mathfrak{J}\mathfrak{A}_{\sigma_1}$. Now, by hypothesis, x does not occur free in $\mathfrak{A}_{\sigma_1}$ (critical condition). Thus, for arbitrary $\mathfrak{x}$, the interpretations $\mathfrak{J}$ and $\mathfrak{J}_x^{\mathfrak{x}}$ coincide for all the variables which occur free in any element of $\mathfrak{A}_{\sigma_1}$.

Thus, by the coincidence theorem (Chap. III, § 3.1), $\mathrm{Mod}\,\mathfrak{J}\mathfrak{A}_{\sigma_1}$ if and only if $\mathrm{Mod}\,\mathfrak{J}_x^{\mathfrak{x}}\mathfrak{A}_{\sigma_1}$. Hence, since $\mathrm{Mod}\,\mathfrak{J}\mathfrak{A}_{\sigma_1}$, we have $\mathrm{Mod}\,\mathfrak{J}_x^{\mathfrak{x}}\mathfrak{A}_{\sigma_1}$. Then, by the soundness of σ_1, $\mathrm{Mod}\,\mathfrak{J}_x^{\mathfrak{x}}\varepsilon_{\sigma_1}$. This is the case for every $\mathfrak{x}$. Hence we have $\mathrm{Mod}\,\mathfrak{J}\bigwedge x\varepsilon_{\sigma_1}$, i.e. $\mathrm{Mod}\,\mathfrak{J}\varepsilon_\sigma$.

The soundness of (S_x^t): Let Mod $\mathfrak{I}\mathfrak{A}_\sigma$. Now the elements of $\mathfrak{A}_\sigma$ are obtained from the elements of $\mathfrak{A}_{\sigma_1}$ by substitution of t for x. Thus, if β is an arbitrary element of $\mathfrak{A}_\sigma$ and α the corresponding element of $\mathfrak{A}_{\sigma_1}$, we have Subst $\alpha\, x\, t\, \beta$. By the substitution theorem (Chap. III, § 3.2), Mod $\mathfrak{I}_x^{\mathfrak{I}(t)}\alpha$ if and only if Mod $\mathfrak{I}\beta$. Hence, since Mod $\mathfrak{I}\beta$, Mod $\mathfrak{I}_x^{\mathfrak{I}(t)}\alpha$. Since this holds for every element α of $\mathfrak{A}_{\sigma_1}$, we have Mod $\mathfrak{I}_x^{\mathfrak{I}(t)} A_{\sigma_1}$. By the soundness of σ_1, this yields Mod $\mathfrak{I}_x^{\mathfrak{I}(t)}\varepsilon_{\sigma_1}$. But Subst $\varepsilon_{\sigma_1}\, x\, t\, \varepsilon_\sigma$. Thus, by the substitution theorem: Mod $\mathfrak{I}_x^{\mathfrak{I}(t)}\varepsilon_{\sigma_1}$ if and only if Mod $\mathfrak{I}\varepsilon_\sigma$. We know that Mod $\mathfrak{I}_x^{\mathfrak{I}(t)}\varepsilon_{\sigma_1}$. Hence Mod $\mathfrak{I}\varepsilon_\sigma$.

The soundness of (E): We need to show that Mod $\mathfrak{I}\, t = t$ for every $\mathfrak{I}$ and every t. But this is trivial, since $\mathfrak{I}(t) = \mathfrak{I}(t)$ always holds.

The soundness of (E_x^t): Let Mod $\mathfrak{I}\mathfrak{A}_\sigma$, i.e. Mod $\mathfrak{I}\mathfrak{A}_{\sigma_1}$ and Mod $\mathfrak{I}\, x = t$, i.e. Mod $\mathfrak{I}\mathfrak{A}_{\sigma_1}$ and $\mathfrak{I}(x) = \mathfrak{I}(t)$. By Mod $\mathfrak{I}\mathfrak{A}_{\sigma_1}$, we have Mod $\mathfrak{I}\varepsilon_{\sigma_1}$. By Subst $\varepsilon_{\sigma_1}\, x\, t\, \varepsilon_\sigma$ and the substitution theorem, Mod $\mathfrak{I}_x^{\mathfrak{I}(t)}\varepsilon_{\sigma_1}$ if and only if Mod $\mathfrak{I}\varepsilon_\sigma$. Now $\mathfrak{I}(t) = \mathfrak{I}(x)$, and hence $\mathfrak{I}_x^{\mathfrak{I}(t)} = \mathfrak{I}_x^{\mathfrak{I}(x)} = \mathfrak{I}$. Hence Mod $\mathfrak{I}\varepsilon_{\sigma_1}$ if and only if Mod $\mathfrak{I}\varepsilon_\sigma$, and hence, since Mod $\mathfrak{I}\varepsilon_{\sigma_1}$, Mod $\mathfrak{I}\varepsilon_\sigma$.

Exercises. 1. Show that the rule

$$(B):\quad \frac{\begin{array}{ll}\Sigma_1 & : \neg(\alpha \wedge \beta)\\ \Sigma_2 & : \neg\alpha\end{array}}{\Sigma_{12} \;:\; \neg\beta}$$

is unsound.

2. Show that the rule

$$(\bar{G}_x):\quad \frac{\Sigma : \alpha}{\Sigma : \bigwedge x\alpha}$$

is not sound. (Note that this is the rule (G_x) without the restriction that x may not occur free in Σ.)

3. Show that the rule

$$(B_y^x):\quad \frac{\Sigma : \bigwedge x\alpha}{\Sigma : \bigwedge y\alpha}$$

is unsound.

4. Using the following two methods, prove that the predicate calculus without the rule (E) is incomplete. Method I: Introduce a modified model relation Mod* by stipulating that, by contrast to the definition of Mod, Mod* $\mathfrak{I}\, t_1 = t_2$ is always false. The predicate calculus without (E) is sound relative to the relation of consequence $\models^*$ based on Mod*. Method II: Define an expression α^* for each expression α by means of the following inductive definition: If α is predicative, then $\alpha^* \equiv \alpha$. If α is an equation, then $\alpha^* \equiv (P \wedge \neg P)$, where P is a fixed no-place predicate variable. $[\neg\alpha]^* \equiv \neg\alpha^*$, $(\alpha \wedge \beta)^* \equiv (\alpha^* \wedge \beta^*)$, $[\bigwedge x\alpha]^* \equiv \bigwedge x\alpha^*$. If a sequent $\alpha_1 \ldots \alpha_r$ is derivable without the rule (E), then so is the sequent $\alpha_1^* \ldots \alpha_r^*$. (For this cf. also Example 3 in § 2.5.)

5. In analogy to the previous exercise, show that the predicate calculus without any one of the rules other than (A) and (E) is incomplete. (For (E) cf. Exercise 4, for (A) cf. § 2, Exercise 4.)

§ 4. Derivable rules

4.1 Introduction. In example 3 of 2.5 we derived the (one-member) sequent $\neg(Px \wedge \neg Px)$. We can also give the corresponding derivation for an arbitrary α instead of for Px:

1)	$(\alpha \wedge \neg \alpha)$	$: (\alpha \wedge \neg \alpha)$	(A)
2)	$(\alpha \wedge \neg \alpha)$	$: \alpha$	(C')1
3)	$(\alpha \wedge \neg \alpha)$	$: \neg \alpha$	(C'')1
4)	$(\alpha \wedge \neg \alpha)$	$: \neg(\alpha \wedge \neg \alpha)$	(X)2,3
5)	$\neg(\alpha \wedge \neg \alpha)$	$: \neg(\alpha \wedge \neg \alpha)$	(A)
6)		$: \neg(\alpha \wedge \neg \alpha)$	(R)4,5

Now let us consider the rule which reads as follows: For an arbitrary expression α, we may write down the (one-member) sequent $\neg(\alpha \wedge \neg \alpha)$ (we also express this briefly as: the rule $\neg(\alpha \wedge \neg \alpha)$). This rule is a rule without premises. It is not one of the rules of the predicate calculus. If we add a new rule to the rules of the predicate calculus, then, in general, it is to be expected that we can derive more sequents with this rule together with the rules of the predicate calculus than with the rules of the predicate calculus alone. However, this is not true of the rule which allows us to write down the sequent $\neg(\alpha \wedge \neg \alpha)$ straight away; for we can achieve the same result with the rules of the predicate calculus by writing down the six-line derivation given above. Thus, the rule $\neg(\alpha \wedge \neg \alpha)$ provides us with nothing new when added to the rules of the predicate calculus. It therefore "does no harm" to add it to these rules as an extra rule.

But why should we add this rule if we do not gain anything new by it? One advantage which the addition of the rule $\neg(\alpha \wedge \neg \alpha)$ to the rules of the predicate calculus yields is that some derivations (in the sense of 2.4) can be abbreviated by the use of the new rule . For if, in the course of a longer derivation, the sequent $\neg(\alpha \wedge \neg \alpha)$ is used, then we should really obtain it, every time, by means of the six-line derivation given above (or by means of some other derivation which uses only the rules of the predicate calculus). If, however, we use the rule $\neg(\alpha \wedge \neg \alpha)$ in addition to the rules of the predicate calculus, then we can obtain the desired sequent in one line and thus save ourselves five lines. Thus, we have an abbreviated derivation, which, however, can easily be transformed into a complete derivation by inserting the derivation of $\neg(\alpha \wedge \neg \alpha)$ given above.

Derivations which use only the rules of predicate logic are usually rather long. Thus, it is very convenient to use additional rules in order to abbreviate these derivations. However, such additional rules must be justified by showing how we can replace an application of them by applications of the actual rules of the predicate calculus. Additional rules for which this is possible are called derivable rules, or derived rules.

As we have said, the rule $\neg(\alpha \wedge \neg\alpha)$ which we have considered is a rule without premises. Now we want to consider an example of a derivable rule with one premise, namely the rule which allows us to pass from the sequent $\alpha\beta$ to the sequent $\neg\beta\neg\alpha$. We formulate this rule (in the notation of the schemata in 2.3) briefly as follows:

$$(*) \qquad \frac{\alpha : \beta}{\neg\beta : \neg\alpha} .$$

In order to show the derivability of this rule, we have to show how we can obtain the sequent $\neg\beta\neg\alpha$ from the sequent $\alpha\beta$ by means of the rules of the predicate calculus. (In doing this we could, in principle, also use those additional rules whose derivability has already been shown.) We can do this as follows:

1)	$\alpha : \beta$	(the sequent from which we want to start)
2)	$\neg\beta : \neg\beta$	(A)
3)	$\neg\beta\alpha : \neg\alpha$	(X) 1,2
4)	$\neg\alpha : \neg\alpha$	(A)
5)	$\neg\beta : \neg\alpha$	(R) 3,4

Thus, we have obtained the required sequent $\neg\beta\neg\alpha$ by using only the rules of predicate logic. The above lines serve as a justification for the rule (*).

The derivable rules are very important when it comes to representing proofs in an abbreviated form. We shall therefore note several derivable rules and justify them in the following pages. In doing this, we shall no longer number the lines of the derivations and shall leave it to the reader to find the preceding lines to which each rule relates. (Cf. the tables on pp. 113-115).

If we use the abbreviation

$$\alpha_1 \wedge \ldots \wedge \alpha_r \rightarrow \alpha \qquad \text{for} \qquad \neg(\alpha_1 \wedge \ldots \wedge \alpha_r \wedge \neg\alpha),$$

then (DdRu') yields the so-called

Deduction theorem. If $\alpha_1, \ldots, \alpha_r \vdash \alpha$, then $\alpha_1 \wedge \ldots \wedge \alpha_r \rightarrow \alpha$.

4.2 Derived rules which can be justified by means of the rules (A), (R), (X) alone[5]

Notation	Name	Rule	Justification	
(SeAt)	Self-assertion rule	$\dfrac{\Sigma\neg\alpha : \alpha}{\Sigma : \alpha}$	$\Sigma\neg\alpha : \alpha$	
			$\alpha : \alpha$	(A)
			$\Sigma : \alpha$	(R)
(SeDe)	Self-denial rule	$\dfrac{\Sigma\alpha : \neg\alpha}{\Sigma : \neg\alpha}$	$\Sigma\alpha : \neg\alpha$	
			$\neg\alpha : \neg\alpha$	(A)
			$\Sigma : \neg\alpha$	(R)

[5] The notations for the derived rules contain two capital letters, whereas the notations for the rules which constitute the predicate calculus contain only one capital letter.

4.2 Derived rules which can be justified by means of the rules (A), (R), (X) alone (contd.)

Notation	Name	Rule	Justification	
(NN)	First double negation rule	$\alpha : \neg\neg\alpha$	$\alpha : \alpha$	(A)
			$\neg\alpha : \neg\alpha$	(A)
			$\alpha\neg\alpha : \neg\neg\alpha$	(X)
			$\alpha : \neg\neg\alpha$	(SeDe)
(NN')	Second double negation rule	$\neg\neg\alpha : \alpha$	$\neg\alpha : \neg\alpha$	(A)
			$\neg\neg\alpha : \neg\neg\alpha$	(A)
			$\neg\neg\alpha\neg\alpha : \alpha$	(X)
			$\neg\neg\alpha : \alpha$	(SeAt)
(CuRu)	Cut rule	$\dfrac{\Sigma_1 : \alpha \qquad \Sigma_2\alpha : \beta}{\Sigma_{12} : \beta}$	$\Sigma_1 : \alpha$	
			$\Sigma_2\alpha : \beta$	
			$\neg\alpha : \neg\alpha$	(A)
			$\Sigma_1\neg\alpha : \beta$	(X)
			$\Sigma_{12} : \beta$	(R)
(CaPo)	First contraposition rule	$\dfrac{\Sigma\alpha : \beta}{\Sigma\neg\beta : \neg\alpha}$	$\Sigma\alpha : \beta$	
			$\neg\beta : \neg\beta$	(A)
			$\Sigma\neg\beta\alpha : \neg\alpha$	(X)
			$\Sigma\neg\beta : \neg\alpha$	(SeDe)
(CaPo')	Second contraposition rule	$\dfrac{\Sigma\alpha : \neg\beta}{\Sigma\beta : \neg\alpha}$	$\Sigma\alpha : \neg\beta$	
			$\beta : \beta$	(A)
			$\Sigma\beta\alpha : \neg\alpha$	(X)
			$\Sigma\beta : \neg\alpha$	(SeDe)
(CaPo'')	Third contraposition rule	$\dfrac{\Sigma\neg\alpha : \beta}{\Sigma\neg\beta : \alpha}$	$\Sigma\neg\alpha : \beta$	
			$\neg\beta : \neg\beta$	(A)
			$\Sigma\neg\beta\neg\alpha : \alpha$	(X)
			$\Sigma\neg\beta : \alpha$	(SeAt)
(CaPo''')	Fourth contraposition rule	$\dfrac{\Sigma\neg\alpha : \neg\beta}{\Sigma\beta : \alpha}$	$\Sigma\neg\alpha : \neg\beta$	
			$\beta : \beta$	(A)
			$\Sigma\beta\neg\alpha : \alpha$	(X)
			$\Sigma\beta : \alpha$	(SeAt)
(XQ)	"Ex contradictione quodlibet (1)"	$\alpha\neg\alpha : \beta$	$\alpha : \alpha$	(A)
			$\neg\alpha : \neg\alpha$	(A)
			$\alpha\neg\alpha : \beta$	(X)

4.3 Derived rules for whose justification the rules (C), (C'), (C'') are needed

Notation	Name	Rule	Justification	
(AEx)	Extended assumption rule	$\alpha\beta : \alpha$	$\alpha : \alpha$	(A)
			$\alpha\beta : \alpha$	(§ 2.6(II))
(CCo)	Commutative rule for conjunction	$(\alpha\wedge\beta) : (\beta\wedge\alpha)$	$(\alpha\wedge\beta) : (\alpha\wedge\beta)$	(A)
			$(\alpha\wedge\beta) : \alpha$	(C')
			$(\alpha\wedge\beta) : \beta$	(C'')
			$(\alpha\wedge\beta) : (\beta\wedge\alpha)$	(C)
(CAs)	First associative rule for conjunction	$((\alpha\wedge\beta)\wedge\gamma) : (\alpha\wedge(\beta\wedge\gamma))$	$((\alpha\wedge\beta)\wedge\gamma) : ((\alpha\wedge\beta)\wedge\gamma)$	(A)
			$((\alpha\wedge\beta)\wedge\gamma) : (\alpha\wedge\beta)$	(C')
			$((\alpha\wedge\beta)\wedge\gamma) : \alpha$	(C')
			$((\alpha\wedge\beta)\wedge\gamma) : \beta$	(C'')
			$((\alpha\wedge\beta)\wedge\gamma) : \gamma$	(C'')
			$((\alpha\wedge\beta)\wedge\gamma) : (\beta\wedge\gamma)$	(C)
			$((\alpha\wedge\beta)\wedge\gamma) : (\alpha\wedge(\beta\wedge\gamma))$	(C)
(CAs')	Second associative rule for conjunction	$(\alpha\wedge(\beta\wedge\gamma)) : ((\alpha\wedge\beta)\wedge\gamma)$	analogously	
(AnDc)	Decomposition of a conjunctive member of the antecedent	$\frac{\Sigma(\alpha\wedge\beta) : \gamma}{\Sigma\alpha\beta : \gamma}$	$\Sigma(\alpha\wedge\beta) : \gamma$	
			$\alpha : \alpha$	(A)
			$\beta : \beta$	(A)
			$\alpha\beta : (\alpha\wedge\beta)$	(C)
			$\Sigma\alpha\beta : \gamma$	(CuRu)
(AnU)	Unification of two members of the antecedent	$\frac{\Sigma\alpha\beta : \gamma}{\Sigma(\alpha\wedge\beta) : \gamma}$	$\Sigma\alpha\beta : \gamma$	
			$(\alpha\wedge\beta) : (\alpha\wedge\beta)$	(A)
			$(\alpha\wedge\beta) : \alpha$	(C')
			$\Sigma(\alpha\wedge\beta)\beta : \gamma$	(CuRu)
			$(\alpha\wedge\beta) : \beta$	(C'')
			$\Sigma(\alpha\wedge\beta) : \gamma$	(CuRu)
(UU)	Simultaneous unification of two members of the antecedent and the succedent	$\frac{\Sigma_1\alpha : \beta \quad \Sigma_2\gamma : \delta}{\Sigma_{12}(\alpha\wedge\gamma) : (\beta\wedge\delta)}$	$\Sigma_1\alpha : \beta$	
			$\Sigma_2\gamma : \delta$	
			$\Sigma_{12}\alpha\gamma : (\beta\wedge\delta)$	(C)
			$\Sigma_{12}(\alpha\wedge\gamma) : (\beta\wedge\delta)$	(AnU)

4.3 Derived rules for whose justification the rules (C), (C'), (C'') are needed (contd.)

Notation	Name	Rule	Justification
(XQ')	"Ex contradictione quodlibet (2)"	$(\alpha\wedge\neg\alpha) : \beta$	$(\alpha\wedge\neg\alpha) : (\alpha\wedge\neg\alpha)$ (A) $(\alpha\wedge\neg\alpha) : \alpha$ (C') $(\alpha\wedge\neg\alpha) : \neg\alpha$ (C'') $(\alpha\wedge\neg\alpha) : \beta$ (X)
(NX)	Rule of no contadiction	$: \neg(\alpha\wedge\neg\alpha)$	$(\alpha\wedge\neg\alpha) : \neg(\alpha\wedge\neg\alpha)$ (XQ') $: \neg(\alpha\wedge\neg\alpha)$ (SeDe) (see also 4.1)
(DdRu)	First deduction rule	$\dfrac{: \neg(\alpha_1\wedge\ldots\wedge\alpha_r\wedge\neg\alpha)}{\alpha_1\ldots\alpha_r : \alpha}$	$: \neg(\alpha_1\wedge\ldots\wedge\alpha_r\wedge\neg\alpha)$ $(\alpha_1\wedge\ldots\wedge\alpha_r\wedge\neg\alpha) : (\alpha_1\wedge\ldots\wedge\alpha_r\wedge\neg\alpha)$ (A) $(\alpha_1\wedge\ldots\wedge\alpha_r\wedge\neg\alpha) : \alpha$ (X) $\alpha_1\ldots\alpha_r\neg\alpha : \alpha$ (AnDc) $\alpha_1\ldots\alpha_r : \alpha$ (SeAt)
(DdRu')	Second deduction rule	$\dfrac{\alpha_1\ldots\alpha_r : \alpha}{: \neg(\alpha_1\wedge\ldots\wedge\alpha_r\wedge\neg\alpha)}$	$\alpha_1\ldots\alpha_r : \alpha$ $\alpha_1\ldots\alpha_r\neg\alpha : \alpha$ (2.6, Rem. II) $(\alpha_1\wedge\ldots\wedge\alpha_r\wedge\neg\alpha) : \alpha$ (AnU) $(\alpha_1\wedge\ldots\wedge\alpha_r\wedge\neg\alpha) : (\alpha_1\wedge\ldots\wedge\alpha_r\wedge\neg\alpha)$ (A) $(\alpha_1\wedge\ldots\wedge\alpha_r\wedge\neg\alpha) : \neg\alpha$ (C'') $(\alpha_1\wedge\ldots\wedge\alpha_r\wedge\neg\alpha) : \neg(\alpha_1\wedge\ldots\wedge\alpha_r\wedge\neg\alpha)$ (X) $: \neg(\alpha_1\wedge\ldots\wedge\alpha_r\wedge\neg\alpha)$ (SeDe)

4.4 Derived rules for whose justification the rules (A), (R), (X), (G), (G_x), (S_x^t) are needed

Notation	Name	Rule		Justification	
(GG_x)	Simultaneous generalisation over x	$\dfrac{\Sigma\alpha : \beta}{\Sigma\bigwedge x\alpha : \bigwedge x\beta}$	if x does not occur free in Σ	$\Sigma\alpha : \beta$	
				$\bigwedge x\alpha : \bigwedge x\alpha$	(A)
				$\bigwedge x\alpha : \alpha$	(G)
				$\Sigma\bigwedge x\alpha : \beta$	(CuRu)
				$\Sigma\bigwedge x\alpha : \bigwedge x\beta$	(G_x)
($ExG_{x,t}$)	Extended generalisation rule	$\dfrac{\Sigma : \bigwedge x\alpha}{\Sigma : \beta}$	if Subst $\alpha\, x\, t\, \beta$	see below	
($SG_{x,y}$)	Substitution and generalisation rule	$\dfrac{\Sigma : \beta}{\Sigma : \bigwedge x\alpha}$	if Subst $\alpha\, x\, y\, \beta$ Subst $\beta\, y\, x\, \alpha$ y not free in Σ	see below	
($ReG_{x,y}$)	Renaming of bound variables in generalisations	$\bigwedge x\alpha : \bigwedge y\beta$	if Subst $\alpha\, x\, y\, \beta$ and Subst $\beta\, y\, x\, \alpha$	$\bigwedge x\alpha : \bigwedge x\alpha$	(A)
				$\bigwedge x\alpha : \alpha$	(G)
				$\bigwedge x\alpha : \bigwedge y\beta$	($SG_{x,y}$)

Justification for ($ExG_{x,t}$). First of all, a preparatory note. Let z be an individual variable which does not occur either in Σ or in β. Supose also that $z \neq x$ and that z does not occur in t. Then, by Chap. II, §5.4, Theorem 6, the substitution of z for x can be carried out in every member of Σ and, conversely, the substitution of x for z can be carried out in the resulting expressions and leads back to the original expressions. If we denote by Σ' the sequent which is obtained from Σ by replacing x by z in each of its members, then we can formulate the last statement symbolically as follows:

(**) $\qquad$ Subst $\Sigma\, x\, z\, \Sigma'$ $\quad$ and $\quad$ Subst $\Sigma'\, z\, x\, \Sigma$.

Now we proceed to the justification of ($ExG_{x,t}$):

$\Sigma : \bigwedge x\alpha$

$\Sigma' : \bigwedge x\alpha$ $\quad$ (S_x^z) $\quad$ because (**) and Subst $\bigwedge x\alpha\, x\, z \bigwedge x\alpha$

$\Sigma' : \alpha$ $\quad$ (G)

$\Sigma' : \beta$	(S^t_x)	by Chap. II, § 5.4, Theorem 8, x does not occur free in Σ', so that, by Theorem 4 there, Subst Σ' x t Σ'; moreover, by hypothesis, Subst α x t β
$\Sigma : \beta$	(S^x_z)	because (**) and because (since z does not occur free in β and therefore does not occur in β), by Chap. II, § 5.4, Theorem 4, Subst β z x β.

Justification for $(SG_{x,y})$. If $x \equiv y$ then, since Subst α x y β and noting Chap. II, § 5.4, Theorems 1 and 4, $\beta \equiv \alpha$. Thus $\Sigma \bigwedge x\alpha$ can be obtained from $\Sigma\beta$ by applying (G_x). Thus we may, for the following, assume that $x \not\equiv y$. As a preparation for the proof, as in the justification of $(ExG_{x,t})$, we introduce an individual variable z which does not occur either in Σ or in $\bigwedge x\alpha$. Moreover, let $z \not\equiv y$. (Clearly also $z \not\equiv x$, since z does not occur in $\bigwedge x\alpha$.)

Then there is a sequent Σ' such that

(***) Subst Σ x z Σ' and Subst Σ' z x Σ .

Now for the justification of $(SG_{x,y})$:

$\Sigma : \beta$		
$\Sigma' : \beta$	(S^z_x)	because (***) and since x does not occur free in β (the latter because Subst α x y β and $x \not\equiv y$)
$\Sigma' : \alpha$	(S^x_y)	Subst Σ' x y Σ', since y does not occur free in Σ'; the latter because Subst Σ x z Σ' and because, by Chap. II, § 5.4, Theorem 7, the variable y occurs free in Σ' if and only if (i) y occurs free in Σ and $y \not\equiv x$, or (ii) x occurs free in Σ and $y \equiv z$. [(i) does not hold since, by hypothesis, y does not occur free in Σ (ii) does not hold, since $z \not\equiv y$.]
$\Sigma' : \bigwedge x\alpha$	(G_x)	in applying (G_x) we use the fact that x does not occur free in Σ'. This follows from Subst Σ x z Σ' and $x \not\equiv z$.
$\Sigma : \bigwedge x\alpha$	(S^x_z)	because (***) and because Subst $\bigwedge x\alpha$ z x $\bigwedge x\alpha$, since, by hypothesis, z does not occur in $\bigwedge x\alpha$ and therefore does not occur free in it.

4.5 Derived rules for whose justification the rules (A), (R), (X), (S_x^t), (E) and (E_x^t) are used

Notation	Name	Rule
(ESy)	Symmetry rule for equality	$t_1 = t_2 : t_2 = t_1$
(ET)	Transitivity rule for equality	$t_1 = t_2\ t_2 = t_3 : t_1 = t_3$
(ER)	First replacement rule for equality	$t_1 = t'_1 \ldots t_r = t'_r\ Pt_1 \ldots t_r : Pt'_1 \ldots t'_r$
(ER')	Second replacement rule for equality	$t_1 = t'_1 \ldots t_r = t'_r : ft_1 \ldots t_r = ft'_1 \ldots t'_r$

Justification.

(ESy): Let x, y, z be pairwise distinct individual variables which are chosen so that none of them occurs in t_1 or t_2. Then we form the derivation:

$$x = z\ x = z\ \text{(A)}$$
$$x = z\ x = y\ y = z\ (E_x^y)$$
$$x = x\ x = y\ y = x\ (S_z^x)$$
$$x = x\ \text{(E)}$$
$$x = y\ y = x\ \text{(CuRu)}$$
$$t_1 = y\ y = t_1\ (S_x^{t_1})$$
$$t_1 = t_2 t_2 = t_1\ (S_y^{t_2})$$

(ET): Let x, y and z be pairwise distinct and chosen so that they do not occur in t_1, t_2 or t_3. We form the derivation:

$$x = y\ x = y\ \text{(A)}$$
$$x = y\ y = z\ x = z\ (E_x^y)$$
$$t_1 = y\ y = z\ t_1 = z\ (S_x^{t_1})$$
$$t_1 = t_2 t_2 = z\ t_1 = z\ (S_y^{t_2})$$
$$t_1 = t_2 t_2 = t_3 t_1 = t_3\ (S_z^{t_3})$$

(ER): Let $x_1,\dots,x_r$ be pairwise distinct and chosen so that they do not occur in any of the terms $t_1,\dots,t_r$, $t'_1,\dots,t'_r$. We form the derivation:

$$\begin{array}{rr}
Px_1\dots x_r\ Px_1\dots x_r & (A)\\
x_r = t'_r\ Px_1\dots x_r\ Px_1\dots t'_r & (E^{t'_r}_{x_r})\\
\dots\dots\dots\dots\dots\dots & \dots\\
x_1 = t'_1\dots x_r = t'_r\ Px_1\dots x_r\ Pt'_1\dots t'_r & (E^{t'_1}_{x_1})\\
t_1 = t'_1\dots x_r = t'_r\ Pt_1\dots x_r\ Pt'_1\dots t'_r & (S^{t_1}_{x_1})\\
\dots\dots\dots\dots\dots\dots & \dots\\
t_1 = t'_1\dots t_r = t'_r\ Pt_1\dots t_r\ Pt'_1\dots t'_r & (S^{t_r}_{x_r})
\end{array}$$

(ER'): Let $x_1,\dots,x_r$, $y_1,\dots,y_r$ be pairwise distinct and chosen so that they do not occur in any of the terms $t_1,\dots,t_r$, $t'_1,\dots,t'_r$. We form the derivation:

$$\begin{array}{rr}
fx_1\dots x_r = fy_1\dots y_r\ fx_1\dots x_r = fy_1\dots y_r & (A)\\
y_r = t'_r\ fx_1\dots x_r = fy_1\dots y_r\ fx_1\dots x_r = fy_1\dots t'_r & (E^{t'_r}_{y_r})\\
\dots\dots\dots\dots\dots\dots & \dots\\
y_1 = t'_1\dots y_r = t'_r\ fx_1\dots x_r = fy_1\dots y_r\ fx_1\dots x_r = ft'_1\dots t'_r & (E^{t'_1}_{y_r})\\
x_1 = t'_1\dots y_r = t'_r\ fx_1\dots x_r = fx_1\dots y_r\ fx_1\dots x_r = ft'_1\dots t'_r & (S^{x_1}_{y_1})\\
\dots\dots\dots\dots\dots\dots & \dots\\
x_1 = t'_1\dots x_r = t'_r\ fx_1\dots x_r = fx_1\dots x_r\ fx_1\dots x_r = ft'_1\dots t'_r & (S^{x_r}_{y_r})\\
fx_1\dots x_r = fx_1\dots x_r & (E)\\
x_1 = t'_1\dots x_r = t'_r\ fx_1\dots x_r = ft'_1\dots t'_r & (CuRu)\\
t_1 = t'_1\dots x_r = t'_r\ ft_1\dots x_r = ft'_1\dots t'_r & (S^{t_1}_{x_1})\\
\dots\dots\dots\dots\dots\dots & \dots\\
t_1 = t'_1\dots t_r = t'_r\ ft_1\dots t_r = ft'_1\dots t'_r & (S^{t_r}_{x_r})
\end{array}$$

Exercises. 1. Prove the derivability of the following rule:

$$\frac{\begin{array}{l}\Sigma_1\alpha : \beta\\ \Sigma_2\alpha : \neg\beta\end{array}}{\Sigma_{12}\ \ \cdot \neg\alpha}$$

2. If, in the system of rules which defines the predicate calculus, we replace the rule (X) by the rules (XQ') and (CuRu), then we obtain an equivalent system. (Two systems of rules are said to be equivalent if the same sequents are derivable from them.)

3. If, in the system of rules which defines the predicate calculus, we replace the rule (R) by the two rules (SeAs) and (SeDe), then we obtain an equivalent system.

4. Under the assumption Subst $\alpha x t \beta$, prove the derivability of the following two rules:

(a) $$\frac{\Sigma : \beta}{\Sigma : \bigwedge x(x = t \rightarrow \alpha)}$$ if x does not occur in t,

(b) $$\frac{\Sigma : \alpha \wedge x = t}{\Sigma : \beta}.$$

5. Let a term be called simple if it consists of an n-place function variable followed by n individual variables ($n \geqslant 0$). Let an equation be called simple if the left-hand side is an individual variable and the right-hand side a simple term. Let a predicative expression $Pt_1 \ldots t_n$ be called simple if the terms $t_1, \ldots, t_n$ are individual variables. Let an expression be called simple if all the atomic expressions which occur in it are simple. Using Exercise 4(a), show that for every expression α we can find a simple expression β such that $\alpha \dashv\vdash \beta$ (i.e. such that $\alpha \vdash \beta$ and $\beta \vdash \alpha$).

6. If P and Q are predicate variables with the same number of places and Σ^* is obtained from Σ by replacing P by Q everywhere in every expression in Σ, and if $\vdash \Sigma$, then $\vdash \Sigma^*$. The same holds for the substitution of function variables.

7. If $\vdash \Sigma$, then there is a derivation of Σ in which a predicate variable or a more-than-no-place function variable occurs only if it occurs in Σ.

8. There is no proof of $\bigwedge x\, x = x$ which contains no free individual variables. Thus the statement in Exercise 7 cannot be extended to no-place function variables.

9. Find a derivation for $\Gamma \bigvee x \bigwedge y\, xy = y$, where x, y are distinct individual variables, xy stands for fxy where f is 2-place, and the sequent Γ contains the three group-theoretic axioms as they are given in mathematical language in Chap. I, §2.3.

§5. Some properties of the concept of derivability

Consistency

5.1 Survey. In 5.2 we bring together a number of simple laws which hold for the relation $\mathfrak{M} \vdash \alpha$. In 5.3 we explain the syntactic concept of consistency with the aid of the concept of derivability. The idea of consistency makes it possible to formulate the statement of the completeness of the predicate calculus in a simpler way which we shall use later as the basis of the actual proof.

5.2 Some properties of the concept of derivability

(1) If $\mathfrak{M} \vdash \alpha$ and $\mathfrak{M} \subset \mathfrak{N}$, then $\mathfrak{N} \vdash \alpha$.

(2) If $\alpha \in \mathfrak{M}$, then $\mathfrak{M} \vdash \alpha$.

(3) If $\mathfrak{M} \vdash \alpha$ and $\mathfrak{N} \cup \{\alpha\} \vdash \beta$, then $\mathfrak{M} \cup \mathfrak{N} \vdash \beta$.

(4) If $\mathfrak{M} \cup \{\neg \alpha\} \vdash \alpha$, then $\mathfrak{M} \vdash \alpha$.

(5) If $\mathfrak{M} \cup \{\alpha\} \vdash \neg \alpha$, then $\mathfrak{M} \vdash \neg \alpha$.

(6) $\mathfrak{M} \cup \{\alpha, \beta\} \vdash \gamma$ if and only if $\mathfrak{M} \cup \{(\alpha \wedge \beta)\} \vdash \gamma$.

(7) If $\mathfrak{M} \cup \{\alpha\} \vdash \beta$ and $\mathfrak{N} \cup \{\neg \alpha\} \vdash \beta$, then $\mathfrak{M} \cup \mathfrak{N} \vdash \beta$.

(8) If $\mathfrak{M} \vdash \alpha$ and $\mathfrak{N} \vdash \neg\alpha$, then $\mathfrak{M} \cup \mathfrak{N} \vdash \beta$ for every β.

(9) If $\mathfrak{M} \vdash (\alpha \wedge \neg\alpha)$, then $\mathfrak{M} \vdash \beta$ for every β.

(10) If $\mathfrak{M} \vdash \bigwedge x\alpha$, then $\mathfrak{M} \vdash \alpha$.

(11) If $\mathfrak{M} \vdash \bigwedge x\alpha$ and Subst $\alpha x t \beta$, then $\mathfrak{M} \vdash \beta$.

(12) If $\mathfrak{M} \vdash \alpha$ and if x does not occur free in $\mathfrak{M}$, then $\mathfrak{M} \vdash \bigwedge x\alpha$.

(13) If $\mathfrak{M} \vdash \beta$ and both Subst $\alpha x y \beta$ and Subst $\beta y x \alpha$ and if y does not occur free in $\mathfrak{M}$, then $\mathfrak{M} \vdash \bigwedge x\alpha$.

(14) If $\alpha \vdash \beta$, then $\bigwedge x\alpha \vdash \bigwedge x\beta$.

Remark. After the proof of the completeness theorem (Chap. V) we shall know that $\mathfrak{M} \vdash \alpha$ if and only if $\mathfrak{M} \models \alpha$. Thus, all the laws given above also hold for the notion of consequence.

Proof. (1) can be obtained immediately from the definition of the relation of derivability: $\mathfrak{M} \vdash \alpha$ means that there are finitely many elements $\alpha_1, \ldots, \alpha_r$ of $\mathfrak{M}$ with $\alpha_1, \ldots, \alpha_r \vdash \alpha$. But these elements are also in $\mathfrak{N}$; hence $\mathfrak{N} \vdash \alpha$.

(2) This is shown by the derivation $\alpha\alpha$(A).

(3) Because of $\mathfrak{M} \vdash \alpha$ there are elements $\alpha_1, \ldots, \alpha_r$ in $\mathfrak{M}$ such that

$$\alpha_1 \ldots \alpha_r \alpha$$

is derivable. Because of 2.6 (III) there are elements $\beta_1, \ldots, \beta_s$ of $\mathfrak{N}$ such that

$$\beta_1 \ldots \beta_s \alpha\beta$$

is derivable. If we write derivations for the above-named sequents one below the other, then we obtain another derivation. This can be extended by the use of the cut rule by adding the sequent

$$\alpha_1 \ldots \alpha_r \beta_1 \ldots \beta_s \beta.$$

The derivation obtained in this way shows that $\mathfrak{M} \cup \mathfrak{N} \vdash \beta$.

Many of the following assertions will be proved using the method we have just applied. However, we do not want to write down all the details of these proofs. We shall limit ourselves to noting the rule (or rules) of inference which we have to use. In this shortened notation, the proof for (3) which we have just given would read as follows (on the left are derivable sequents, on the right explanations):

$\alpha_1 \ldots \alpha_r \alpha$ $(\alpha_1, \ldots, \alpha_r \in \mathfrak{M})$
$\beta_1 \ldots \beta_s \alpha\, \beta$ $(\beta_1, \ldots, \beta_s \in \mathfrak{M};$ § 2.6 (III))
$\alpha_1 \ldots \alpha_1 \beta_1 \ldots \beta_s \beta$ (CuRu)

(4) $\alpha_1 \ldots \alpha_r \neg \alpha\, \alpha$ $(\alpha_1, \ldots, \alpha_r \in \mathfrak{M};$ § 2.6 (III))
$\alpha_1 \ldots \alpha_r\, \alpha$ (SeAt)

(5) $\alpha_1 \ldots \alpha_r \alpha \neg \alpha$ $(\alpha_1, \ldots, \alpha_r \in \mathfrak{M};$ § 2.6 (III))
$\alpha_1 \ldots \alpha_r \neg \alpha$ (SeDe)

(6) $\alpha_1 \ldots \alpha_r \alpha\beta\, \gamma$ $(\alpha_1, \ldots, \alpha_r \in \mathfrak{M};$ § 2.6 (III))
$\alpha_1 \ldots \alpha_r (\alpha \wedge \beta)\, \gamma$ (AnU)

and:

$\alpha_1 \ldots \alpha_r (\alpha \wedge \beta)\, \gamma$ $(\alpha_1, \ldots, \alpha_r \in \mathfrak{M};$ § 2.6 (III))
$\alpha_1 \ldots \alpha_r \alpha\beta\, \gamma$ (AnDc)

(7) $\alpha_1 \ldots \alpha_r \alpha\, \beta$ $(\alpha_1, \ldots, \alpha_r \in \mathfrak{M};$ § 2.6 (III))
$\beta_1 \ldots \beta_s \neg \alpha\, \beta$ $(\beta_1, \ldots, \beta_s \in \mathfrak{N};$ § 2.6 (III))
$\alpha_1 \ldots \alpha_r \beta_1 \ldots \beta_s\, \beta$ (R)

(8) $\alpha_1 \ldots \alpha_r\, \alpha$ $(\alpha_1, \ldots \alpha_r \in \mathfrak{M})$
$\beta_1 \ldots \beta_s \neg \alpha$ $(\beta_1, \ldots, \beta_s \in \mathfrak{N})$
$\alpha_1 \ldots \alpha_r \beta_1 \ldots \beta_s\, \beta$ (X)

(9) $\alpha_1 \ldots \alpha_r\, (\alpha \wedge \neg \alpha)$ $(\alpha_1, \ldots, \alpha_r \in \mathfrak{M})$
$(\alpha \wedge \neg \alpha)\, \beta$ (XQ')
$\alpha_1 \ldots \alpha_r\, \beta$ (CuRu)

(10) $\alpha_1 \ldots \alpha_r \bigwedge x \alpha$ $(\alpha_1, \ldots, \alpha_r \in \mathfrak{M})$
$\alpha_1 \ldots \alpha_r\, \alpha$ (G)

(11) $\alpha_1 \ldots \alpha_r \bigwedge x \alpha$ $(\alpha_1, \ldots, \alpha_r \in \mathfrak{M})$
$\alpha_1 \ldots \alpha_r\, \beta$ $(\mathrm{ExG}_{x,t})$

(12) $\alpha_1 \ldots \alpha_r\, \alpha$ $(\alpha_1, \ldots, \alpha_r \in \mathfrak{M})$
$\alpha_1 \ldots \alpha_r \bigwedge x \alpha$ (G_x)

(13) $\alpha_1 \ldots \alpha_r\, \beta$ $(\alpha_1, \ldots, \alpha_r \in \mathfrak{M})$
$\alpha_1 \ldots \alpha_r \bigwedge x \alpha$ $(\mathrm{SG}_{x,y})$

(14) $\alpha\, \beta$
$\bigwedge x \alpha \bigwedge x \alpha$ (A)
$\bigwedge x \alpha\, \alpha$ (G)
$\bigwedge x \alpha\, \beta$ (CuRu)
$\bigwedge x \alpha \bigwedge x \beta$ (G_x)

5.3 Consistent and inconsistent sets of expressions. These syntactical concepts are introduced by:

Definition 1. The set $\mathfrak{M}$ of expressions is consistent if and only if there is an expression α which is not derivable from $\mathfrak{M}$.

Definition 2. The set $\mathfrak{M}$ of expressions is inconsistent if and only if every expression α is derivable from $\mathfrak{M}$.

Thus, a set of expressions $\mathfrak{M}$ is inconsistent if and only if $\mathfrak{M}$ is not consistent. Moreover, the following laws hold:

(15) If $\mathfrak{N}$ is inconsistent and $\mathfrak{N} \subset \mathfrak{M}$, then $\mathfrak{M}$ is also inconsistent.

(16) If $\mathfrak{M}$ is consistent and $\mathfrak{N} \subset \mathfrak{M}$, then $\mathfrak{N}$ is consistent.

(17) $\mathfrak{M}$ is inconsistent if and only if there is an expression α such that $\mathfrak{M} \vdash (\alpha \wedge \neg \alpha)$.

(18) $\mathfrak{M}$ is inconsistent if and only if there is an expression α such that $\mathfrak{M} \vdash \alpha$ and $\mathfrak{M} \vdash \neg \alpha$.

(19) $\mathfrak{M} \cup \{\alpha\}$ is inconsistent if and only if $\mathfrak{M} \vdash \neg \alpha$.

(20) $\mathfrak{M} \cup \{\neg \alpha\}$ is inconsistent if and only if $\mathfrak{M} \vdash \alpha$.

(21) $\mathfrak{M}$ is inconsistent if and only if both $\mathfrak{M} \cup \{\alpha\}$ and $\mathfrak{M} \cup \{\neg \alpha\}$ are inconsistent.

Proof.

(15) follows from (1).

(16) is the contraposition of (15).

(17) (a) If $\mathfrak{M}$ is inconsistent, then every expression, in particular $(\alpha \wedge \neg \alpha)$ for an arbitrary α, is derivable from $\mathfrak{M}$.

(b) If $\mathfrak{M} \vdash (\alpha \wedge \neg \alpha)$ for an expression α, then, by (9), $\mathfrak{M} \vdash \beta$ for every β, hence $\mathfrak{M}$ is inconsistent.

(18) (a) If $\mathfrak{M}$ is inconsistent, then every expression is derivable from $\mathfrak{M}$; hence, so are α and $\neg \alpha$ for an arbitrary α.

(b) If $\mathfrak{M} \vdash \alpha$ and $\mathfrak{M} \vdash \neg \alpha$ for an expression α, then, by (8), $\mathfrak{M} \vdash \beta$ for an arbitrary expression β; hence, $\mathfrak{M}$ is inconsistent.

(19) (a) If $\mathfrak{M} \cup \{\alpha\}$ is inconsistent, then every expression is derivable from $\mathfrak{M} \cup \{\alpha\}$. Hence, $\mathfrak{M} \cup \{\alpha\} \vdash \neg \alpha$. Then, by (5), $\mathfrak{M} \vdash \neg \alpha$.

(b) If $\mathfrak{M} \vdash \neg \alpha$, then, by (1), $\mathfrak{M} \cup \{\alpha\} \vdash \neg \alpha$. By (2), $\mathfrak{M} \cup \{\alpha\} \vdash \alpha$. Hence, by (18), $\mathfrak{M} \cup \{\alpha\}$ is inconsistent.

(20) (a) If $\mathfrak{M} \cup \{\neg \alpha\}$ is inconsistent, then every expression is derivable from $\mathfrak{M} \cup \{\neg \alpha\}$. Hence, $\mathfrak{M} \cup \{\neg \alpha\} \vdash \alpha$. Then, by (4), $\mathfrak{M} \vdash \alpha$.

(b) If $\mathfrak{M} \vdash \alpha$, then, by (1), $\mathfrak{M} \cup \{\neg \alpha\} \vdash \alpha$. By (2), $\mathfrak{M} \cup \{\neg \alpha\} \vdash \neg \alpha$. Hence, by (18), $\mathfrak{M} \cup \{\neg \alpha\}$ is inconsistent.

(21) (a) If $\mathfrak{M}$ is inconsistent, then, by (15), so are $\mathfrak{M} \cup \{\alpha\}$ and $\mathfrak{M} \cup \{\neg \alpha\}$.

(b) If $\mathfrak{M} \cup \{\alpha\}$ and $\mathfrak{M} \cup \{\neg \alpha\}$ are inconsistent, then, by (19) and (20), $\mathfrak{M} \vdash \neg \alpha$ and $\mathfrak{M} \vdash \alpha$; hence, by (18), $\mathfrak{M}$ is also inconsistent.

We now prove two important lemmas (for satisfiability cf. Chap. III, § 2.5):

Lemma 1. If the predicate calculus is sound, then every satisfiable set of expressions is consistent.

Lemma 2. If every consistent set of expressions is satisfiable, then the predicate calculus is complete.

Since the soundness of the predicate calculus has already been proved, Lemma 1 yields the

Theorem. If any set of expressions is satisfiable, then it is also consistent.

In Chap. V, § 4.,5, we shall prove that every consistent set of expressions is satisfiable; by Lemma 2, the completeness of the predicate calculus follows from this. In particular, we shall then have proved the equivalence of the semantic statement $\mathfrak{M}$ is satisfiable with the syntactic statement $\mathfrak{M}$ is consistent.

P r o o f of L e m m a 1. Let $\mathfrak{M}$ be satisfiable. Then there is a domain of individuals ω and an interpretation $\mathfrak{J}$ over ω such that $\mathrm{Mod}_\omega \mathfrak{J}\mathfrak{M}$. Let α be an arbitrary expression. Then $\mathrm{Mod}_\omega \mathfrak{J} \neg \alpha$ if and only if not $\mathrm{Mod}_\omega \mathfrak{J}\alpha$. Thus, at least one of the expressions α, $\neg \alpha$ is not valid in $\mathfrak{J}$. Hence, there is an expression β which is not valid in $\mathfrak{J}$. We assert that β is not derivable from $\mathfrak{M}$ (which proves the asserted consistency of $\mathfrak{M}$). For otherwise we should have $\mathfrak{M} \vdash \beta$, and hence $\mathfrak{M} \models \beta$, since we are assuming the soundness of the predicate calculus. But then, since $\mathrm{Mod}_\omega \mathfrak{J}\mathfrak{M}$, we should also have $\mathrm{Mod}_\omega \mathfrak{J}\beta$.

P r o o f of L e m m a 2. In order to prove completeness, we start from arbitrary $\mathfrak{M}$, α for which we assume that $\mathfrak{M} \models \alpha$. We then have to show that $\mathfrak{M} \vdash \alpha$. Since $\mathfrak{M} \models \alpha$, every model of $\mathfrak{M}$ is also a model of α and is thus not a model of $\neg \alpha$. Thus, $\mathfrak{M} \cup \{\neg \alpha\}$ has no models and is therefore not satisfiable. In Lemma 2 we assume that every consistent set is satisfiable. Thus, $\mathfrak{M} \cup \{\neg \alpha\}$ is not consistent and is therefore inconsistent. The assertion $\mathfrak{M} \vdash \alpha$ then follows from (20).

Exercises. 1. Show that the properties of the relation of derivability given in 5.2 also hold for the relation of consequence. (Prove this directly, not by using the equivalence of $\vdash$ and $\vDash$ which is proved in Chap. V.)

2. Show that the set $\{\alpha_1, \alpha_2\}$ is inconsistent (in the following let x and y be distinct individual variables):

$$\alpha_1 \equiv \bigwedge x \neg Rxx, \qquad \alpha_2 \equiv \neg \bigwedge x \neg \bigwedge y Rxy .$$

3. (a) $\{\alpha\}$ is inconsistent if and only if $\{\bigvee x\alpha\}$ is inconsistent. (b) Refute the assertion that $\{\bigvee x\alpha, \bigvee x\beta\}$ is inconsistent whenever $\{\alpha, \beta\}$ is.

4. Let $\mathfrak{M}$ be a set of expressions. We define a relation $\sim$ on the set of all expressions as follows:

$$\alpha \sim \beta \text{ if and only if } \mathfrak{M} \cup \{\alpha\} \vdash \beta \text{ and } \mathfrak{M} \cup \{\beta\} \vdash \alpha .$$

Show that:

(a) $\sim$ is an equivalence relation.

(b) If $\alpha \sim \beta$ then $\neg\alpha \sim \neg\beta$.

(c) If $\alpha_1 \sim \beta_1$ and $\alpha_2 \sim \beta_2$, then $(\alpha_1 \wedge \alpha_2) \sim (\beta_1 \wedge \beta_2)$.

5. (for $\mathfrak{M}$ and $\sim$ cf. Exercise 4.) Let $\bar{\alpha}$ be the equivalence class of α under $\sim$. Let L be the set of all $\bar{\alpha}$. Show that the definitions $\bar{\alpha}' = \overline{\neg\alpha}$ and $\bar{\alpha} \cap \bar{\beta} = \overline{(\alpha \wedge \beta)}$ are independent of the representatives chosen. Also, put $0 = \overline{x \neq x}$ and $1 = \overline{x = x}$, where x is a fixed individual variable. Show that L is a Boolean algebra relative to $'$, $\cap$, 0, 1. L is called the Lindenbaum algebra of $\mathfrak{M}$. Show further that : (a) $\mathfrak{M} \vdash \alpha$ if and only if $\bar{\alpha} = 1$, $\mathfrak{M} \vdash \neg\alpha$ if and only if $\bar{\alpha} = 0$; (b) $0 = 1$ if and only if $\mathfrak{M}$ is inconsistent; (c) it is not in general true that: If $\alpha \sim \beta$ then $\bigwedge x\alpha \sim \bigwedge x\beta$.

§6. The decidability of propositional derivability[6]

6.1 Propositional derivability. One can ask whether there is an algorithm by means of which we can, for given arbitrary expressions, decide whether or not $\alpha_1, \ldots, \alpha_r \vdash \alpha$. Church has shown that there is no such algorithm; this fact is called the undecidability of predicate logic. In order to prove the undecidability of predicate logic, it is necessary to define precisely the concept of an algorithm. We cannot go into this within the scope of this book.

The relation $\vdash$ of derivability is defined by means of the rules of predicate logic. We divide these rules (cf. the schematic representation of Chap. IV, § 2.3) into two groups:

(a) The propositional rules (A), (C), (C'), (C''), (R), (X), and

(b) the rules specific to predicate logic - (G), (G_x), (S_x^t), (E) and (E_x^t).

We call a sequent $\alpha_1 \ldots \alpha_r \alpha$ propositionally derivable if it is possible to derive this sequent by means of the propositional rules alone. $\alpha_1, \ldots, \alpha_r \vdash_P \alpha$ is to mean that $\alpha_1 \ldots \alpha_r \alpha$ is propositionally derivable.

[6] This section may be omitted at a first reading.

For propositional derivability $\vdash_P$, by contrast to the situation for the usual derivability $\vdash$ of predicate logic, there is a procedure by means of which, for given arbitrary expressions $\alpha_1, \ldots, \alpha_r, \alpha$, we can decide in finitely many steps whether or not $\alpha_1, \ldots, \alpha_r \vdash_P \alpha$. We shall prove this in this section.

In order to do this we introduce, in 6.2, the concept of a tautology, using the matrices for negation and conjunction which we defined in Chap. I, § 6.2.

We have already spoken of tautologies in Chap. I, § 6, Exercises 2 and 3. There we were dealing with statements or with their structure formulas. These structure formulas can also be understood as expressions of predicate logic which contain no quantifiers and in which only no-place predicate variables occur. Here, we take arbitrary expressions of predicate logic as a basis.

In 6.5, 6.6, 6.7 we prove

Lemma 1. $\alpha_1, \ldots, \alpha_r \vdash_P \alpha$ if and only if $\neg(\alpha_1 \wedge \ldots \wedge \alpha_r \wedge \neg\alpha)$ is a tautology.

The decision procedure for $\vdash_P$ which we give in 6.4 is based on this lemma.

6.2 Tautologies. Let $\{T, F\}$ be a set with two elements which is to remain fixed during the following ("T" reminds us of truth, "F" of falsehood).

We define two mappings N and C (reminiscent of negation and conjunction) by laying down:

$$(1) \quad \begin{cases} N(T) = F, \quad N(F) = T, \\ C(T, T) = T, \quad C(T, F) = C(F, T) = C(F, F) = F. \end{cases}$$

These mappings correspond to the logical matrices for negation and conjunction.

A function V which is defined for all expressions and maps every expression α onto an element $V(\alpha)$ of the set $\{T, F\}$ is to be called a valuation if it satisfies the following two conditions:

$$(2) \quad \begin{cases} V(\neg\alpha) = N(V(\alpha))^{7} \\ V(\alpha \wedge \beta) = C(V(\alpha), V(\beta)). \end{cases}$$

[7] Strictly speaking, we should distinguish between the names "(" and ")" for brackets (cf. Chap. II, § 1.2) and the brackets which are usually used in connection with function symbols. Thus, we could, for example, write: $V[\neg\alpha] = N[V[\alpha]]$ and, for the next line: $V[(\alpha \wedge \beta)] = C[V[\alpha], V[\beta]]$. However, we shall continue to use the simpler notation of the text above.

Let an expression α be called a <u>tautology</u> if

$$V(\alpha) = T$$

<u>for every valuation</u> V.

<u>6.3 Assignments.</u>. Let $\mathfrak{A}$ be the set of all expressions which are either atomic or generalisations. By an <u>assignment</u> $\mathfrak{A}$ we mean a mapping from $\mathfrak{A}$ into the set $\{T, F\}$. By Chap. II, § 3.2, there is exactly one mapping V of the set of all expressions into the set $\{T, F\}$ which satisfies:

(a) $V(\alpha) = A(\alpha)$ for every atomic expression α

(b) $V(\neg\alpha) = N(V(\alpha))$

(c) $V(\alpha \wedge \beta) = C(V(\alpha), V(\beta))$

(d) $V(\bigwedge x\alpha) = A(\bigwedge x\alpha)$.

By, (b) and (c), V is a valuation. Let V be called the <u>completion</u> of A.

If $\mathfrak{M}$ is a <u>nonempty finite</u> subset of $\mathfrak{A}$ and A_0 is a mapping from $\mathfrak{M}$ into the set $\{T, F\}$, then we call A_0 a <u>restricted assignment</u> or, more precisely, an $\mathfrak{M}$-<u>restricted assignment</u>. Then a valuation V is called an <u>extension</u> of A_0 if $V(\alpha) = A_0(\alpha)$ for the elements $\alpha \in \mathfrak{M}$. Clearly, for every restricted assignment A_0 there is at least one extension V.

We associate with every expression α a nonempty finite subset $\mathfrak{C}_\alpha$ of $\mathfrak{A}$. Let $\mathfrak{C}_\alpha$ be called the set of the <u>constituents</u> of α. $\mathfrak{C}_\alpha$ is defined by induction on the structure of the expressions, as follows:

(a') If α is atomic, then $\mathfrak{C}_\alpha = \{\alpha\}$.

(b') If $\alpha \equiv \neg\beta$, then $\mathfrak{C}_\alpha = \mathfrak{C}_\beta$.

(c') If $\alpha \equiv (\alpha \wedge \beta)$, then $\mathfrak{C}_\alpha = \mathfrak{C}_\alpha \cup \mathfrak{C}_\beta$.

(d') If $\alpha \equiv \bigwedge x\beta$, then $\mathfrak{C}_\alpha = \{\alpha\}$.

Example. For $\alpha \equiv (\bigwedge x(Px \wedge Qy) \wedge \neg Qx)$, $\mathfrak{C}_\alpha = \{\bigwedge x(Px \wedge Qy), Qx\}$.

Clearly, $\mathfrak{C}_\alpha$ can be determined by a finite procedure for every α.

<u>Lemma 2. If A_0 is a $\mathfrak{C}_\alpha$-restricted assignment, then all extensions V of A_0 have the same value $V(\alpha)$ at α. This value can computed by means of a finite procedure if A_0 is known.</u>

Proof. We prove the lemma by induction on the structure of the expressions.

(a'') If α is atomic, then $\mathfrak{C}_\alpha = \{\alpha\}$. Then we have $V(\alpha) = A_0(\alpha)$, since V is an extension of A_0.

(b'') If $\alpha \equiv \neg\beta$ then, by (b'), $\mathfrak{C}_\alpha = \mathfrak{C}_\beta$. Thus, A_0 is also a $\mathfrak{C}_\beta$-restricted assignment. By induction hypothesis, all extensions V of A_0 have the same value $V(\beta)$, which is an element U of $\{T, F\}$ which can be effectively determined. By (b), $V(\alpha) = N(V(\beta)) = N(U)$. Thus, the assertion has been proved for α.

(c'') If $\alpha \equiv (\beta \wedge \gamma)$ then, by (c') $\mathfrak{C}_\alpha = \mathfrak{C}_\beta \cup \mathfrak{C}_\gamma$. If we consider the $\mathfrak{C}_\alpha$-restricted assignment A_0 only for the elements of $\mathfrak{C}_\beta$ or only for the elements of $\mathfrak{C}_\gamma$, then we obtain a $\mathfrak{C}_\beta$-restricted assignment A_0' and a $\mathfrak{C}_\gamma$-restricted assignment A_0''. By induction hypothesis, all extensions V of A_0', and therefore, in particular, all extensions V of A_0, have the same value $V(\beta)$, which is an element U' of $\{T, F\}$ and can be effectively determined. Correspondingly, all extensions V of A_0 have the same value $V(\gamma)$, which is an element U'' of $\{T, F\}$ and can be effectively determined. By (c), $V(\alpha) = C(U', U'')$, which proves the assertion for α.

(d'') If α is a generalisation, we prove the assertion as in (a'').

6.4 Decidability of the property of being a tautology. We now want to prove that we can decide in finitely many steps whether or not a given expression is a tautology. Then, by Lemma 1 in 6.1, the relation $\vdash_P$ is also decidable.

Let α be a given expression. First of all, we form $\mathfrak{C}_\alpha$. If $\mathfrak{C}_\alpha$ has r elements, then there are 2^r $\mathfrak{C}_\alpha$-restricted assignments $A_1, \ldots, A_{2^r}$. These can be determined effectively by forming all the mappings from the elements of $\mathfrak{C}_\alpha$ into the set $\{T, F\}$.

By Lemma 2 in 6.3, for every j all the extensions V of A_j have the same value $V(\alpha)$; let this be called V_j. Then α is a tautology if and only if

$$V_1 = V_2 = \ldots = V_{2^r} = T.$$

For:

(a) If α is a tautology, then $V(\alpha) = T$ for every V. For every V_j there is, by definition, at least one V with $V_j = V(\alpha)$. Hence every V_j is equal to T.

(b) If α is not a tautology, then there is a V with $V(\alpha) = F$. If we consider V only for the elements of $\mathfrak{C}_\alpha$, then we obtain a $\mathfrak{C}_\alpha$-restricted assignment A_0. A_0 must be the same as one of the A_j. V is an extension of A_j. Hence

$$V_j = V(\alpha) = F.$$

In practice, given an expression α, the decision procedure can most easily be carried out by replacing the constituents of α by proposition variables and then proceeding as in Chap. I, § 6, Exercise 2.

6.5 Proof of Lemma 1, first part. We want to show that $\neg(\alpha_1 \wedge \ldots \wedge \alpha_r \wedge \neg\alpha)$ is a tautology if $\alpha_1, \ldots, \alpha_r \vdash_A \alpha$. We call $\neg(\alpha_1 \wedge \ldots \wedge \neg\alpha)$ the expression corresponding to the sequent $\alpha_1 \ldots \alpha_r \alpha$. It suffices to show that, in a propositional derivation (i.e. in a derivation in which only propositional rules are used), the expressions corresponding to the individual rows are always tautologies. In order to do this, we need only prove that:

(i) The expression $\neg(\alpha \wedge \neg\alpha)$ corresponding to the sequent $\alpha\alpha$ is a tautology.

(ii) If we apply one of the rules (C), (C'), (C''), (R), (X) to rows whose corresponding expressions are tautologies, then we obtain a row with the same property.

to (i): $$V(\neg(\alpha \wedge \neg\alpha)) = N(C(V(\alpha), V(\neg\alpha))) = N(C(V(\alpha), N(V(\alpha)))) = N(F) = T.$$

to (ii): We restrict ourselves to considering a special case of (C), and leave the general case and the investigation of the other rules to the reader.

Suppose that (C) leads from the rows $\alpha_1 \ldots \alpha_r \alpha$ and $\beta_1 \ldots \beta_s \beta$ to the row

$$\alpha_1 \ldots \alpha_r \beta_1 \ldots \beta_s (\alpha \wedge \beta).$$

By hypothesis,

$$\neg(\alpha_1 \wedge \ldots \wedge \alpha_r \wedge \neg\alpha) \quad \text{and} \quad \neg(\beta_1 \wedge \ldots \wedge \beta_s \wedge \neg\beta)$$

are tautologies. We have to show that the same holds of $\neg(\alpha_1 \wedge \ldots \wedge \alpha_r \wedge \beta_1 \wedge \ldots \wedge \beta_s \wedge \neg(\alpha \wedge \beta))$. Let V be an arbitrary valuation. We have to show that

$$V(\alpha_1 \wedge \ldots \wedge \alpha_r \wedge \beta_1 \wedge \ldots \wedge \beta_s \wedge \neg(\alpha \wedge \beta)) = F.$$

We prove this indirectly by assuming that, contrary to assertion,

$$V(\alpha_1 \wedge \ldots \wedge \alpha_r \wedge \beta_1 \wedge \ldots \wedge \beta_s \wedge \neg(\alpha \wedge \beta)) = T.$$

But we have the relation

$$V(\alpha_1 \wedge \ldots \wedge \alpha_r \wedge \beta_1 \wedge \ldots \wedge \beta_s \wedge \neg(\alpha \wedge \beta)) = C(V(\alpha_1), V(\alpha_2 \wedge \ldots \wedge \beta_s \wedge \neg(\alpha \wedge \beta))).$$

Hence, $V(\alpha_1) = T$ and $V(\alpha_2 \wedge \ldots \wedge \beta_s \wedge \neg(\alpha \wedge \beta)) = T$. By proceeding further in the same way, we obtain

(3) $$V(\alpha_1) = V(\alpha_2) = \ldots = V(\alpha_r) = V(\beta_1) = \ldots = V(\beta_s) = T, \quad V(\alpha \wedge \beta) = F.$$

Since $\neg(\alpha_1 \wedge \ldots \wedge \alpha_r \wedge \neg\alpha)$ is a tautology, we have $V(\alpha_1 \wedge \ldots \wedge \alpha_r \wedge \neg\alpha) = F$. But we also have

$$V(\alpha_1 \wedge \ldots \wedge \alpha_r \wedge \neg \alpha) = C(V(\alpha_1), V(\alpha_2 \wedge \ldots \wedge \alpha_r \wedge \neg \alpha))$$
$$= C(T, V(\alpha_2 \wedge \ldots \wedge \alpha_r \wedge \neg \alpha)).$$

But this can only be equal to F if $V(\alpha_2 \wedge \ldots \wedge \alpha_r \wedge \neg \alpha) = F$. Thus, finally, we obtain $V(\neg \alpha) = F$ and hence $V(\alpha) = T$. Correspondingly we find, by starting from the tautology $\neg(\beta_1 \wedge \ldots \wedge \beta_s \wedge \neg \beta)$, that $V(\beta) = T$. Thus we have $V(\alpha \wedge \beta) = C(V(\alpha), V(\beta)) = C(T, T) = T$, which contradicts (3). - This contradiction shows that (C) has the asserted property.

6.6 Proof of Lemma 1, second part. Now we assume, conversely, that $\neg(\alpha_1 \wedge \ldots \wedge \alpha_r \wedge \neg \alpha)$ is a tautology, and we have to show that $\alpha_1, \ldots, \alpha_r \vdash_P \alpha$. Further on, we shall prove that $\vdash_P \alpha$ always holds if α is a tautology. If we apply this result to the tautology $\neg(\alpha_1 \wedge \ldots \wedge \alpha_r \wedge \neg \alpha)$, then we obtain $\vdash_P \neg(\alpha_1 \wedge \ldots \wedge \alpha_r \wedge \neg \alpha)$. By (DdRu) (cf. 4.3), this yields $\alpha_1, \ldots, \alpha_r \vdash_P \alpha$, q.e.d. In 4.3, the deduction theorem was stated for the concept $\vdash$ of derivation. However, only propositional rules were used for its justification; hence, the deduction theorem also holds for $\vdash_P$.

In order to complete the proof, we have yet to show that $\vdash_P \alpha$ for every tautology α.

First of all, we introduce the notation

$$\alpha^T \equiv \alpha, \quad \alpha^F \equiv \neg \alpha.$$

In 6.7, we prove

Lemma 3. For every valuation V: If $\mathfrak{C}_\alpha = \{\alpha_1, \ldots, \alpha_r\}$, then

$$\vdash_P \alpha_1^{V(\alpha_1)} \ldots \alpha_r^{V(\alpha_r)} \alpha^{V(\alpha)}.$$

Let α be a tautology. We want to show by means of Lemma 3 that $\vdash_P \alpha$. Since $V(\alpha) = T$ for every V, we know by the lemma that

$$\vdash_P \alpha_1^{V(\alpha_1)} \ldots \alpha_r^{V(\alpha_r)} \alpha$$

for every valuation V. We shall eliminate the antecedent

$$\alpha_1^{V(\alpha_1)} \ldots \alpha_r^{V(\alpha_r)}$$

in this assertion step by step.

Let a be an arbitrary $\mathfrak{C}_\alpha$-restricted assignment. If V is an extension of a (6.3), then $a(\alpha_j) = V(\alpha_j)$ for every $j = 1, \ldots, r$. Thus we have

$$(*) \qquad \vdash_P \alpha_1^{a(\alpha_1)} \ldots \alpha_r^{a(\alpha_r)} \alpha .$$

Let $a'(\alpha_1) = a(\alpha_1), \ldots, a'(\alpha_{r-1}) = a(\alpha_{r-1})$, but $a'(\alpha_r) = N(a(\alpha_r))$. Since (*) holds for arbitrary restricted assignments, we have also for a':

$$(**) \qquad \vdash_P \alpha_1^{a'(\alpha_1)} \ldots \alpha_r^{a'(\alpha_r)} \alpha .$$

Now, noting that, of the two expressions $\alpha_r^{a(\alpha_r)}$, $\alpha_r^{a'(\alpha_r)}$, one is the negation of the other, and that the other members of the antecedents of the sequents in (*) and (**) are the same, we see that, by means of (R), (*) and (**) yield the assertion

$$\vdash_P \alpha_1^{a(\alpha_1)} \ldots \alpha_{r-1}^{a(\alpha_{r-1})} \alpha .$$

By repeated application of this procedure, we finally obtain

$$\vdash_P \alpha ,$$

q.e.d.

<u>6.7 Proof of Lemma 3.</u> We carry out the proof by induction on the structure of α.

(a) If α is atomic, then $\mathfrak{C}_\alpha = \{\alpha\}$. Thus, we need to show that

$$\vdash_P \alpha^{V(\alpha)} \alpha^{V(\alpha)}$$

for every valuation V. But this can be shown at once by means of rule (A).

(b) Let $\alpha \equiv \neg \beta$. Then $\mathfrak{C}_\alpha = \mathfrak{C}_\beta$. Let $\mathfrak{C}_\alpha = \{\alpha_1, \ldots, \alpha_r\}$. By induction hypothesis, we have

$$\vdash_P \alpha_1^{V(\alpha_1)} \ldots \alpha_r^{V(\alpha_r)} \beta^{V(\beta)} .$$

We distinguish two cases:

Case 1: $V(\beta) = T$, and hence $V(\alpha) = F$. Then $\beta^{V(\beta)} \equiv \beta$, so that

$$\vdash_P \alpha_1^{V(\alpha_1)} \ldots \alpha_r^{V(\alpha_r)} \beta .$$

Then, by (NN) and (CuRu) (cf. 4.2), we obtain

$$\vdash_P \alpha_1^{V(\alpha_1)} \ldots \alpha_r^{V(\alpha_r)} \neg\neg \beta.$$

But $\neg\neg \beta \equiv \neg \alpha \equiv \alpha^F \equiv \alpha^{V(\alpha)}$, which proves the assertion.

Case 2: $V(\beta) = F$, and hence $V(\alpha) = T$. Then $\beta^{V(\beta)} \equiv \neg \beta \equiv \alpha$, so that

$$\vdash_P \alpha_1^{V(\alpha_1)} \ldots \alpha_r^{V(\alpha_r)} \alpha,$$

which is the required result.

(c) Let $\alpha \equiv (\beta \wedge \gamma)$. Then $\mathfrak{C}_\alpha = \mathfrak{C}_\beta \cup \mathfrak{C}_\gamma$. Let $\mathfrak{C}_\beta = \{\beta_1, \ldots, \beta_r\}$ and $\mathfrak{C}_\gamma = \{\gamma_1, \ldots, \gamma_s\}$. Then $\mathfrak{C}_\alpha = \{\beta_1, \ldots, \beta_r, \gamma_1, \ldots, \gamma_s\}$ (where, in the set on the right, some elements may have been written down twice; however, this does no harm here and in the following text). By induction hypothesis, we have

$$\vdash_P \beta_1^{V(\beta_1)} \ldots \beta_r^{V(\beta_r)} \beta \quad \text{and} \quad \vdash_P \gamma_1^{V(\gamma_1)} \ldots \gamma_s^{V(\gamma_s)} \gamma .$$

Then, by (C), we obtain

$$\vdash_P \beta_1^{V(\beta_1)} \ldots \beta_r^{V(\beta_r)} \gamma_1^{V(\gamma_1)} \ldots \gamma_s^{V(\gamma_s)} (\beta \wedge \gamma) .$$

But $\alpha^{V(\alpha)} \equiv \alpha^T \equiv \alpha \equiv (\beta \wedge \gamma)$, which proves the assertion.

Case 2: $V(\beta) = F$ or (vel) $V(\gamma) = F$. We restrict ourselves to the case in which $V(\beta) = F$, hence $V(\alpha) = V(\beta \wedge \gamma) = F$. Then $\beta^{V(\beta)} \equiv \neg \beta$, so that

$$\vdash_P \beta_1^{V(\beta_1)} \ldots \beta_r^{V(\beta_r)} \neg \beta .$$

Then, by (A), (C'), (CaPo), (CuRu), we obtain

$$\vdash_P \beta_1^{V(\beta_1)} \ldots \beta_r^{V(\beta_r)} \neg (\beta \wedge \gamma) .$$

This is the required assertion, since $\alpha^{V(\alpha)} \equiv (\beta \wedge \gamma)^F \equiv \neg (\beta \wedge \gamma)$.

(d) If α is a generalisation $\bigwedge x\beta$, then we prove the assertion as in (a).

<u>Exercise.</u> Decide whether

(a) $(\alpha \wedge \neg(\alpha \wedge \beta)) \vdash_P \neg(\neg \alpha \wedge \neg(\alpha \wedge \neg \beta))$

(b) $\neg(\neg \alpha \wedge \neg(\alpha \wedge \beta)) \vdash_P (\alpha \wedge \neg(\alpha \wedge \neg \beta))$

(c) $(\bigwedge x(\alpha \wedge \beta) \wedge \bigwedge x\beta) \vdash_P \bigwedge x\beta$

(d) $(\bigwedge x(\alpha \wedge \beta) \wedge \bigwedge x\alpha) \vdash_P \bigwedge x\beta$.

V. Gödel's Completeness Theorem

§ 1. Isomorphisms of expressions

In order to prove the completeness theorem (cf. § 2), we use as a technical aid a mapping of the set of all expressions into itself, which we can regard as a sort of isomorphism. In this section, we shall bring together the theorems which we need about such isomorphisms; we shall treat the notion of an isomorphism only insofar as we need it for the following.

1.1 Definition of isomorphisms. We start from a mapping Φ which is defined for every individual variable and maps every individual variable x into another individual variable x . We assume in addition that Φ is 1-1, i.e. that

(*) $$\text{if } x^{\Phi} \equiv y^{\Phi}, \text{ then } x \equiv y .$$

However, we do not require that every individual variable should be the image under Φ of some individual variable.

With the aid of such a mapping Φ, we can assign to every expression α an image α^{Φ}, which we obtain by replacing each individual variable which occurs in α by its Φ-image. We shall call such a mapping of the set of all expressions into itself an isomorphism. If $\mathfrak{M}$ is a set of expressions, then let $\mathfrak{M}^{\Phi}$ be the set of the Φ-images of the elements of $\mathfrak{M}$.

We can replace the above globally-given definition of α^{Φ} by an inductive definition. For this purpose, we first of all assign to every term t a term t^{Φ}: If t is an individual variable, then t^{Φ} is already defined. For the compound term $t \equiv f t_1 \ldots t_r$, let $t^{\Phi} \equiv f t_1^{\Phi} \ldots t_r^{\Phi}$. Furthermore, we define:

$$[P t_1 \ldots t_r]^{\Phi} \equiv P t_1^{\Phi} \ldots t_r^{\Phi}$$

$$[t_1 = t_2]^{\Phi} \equiv t_1^{\Phi} = t_2^{\Phi}$$

$$[\neg \alpha]^{\Phi} \equiv \neg \alpha^{\Phi}$$

$$(\alpha \wedge \beta)^{\Phi} \equiv (\alpha^{\Phi} \wedge \beta^{\Phi})$$

$$[\bigwedge x \alpha]^{\Phi} \equiv \bigwedge x^{\Phi} \alpha^{\Phi}$$

Finally, we want to associate with every 1-1 mapping Φ of the set of individual variables into itself and every interpretation $\mathfrak{I}$ over a domain of individuals an interpretation $\mathfrak{I}^{\Phi}$ over the same domain of individuals. Now let

$$\mathfrak{I}^{\Phi}(x) = \mathfrak{I}(x^{\Phi}) \quad \text{for every individual variable } x,$$
$$\mathfrak{I}^{\Phi}(f) = \mathfrak{I}(f) \quad \text{for every } f \text{ which is at least one-place,}$$
$$\mathfrak{I}^{\Phi}(P) = \mathfrak{I}(P) \quad \text{for every predicate variable } P.$$

It is easy to show that $\mathfrak{I}(t^{\Phi}) = \mathfrak{I}^{\Phi}(t)$ for every term t: This follows from the definition of $\mathfrak{I}^{\Phi}$ if t is an individual variable. If t is a compound term $t \equiv ft_1 \ldots t_r$ and if the assertion holds for $t_1, \ldots, t_r$, then we have:

$$\begin{aligned}
\mathfrak{I}([ft_1 \ldots t_r]^{\Phi}) &= \mathfrak{I}(ft_1^{\Phi} \ldots t_r^{\Phi}) \\
&= \mathfrak{I}(f)(\mathfrak{I}(t_1^{\Phi}), \ldots, \mathfrak{I}(t_r^{\Phi})) \\
&= \mathfrak{I}^{\Phi}(f)(\mathfrak{I}^{\Phi}(t_1), \ldots, \mathfrak{I}^{\Phi}(t_r)) \\
&= \mathfrak{I}^{\Phi}(ft_1 \ldots t_r).
\end{aligned}$$

1.2 Local invertibility. We have already said above that not every individual variable has to be the Φ-image of some individual variable. Thus, it is not true to say that, for every isomorphism Φ, there is an isomorphism Ψ which is the "global inversion" of Φ in the sense that $x^{\Phi\Psi} \equiv x$ for every x (and consequently $\alpha^{\Phi\Psi} \equiv \alpha$ for every α). For if, given an isomorphism Φ, there were an isomorphism Ψ such that $x^{\Phi\Psi} \equiv x$ for every x and if y is an arbitrary individual variable, then we should have $y^{\Psi\Phi\Psi} \equiv y^{\Psi}$, and hence $y^{\Psi\Phi} \equiv y$ since Ψ is 1-1; thus, y would be the Φ-image of y^{Ψ}.

However, in place of the "global inversion", which does not, in general exist, we always have at least a "local inversion" in the following sense: Let Φ be an arbitrary isomorphism and $\mathfrak{S}$ a finite set of individual variables. Then there is always an isomorphism Ψ such that $x^{\Phi\Psi} \equiv x$ for all $x \in \mathfrak{S}$.

In order to show this, we consider first the set $\mathfrak{T}$ of the Φ-images of $\mathfrak{S}$. We define Ψ first of all for the elements of $\mathfrak{T}$. If $z \in \mathfrak{T}$, then there is an $x \in \mathfrak{S}$ with $x^{\Phi} \equiv z$. Since Φ is 1-1, x is uniquely determined. We put z^{Ψ} equal to x. For the elements of $\mathfrak{T}$, Ψ is 1-1: let $z_1, z_2 \in \mathfrak{T}$ and $z_1^{\Psi} \equiv z_2^{\Psi}$. Then there are elements $x_1, x_2 \in \mathfrak{S}$ such that $x_1^{\Phi} \equiv z_1$, $x_2^{\Phi} \equiv z_2$. Then $z_1^{\Psi} \equiv x_1$, $z_2^{\Psi} \equiv x_2$, hence $x_1 \equiv x_2$, and therefore $z_1 \equiv x_1^{\Phi} \equiv x_2^{\Phi} \equiv z_2$. Moreover, clearly $x^{\Phi\Psi} \equiv x$ for all elements x of $\mathfrak{S}$.

Now we have the task of extending the function Ψ, which, up till now, has been defined only for the elements of $\mathfrak{T}$ and is 1-1 on $\mathfrak{T}$, to a 1-1 mapping which is defined for all individual variables. We can do this by, for example, enumerating both the infinitely many individual variables which do not belong to $\mathfrak{T}$ and the infinitely many individual variables which do not belong to $\mathfrak{S}$, and then, for every k, mapping the k-th individual variable of the first sort in this enumeration onto the k-th individual variable of the second sort in this enumeration.

1.3 Theorems about isomorphisms

Theorem 1. x occurs free in α if and only if x^Φ occurs free in α^Φ.

Theorem 2. If Subst $\alpha x t \beta$, then Subst $\alpha^\Phi x^\Phi t^\Phi \beta^\Phi$.

Theorem 3. $\mathfrak{M} \vdash \alpha$ if and only if $\mathfrak{M}^\Phi \vdash \alpha^\Phi$.

Theorem 4. $\mathfrak{M}$ is consistent if and only if $\mathfrak{M}^\Phi$ is consistent.

Theorem 5. Mod $\mathfrak{J}\alpha^\Phi$ if and only if Mod $\mathfrak{J}^\Phi\alpha$.

Theorem 6. If $\mathfrak{M}^\Phi$ is satisfiable, then $\mathfrak{M}$ is also satisfiable.

1.4 Proof of the theorems stated

To Theorem 1. (1) We show first of all that x^Φ occurs free in α^Φ if x occurs free in α. We prove the assertion for the expression α under the assumption that it has already been proved for all shorter expressions. We assume that x occurs free in α.

(a) If α is atomic and x occurs free in α, then x occurs in α, hence x^Φ in α^Φ, where α^Φ is also atomic. Hence x^Φ occurs free in α^Φ.

(b) If α is a negation, i.e. $\alpha \equiv \neg\beta$ and if x occurs free in α, then x also occurs free in β. β is shorter than α. By hypothesis, x^Φ occurs free in β^Φ, and therefore also in $\neg\beta^\Phi$. But $\neg\beta^\Phi \equiv [\neg\beta]^\Phi \equiv \alpha^\Phi$.

(c) We treat the case in which α is a conjunction in the same way.

(d) If α is a generalisation, i.e. $\alpha \equiv \bigwedge u\beta$ and if x occurs free in α, then $x \neq u$ and x occurs free in β. β is shorter than α. By hypothesis, x^Φ occurs free in β^Φ. Since Φ is 1-1, $x^\Phi \neq u^\Phi$ because $x \neq u$. Hence x^Φ occurs free in $\bigwedge u^\Phi\beta^\Phi$. But $\bigwedge u^\Phi\beta^\Phi \equiv$ $\equiv [\bigwedge u\beta]^\Phi \equiv \alpha^\Phi$.

(2) We now need only show that x occurs free in α if x^Φ occurs free in α^Φ. Consider the set $\mathfrak{S}$ which contains the individual variables which occur in α together with x. $\mathfrak{S}$ is finite. Hence there is a local inversion Ψ of Φ with $x^{\Phi\Psi} \equiv x$ for all $x \in \mathfrak{S}$. From this, it follows that $\alpha^{\Phi\Psi} \equiv \alpha$. Part (1), which we have already proved, applied to x^Φ, α^Φ and Ψ (instead of x, α and Φ) then provides us with the assertion (2).

T o T h e o r e m 2 . We prove the assertion for α under the assumption that it holds for all expressions which are shorter than α.

(a) If α is a predicative expression, i.e. $\alpha \equiv P t_1 \ldots t_r$, then $\beta \equiv P \Delta^t_x t_1 \ldots \Delta^t_x t_r$. It follows that $\alpha^\Phi \equiv P t_1^\Phi \ldots t_r^\Phi$ and $\beta \equiv P[\Delta^t_x t_1]^\Phi \ldots [\Delta^t_x t_r]^\Phi$. Subst $\alpha^\Phi x^\Phi t^\Phi \beta^\Phi$ states that $\beta \equiv P \Delta^{t^\Phi}_{x^\Phi} t_1^\Phi \ldots \Delta^{t^\Phi}_{x^\Phi} t_r^\Phi$. Thus, it suffices to show that, for every j,

$$\Delta^{t^\Phi}_{x^\Phi} t_j^\Phi \equiv [\Delta^t_x t_j]^\Phi .$$

We show this by induction on the structure of t_j.

If t_j is an individual variable, i.e. $t_j \equiv u$, then we have

$$\Delta^{t^\Phi}_{x^\Phi} u^\Phi \equiv \begin{cases} t^\Phi \text{ if } x^\Phi \equiv u^\Phi; \text{ and hence also if } x \equiv u \\ u^\Phi \text{ if } x^\Phi \not\equiv u^\Phi; \text{ and hence, by } (*), \text{ also if } x \not\equiv u . \end{cases}$$

Moreover,

$$\Delta^t_x u \equiv \begin{cases} t \text{ if } x \equiv u \\ u \text{ if } x \not\equiv u, \end{cases} \qquad \text{and hence} \quad [\Delta^t_x u]^\Phi \equiv \begin{cases} t^\Phi \text{ if } x \equiv u \\ u^\Phi \text{ if } x \not\equiv u. \end{cases}$$

If t_j is compound, i.e. $t_j \equiv f t_1' \ldots t_s'$, then, if we assume that the assertion has already been proved for the terms t_k', we obtain:

$$\begin{aligned} \Delta^{t^\varphi}_{x^\varphi} t_j^\varphi &\equiv \Delta^{t^\varphi}_{x^\varphi} [f t_1' \ldots t_s']^\varphi \equiv \Delta^{t^\varphi}_{x^\varphi} f t_1'^\Phi \ldots t_s'^\Phi \equiv f \Delta^{t^\varphi}_{x^\varphi} t_1'^\Phi \ldots \Delta^{t^\varphi}_{x^\varphi} t_s'^\Phi \\ &\equiv f[\Delta^t_x t_1']^\varphi \ldots [\Delta^t_x t_s']^\varphi \equiv [f \Delta^t_x t_1' \ldots \Delta^t_x t_s']^\varphi \\ &\equiv [\Delta^t_x f t_1' \ldots t_s']^\varphi \equiv [\Delta^t_x t_j]^\varphi . \end{aligned}$$

(b) If α is an equation, then we proceed as in (a).

(c) If α is a negation, then there is a shorter α_1 and a β_1 such that

$$\alpha \equiv \neg \alpha_1, \quad \beta \equiv \neg \beta_1 \quad \text{and} \quad \text{Subst } \alpha_1 x t \beta_1 .$$

By hypothesis, we have Subst $\alpha_1^\Phi x^\Phi t^\Phi \beta_1^\Phi$. From this it follows that Subst $\neg \alpha_1^\Phi x^\Phi t^\Phi \neg \beta_1^\Phi$, i.e. Subst $\alpha^\Phi x^\Phi t^\Phi \beta^\Phi$.

(d) If α is a conjunction, we proceed similarly.

(e) If α is a generalisation, i.e. $\alpha \equiv \bigwedge u\alpha_1$, then we distinguish two cases:

Case 1. x does not occur free in α. Then $\beta \equiv \alpha$. Moreover, by Theorem 1, x^Φ does not occur free in α^Φ. Hence Subst $\alpha^\Phi x^\Phi t^\Phi \alpha^\Phi$, i.e. Subst $\alpha^\Phi x^\Phi t^\Phi \beta^\Phi$.

Case 2. x occurs free in α, i.e. in $\bigwedge u\alpha_1$. Since Subst $\alpha x t \beta$, u does not occur in t and there is a β_1 with Subst $\alpha_1 x t \beta_1$ and $\beta \equiv \bigwedge u\beta_1$. By hypothesis, Subst $\alpha_1^\Phi x^\Phi t^\Phi \beta_1^\Phi$. By Theorem 1, x^Φ occurs free in α_1^Φ because x occurs free in α_1. Since u does not occur in t, u^Φ does not occur in t^Φ. Hence we have Subst $\bigwedge u^\Phi \alpha_1^\Phi x^\Phi t^\Phi \bigwedge u^\Phi \beta_1^\Phi$, i.e. Subst $\alpha^\Phi x^\Phi t^\Phi \beta^\Phi$.

To Theorem 3. (1) First of all, we assume that $\mathfrak{M} \vdash \alpha$ and show that $\mathfrak{M}^\Phi \vdash \alpha^\Phi$. Since $\mathfrak{M} \vdash \alpha$, there are finitely many elements $\alpha_1, \ldots, \alpha_n \in \mathfrak{M}$ such that $\alpha_1 \ldots \alpha_n \alpha$ is derivable. It suffices to show that $\alpha_1^\Phi \ldots \alpha_n^\Phi \alpha^\Phi$ is derivable. In fact we show quite generally that, if a sequent σ is derivable, then the sequent σ^Φ is always also derivable. In order to do this, it clearly suffices to prove that:

(*) If we can write down σ by means of (A) or (E), then the same holds for σ^Φ. But this is trivial.

(**) If σ can be obtained from σ_1 or from σ_1 and σ_2 by means of one of the rules (C), (C'), (C''), (R), (X), (G), (G_x), (S_x^t), (E_x^t), then σ^Φ can be obtained from σ_1^Φ or from σ_1^Φ and σ_2^Φ (as the case may be) by means of the corresponding rule (C), (C'), (C''), (R), (X), (G), (G_x), (S_x^t), (E_x^t). As typical examples, we treat the rules (C), (G_x), (S_x^t).

<u>to</u> (C): We have (for this and the following cf. the representation of the rules given in Chap. IV, § 2.3)

$$\begin{array}{lcl} \sigma_1 \equiv \Sigma_1 \alpha & & \sigma_1^\Phi \equiv \Sigma_1^\Phi \alpha^\Phi \\ \sigma_2 \equiv \Sigma_2 \beta & \text{and hence} & \sigma_2^\Phi \equiv \Sigma_2^\Phi \beta^\Phi \\ \sigma \equiv \Sigma_{12}(\alpha \wedge \beta), & & \sigma^\Phi \equiv \Sigma_{12}^\Phi (\alpha \wedge \beta)^\Phi . \end{array}$$

From this we can see immediately that σ^Φ can be obtained from σ_1^Φ and σ_2^Φ by means of (C).

<u>to</u> (G_x): We have

$$\begin{array}{lcl} \sigma_1 \equiv \Sigma\alpha & \text{and hence} & \sigma_1^\Phi \equiv \Sigma^\Phi \alpha^\Phi \\ \sigma \equiv \Sigma \bigwedge x\alpha, & & \sigma^\Phi \equiv \Sigma^\Phi \bigwedge x^\Phi \alpha^\Phi . \end{array}$$

Since σ is obtained from σ_1 by means of (G_x), the critical condition, namely that x does not occur free in Σ, must be fulfilled. Then by Theorem 1, x^Φ does not occur free in Σ^Φ. Hence, we can obtain σ^Φ from σ_1^Φ by means of (G_{x^Φ}).

<u>to</u> (S_x^t): We have

$$\sigma_1 \equiv \Sigma\alpha \qquad\qquad \sigma_1^\Phi \equiv \Sigma^\Phi\alpha^\Phi$$
and hence
$$\sigma \equiv \Sigma'\beta, \qquad\qquad \sigma^\Phi \equiv \Sigma'^\Phi\beta^\Phi.$$

But Subst $\Sigma\alpha x t \Sigma'\beta$. It then follows by Theorem 2 that

$$\text{Subst } \Sigma^\Phi\alpha^\Phi x^\Phi t^\Phi \Sigma'^\Phi\beta^\Phi .$$

Hence σ^Φ can be obtained from σ_1^Φ by means of $(S_{x^\Phi}^{t^\Phi})$.

(2) Now we have to show, conversely, that $\mathfrak{M} \vdash \alpha$ follows from $\mathfrak{M}^\Phi \vdash \alpha^\Phi$. Let us assume that $\mathfrak{M}^\Phi \vdash \alpha^\Phi$. Then there is a finite subset $\mathfrak{E}$ of $\mathfrak{M}$ such that $\mathfrak{E}^\Phi \vdash \alpha^\Phi$. Let $\mathfrak{S}$ be the set of those individual variables which occur in the expressions in $\mathfrak{E}$ or in α. $\mathfrak{S}$ is finite. Hence there is an isomorphism Ψ with the property that $x^{\Phi\Psi} \equiv x$ for all $x \in \mathfrak{S}$. By (1), we obtain from $\mathfrak{E}^\Phi \vdash \alpha^\Phi$ the statement $\mathfrak{E}^{\Phi\Psi} \vdash \alpha^{\Phi\Psi}$, and this is the same as $\mathfrak{E} \vdash \alpha$. A fortiori, $\mathfrak{M} \vdash \alpha$, q.e.d.

T o T h e o r e m 4 . We prove here only the somewhat less trivial half of the equivalence. We prove it indirectly, starting with the assumption that $\mathfrak{M}^\Phi$ is inconsistent. We then have to show that $\mathfrak{M}$ is also inconsistent. The fact that $\mathfrak{M}^\Phi$ is inconsistent means that every expression is derivable from $\mathfrak{M}$. Let α be an arbitrary expression. Then we have $\mathfrak{M}^\Phi \vdash \alpha^\Phi$. But then, by Theorem 3, $\mathfrak{M} \vdash \alpha$. Since this holds for every α, $\mathfrak{M}$ is inconsistent.

T o T h e o r e m 5 . It suffices to show that the assertion of Theorem 5 holds for α and $\mathfrak{J}$, assuming that it holds for any expression β which is shorter than α and for any interpretation $\mathfrak{K}$. According to the form of α, we have the following cases:

(a) α is a predicative expression, i.e. $\alpha \equiv P t_1 \ldots t_r$. Using the fact that $\mathfrak{J}(t^\Phi) = \mathfrak{J}^\Phi(t)$ (see the end of 1.1), we have:

$$\begin{aligned} \text{Mod } \mathfrak{J}[P t_1 \ldots t_r]^\Phi & \text{ iff Mod } \mathfrak{J} P t_1^\Phi \ldots t_r^\Phi \\ & \text{ iff } \mathfrak{J}(P) \text{ fits } \mathfrak{J}(t_1^\Phi), \ldots, \mathfrak{J}(t_r^\Phi) \\ & \text{ iff } \mathfrak{J}^\Phi(P) \text{ fits } \mathfrak{J}^\Phi(t_1), \ldots, \mathfrak{J}^\Phi(t_r) \\ & \text{ iff Mod } \mathfrak{J}^\Phi P t_1 \ldots t_r . \end{aligned}$$

(b) α is an equation. We proceed as in (a).

(c) α is a negation, i.e. $\alpha \equiv \neg\beta$. β is shorter than α. Hence we have

$$\begin{aligned}
\text{Mod}\,\mathfrak{J}\alpha^{\Phi} \quad &\text{iff Mod}\,\mathfrak{J}\neg\beta^{\Phi} \\
&\text{iff not Mod}\,\mathfrak{J}\beta^{\Phi} \\
&\text{iff not Mod}\,\mathfrak{J}^{\Phi}\beta \\
&\text{iff Mod}\,\mathfrak{J}^{\Phi}\neg\beta \\
&\text{iff Mod}\,\mathfrak{J}^{\Phi}\alpha .
\end{aligned}$$

(d) α is a conjunction. We proceed as in (c).

(e) α is a generalisation, i.e. $\alpha \equiv \bigwedge x\alpha_1$. α_1 is shorter than α. We have:

$$\begin{aligned}
\text{Mod}\,\mathfrak{J}\alpha^{\Phi} \quad &\text{iff Mod}\,\mathfrak{J}\bigwedge x^{\Phi}\alpha_1{}^{\Phi} \\
&\text{iff for all } \mathfrak{x}\text{: Mod}\,\mathfrak{J}^{\mathfrak{x}}_{x^{\Phi}}\alpha_1{}^{\Phi} \\
&\text{iff for all } \mathfrak{x}\text{: Mod}\,[\mathfrak{J}^{\mathfrak{x}}_{x^{\Phi}}]^{\Phi}\alpha_1 \quad \text{(condition on } \alpha_1\text{)} \\
&\text{iff for all } \mathfrak{x}\text{: Mod}\,\mathfrak{J}^{\Phi\mathfrak{x}}_{x}\alpha_1 \quad \text{(see below)} \\
&\text{iff Mod}\,\mathfrak{J}^{\Phi}\bigwedge x\alpha_1 \\
&\text{iff Mod}\,\mathfrak{J}^{\Phi}\alpha .
\end{aligned}$$

We have yet to show that $[\mathfrak{J}^{\mathfrak{x}}_{x^{\Phi}}]^{\Phi} = \mathfrak{J}^{\Phi\mathfrak{x}}_{x}$ always holds. In order to do this, we have only to show that the interpretations on the left and the right give the same image for every individual variable. First of all, for the variable x we have:

$$[\mathfrak{J}^{\mathfrak{x}}_{x^{\Phi}}]^{\Phi}(x) = \mathfrak{J}^{\mathfrak{x}}_{x^{\Phi}}(x^{\Phi}) = \mathfrak{x}, \quad \text{and} \quad \mathfrak{J}^{\Phi\mathfrak{x}}_{x}(x) = \mathfrak{x} .$$

If y is an individual variable which is different from x, then

$$[\mathfrak{J}^{\mathfrak{x}}_{x^{\Phi}}]^{\Phi}(y) = \mathfrak{J}^{\mathfrak{x}}_{x^{\Phi}}(y^{\Phi}) = \mathfrak{J}(y^{\Phi}) \quad (\text{since } y^{\Phi} \not\equiv x^{\Phi}), \text{ and } \mathfrak{J}^{\Phi\mathfrak{x}}_{x}(y) = \mathfrak{J}^{\Phi}(y) = \mathfrak{J}(y^{\Phi}) .$$

To Theorem 6. Let $\mathfrak{M}^{\Phi}$ be satisfiable. This means that there is an interpretation $\mathfrak{J}$ over a suitable domain of individuals such that Mod $\mathfrak{J}\mathfrak{M}^{\Phi}$. Theorem 5 shows us immediately that Mod $\mathfrak{J}^{\Phi}\mathfrak{M}$. But this shows that $\mathfrak{M}$ is satisfiable.

Exercises. 1. Show that 1.3, Theorem 4 and the inverse of Theorem 6 do not hold if we do not assume that Φ is 1-1.

2. Prove the inverse of 1.3, Theorem 6.

3. Formulate and prove theorems about homomorphisms which map predicate variables and (more-than-no-place) function variables.

§2. Sketch of the proof of the completeness theorem

2.1 Sketch of the proof. We want to prove (cf. Chap. IV, §1) the

Completeness theorem. If any expression α follows from a set $\mathfrak{M}$ of expressions, then α is derivable from $\mathfrak{M}$; i.e., expressed briefly: If $\mathfrak{M} \models \alpha$, then $\mathfrak{M} \vdash \alpha$.

The completeness of a predicate calculus was first shown by Gödel in 1930.

In Chap. IV, §5.3, Lemma 2 we showed that the completeness theorem can be reduced to the statement:

If a set of expressions $\mathfrak{M}$ is consistent, then $\mathfrak{M}$ is satisfiable. We shall prove this statement in §3 and §4. The proof consists of the following steps:

(1) $\mathfrak{M}$ is consistent (this is the assumption).

The set of expressions $\mathfrak{M}$ is mapped onto a set of expressions $\mathfrak{M}^{\Phi_0}$ by a special isomorphism Φ_0 which we shall define in 2.2. Then, by 1.3, Theorem 4, we have:

(2) $\mathfrak{M}^{\Phi_0}$ is consistent.

Now the set $\mathfrak{M}^{\Phi_0}$ is embedded in a set $\mathfrak{M}^*$ by a "maximalisation process" (§3). We talk of an "embedding" in order to indicate that $\mathfrak{M}^{\Phi_0} \subset \mathfrak{M}^*$. The maximalisation is carried out in such a way that we can show that:

(3) $\mathfrak{M}^*$ is consistent.

The sets of expressions $\mathfrak{M}$, $\mathfrak{M}^{\Phi_0}$ and $\mathfrak{M}^*$ which we have considered are consistent. Consistency is a syntactic property. Up to this point, the proof has been a purely syntactic one; now we make the telling transition to the realm of semantics. We show that there is a domain of individuals and an interpretation in which $\mathfrak{M}^*$ is valid (§4). Thus, we have:

(4) $\mathfrak{M}^*$ is satisfiable.

Now we must find the way back to $\mathfrak{M}$ via $\mathfrak{M}^{\Phi_0}$. Since, because of the embedding, $\mathfrak{M}^{\Phi_0} \subset \mathfrak{M}^*$, we have trivially:

(5) $\mathfrak{M}^{\Phi_0}$ is satisfiable.

By 1.3, Theorem 6, this provides us with:

(6) $\mathfrak{M}$ is satisfiable (this is the assertion).

In 2.2, we define the isomorphism Φ_0 which is used. In § 3, we carry out the maximalisation process and prove some lemmas about $\mathfrak{M}^*$ which are necessary for the proof of the satisfiability of $\mathfrak{M}^*$ in § 4.

2.2 Definition of the isomorphism Φ_0 of expressions. We define Φ_0 by requiring that x^{Φ_0} should be the individual variable whose index is twice the index of x. The mapping Φ_0 is clearly 1-1 and is thus an isomorphism. It is exactly those individual variables with even indices which appear as images of Φ_0. Hence, none of the individual variables which occur in the expressions α^{Φ_0} have odd indices.

§ 3. The process of maximalisation

3.1 Survey. The set $\mathfrak{M}^{\Phi_0}$, is, as we have seen in the last section, a consistent set of expressions. Moreover, no individual variable with an odd index occurs in any of the expressions of $\mathfrak{M}^{\Phi_0}$. These two properties of $\mathfrak{M}^{\Phi_0}$ are the only ones of which we shall make use here.

The set $\mathfrak{M}^{\Phi_0}$ can be extended by adding to it expressions which do not belong to it (e.g. an expression $x = x$, where x is an individual variable with an odd index), but which are such that the extended set is still consistent. In other words, $\mathfrak{M}^{\Phi_0}$ is not "maximal consistent". Here, we define the maximal consistency of a set $\mathfrak{M}$ of expressions by the two requirements:

(*) $\mathfrak{M}$ is consistent.
(**) If α is an arbitrary expression which does not belong to $\mathfrak{M}$, then $\mathfrak{M} \cup \{\alpha\}$ is inconsistent.

We want to embed $\mathfrak{M}^{\Phi_0}$ in a maximal consistent set $\mathfrak{M}^*$. In order to do this, we define a sequence $\mathfrak{M}_0, \mathfrak{M}_1, \mathfrak{M}_2, \ldots$ of sets of expressions, for which we prove (simultaneously with the definition) the following property:

(***) There are infinitely many individual variables which do not occur in any of the expressions of $\mathfrak{M}_j$.

We define the $\mathfrak{M}_j$'s and $\mathfrak{M}^*$ in 3.2. In 3.3 we prove that $\mathfrak{M}^*$ is maximal consistent. In 3.4 we obtain some consequences of the maximal consistency of $\mathfrak{M}^*$.

3.2 Definition of the sets $\mathfrak{M}_j$ and $\mathfrak{M}^*$ of expressions. There are countably many terms and countably many expressions, as we have already seen in Chap. II, § 1.6. In the following, we take as a basis a fixed denumeration $t_0, t_1, t_2, \ldots$ of all the terms and a fixed denumeration $\alpha_0, \alpha_1, \alpha_2, \ldots$ of all the expressions. The maximalisation process depends on the denumerations we have chosen.

First of all we put:

(0) $\mathfrak{M}_0 = \mathfrak{M}^{\Phi_0}$.

Φ_0 was so chosen that $\mathfrak{M}_0$ has the property (***).

In preparation for the definition of $\mathfrak{M}_{j+1}$, we define the set $\mathfrak{S}_j$ of expressions and the expression β_j (the latter only in the case that α_j is a generalisation).

By the induction hypothesis (***), there are infinitely many individual variables which occur neither in any of the expressions of $\mathfrak{M}_j$, nor in α_j, nor in t_j. In the natural order, let these be $y_0, y_1, y_2, \ldots$. Let $\mathfrak{S}_j$ be the set of all expressions $y_{2k} = t_j$ $(k = 0, 1, 2, \ldots)$. Suppose that α_j is a generalisation, i.e. $\alpha_j \equiv \bigwedge x\, \gamma_j$. y_1 does not occur in γ_j. Hence, by Chap. II, § 5.4, Theorem 6, there is exactly one expression β_j such that Subst $\gamma_j\, x\, y_1\, \beta_j$ and, conversely, Subst $\beta_j\, y_1\, x\, \gamma_j$.

Now we define:

$$(1) \qquad \mathfrak{M}_{j+1} = \begin{cases} \mathfrak{M}_j \cup \mathfrak{S}_j \cup \{\alpha_j\} & \underline{\text{if}}\ \ \mathfrak{M}_j \cup \{\alpha_j\}\ \ \underline{\text{is consistent (case 1)}}, \\ \mathfrak{M}_j \cup \mathfrak{S}_j & \underline{\text{if}}\ \ \mathfrak{M}_j \cup \{\alpha_j\}\ \ \underline{\text{is inconsistent and}}\ \alpha_j\ \underline{\text{is not a}} \\ & \qquad \underline{\text{generalisation (case 2)}}, \\ \mathfrak{M}_j \cup \mathfrak{S}_j \cup \{\neg\beta_j\} & \underline{\text{if}}\ \ \mathfrak{M}_j \cup \{\alpha_j\}\ \ \underline{\text{is inconsistent and}}\ \alpha_j\ \underline{\text{is a}} \\ & \qquad \underline{\text{generalisation (case 3)}}. \end{cases}$$

The variables $y_3, y_5, y_7, \ldots$ do not occur in $\mathfrak{M}_{j+1}$. Hence, (***) also holds for $\mathfrak{M}_{j+1}$.

By construction, we have $\mathfrak{M}_0 \subset \mathfrak{M}_1 \subset \mathfrak{M}_2 \subset \ldots$. We put:

$$(2) \qquad \mathfrak{M}^* = \bigcup_{j=0}^{\infty} \mathfrak{M}_j .$$

3.3 The maximal consistency of $\mathfrak{M}^*$. First of all, we prove by induction on j that every $\mathfrak{M}_j$ is consistent.

$\mathfrak{M}_0 = \mathfrak{M}^{\Phi_0}$ is consistent by 3.1.

By induction hypothesis, $\mathfrak{M}_j$ is consistent. In order to prove the consistency of $\mathfrak{M}_{j+1}$, we distinguish the three cases of the definition:

Case 1. If $\mathfrak{M}_j \cup \mathfrak{S}_j \cup \{\alpha_j\}$ were inconsistent, then we should have $\mathfrak{M}_j \cup \mathfrak{S}_j \cup \{\alpha_j\} \vdash P \wedge \neg P$ (for a no-place predicate variable P). Thus, there would be expressions $\mu_0, \ldots, \mu_r$ in $\mathfrak{M}_j$ and individual variables $y_{2k_0}, \ldots, y_{2k_s}$ such that

$$\vdash \mu_0 \ldots \mu_r \quad y_{2k_0} = t_j \ldots y_{2k_s} = t_j \quad \alpha_j : P \wedge \neg P .$$

Since y_{2k_0} appears only in the position indicated, $(S_{y_{2k_0}}^{t_j})$ gives us

$$\vdash \mu_0 \ldots \mu_r \quad t_j = t_j \ldots y_{2k_s} = t_j \quad \alpha_j : P \wedge \neg P ,$$

which, with (E) and (CuRu), yields

$$\vdash \mu_0 \ldots \mu_r \quad y_{2k_1} = t_j \ldots y_{2k_s} = t_j \quad \alpha_j : P \wedge \neg P .$$

By repeating this procedure, we finally obtain

$$\vdash \mu_0 \ldots \mu_r \quad \alpha_j : P \wedge \neg P .$$

But, by Chap. IV, § 5.3 (17), this means that $\mathfrak{M}_j \cup \{\alpha_j\}$ is inconsistent, which contradicts the assumption for case 1.

Case 2. If $\mathfrak{M}_j \cup \mathfrak{S}_j$ were inconsistent, then $\mathfrak{M}_j$ would also be inconsistent; we show this as in case 1.

Case 3. If $\mathfrak{M}_j \cup \mathfrak{S}_j \cup \{\neg \beta_j\}$ were inconsistent, then we should have (cf. case 1):

$$\vdash \mu_0 \ldots \mu_r \quad y_{2k_0} = t_j \ldots y_{2k_s} = t_j \quad \neg \beta_j : \bigwedge x \, \gamma_j .$$

As in case 1, we could then obtain

$$\vdash \mu_0 \ldots \mu_r \quad \neg \beta_j : \bigwedge x \, \gamma_j .$$

It would then follow that

$$\vdash \mu_0 \ldots \mu_r \neg \bigwedge x\gamma_j : \beta_j \qquad \text{(CaPo'')}$$

$$\vdash \mu_0 \ldots \mu_r \neg \bigwedge x\gamma_j : \bigwedge y_1\beta_j \qquad (G_{y_1})\ (y_1 \text{ does not occur free in the antecedent})$$

$$\vdash \bigwedge y_1\beta_j : \bigwedge x\gamma_j \qquad (\text{ReG}_{y_1, x})$$

$$\vdash \mu_0 \ldots \mu_r \neg \bigwedge x\gamma_j : \bigwedge x\gamma_j \qquad \text{(CuRu)}$$

$$\vdash \mu_0 \ldots \mu_r : \bigwedge x\gamma_j \qquad \text{(SeAs)}.$$

If we note that $\bigwedge x\gamma_j \equiv \alpha_j$, then this derivability relation shows that $\mathfrak{M}_j \vdash \alpha_j$. In case 3, $\mathfrak{M}_j \cup \{\alpha_j\}$ is inconsistent; hence $\mathfrak{M}_j \vdash \neg\alpha_j$, by Chap. IV, § 5.3 (19). Now Chap. IV, § 5.3 (21) shows that $\mathfrak{M}_j$ is inconsistent, which contradicts the induction hypothesis.

Given the consistency of the sets $\mathfrak{M}_j$, which we have just proved, we can prove the consistency of $\mathfrak{M}^*$ as follows: If $\mathfrak{M}^*$ were inconsistent, then we should have $\mathfrak{M}^* \vdash (\alpha \wedge \neg\alpha)$ for an arbitrary expression α. Then there would be finitely many elements $\gamma_1, \ldots, \gamma_n \in \mathfrak{M}^*$ with $\gamma_1, \ldots, \gamma_n \vdash (\alpha \wedge \neg\alpha)$. Since γ_j belongs to $\mathfrak{M}^*$, there is an $\mathfrak{M}_{k_j}$ such that $\gamma_j \in \mathfrak{M}_{k_j}$. Let k be the largest of the indices $k_1, \ldots, k_n$. Since the sequence of the sets $\mathfrak{M}_j$ is a non-decreasing one, $\mathfrak{M}_{k_j} \subset \mathfrak{M}_k$ for each $j = 1, \ldots, n$. Thus, each of $\gamma_1, \ldots, \gamma_n$ is an element of $\mathfrak{M}_k$. It follows that $\mathfrak{M}_k \vdash (\alpha \wedge \neg\alpha)$. But then $\mathfrak{M}_k$ would be inconsistent, whereas we have proved that it is consistent.

In order to show the maximal consistency of $\mathfrak{M}^*$, we take an arbitrary expression α which does not belong to $\mathfrak{M}^*$. We must show that $\mathfrak{M}^* \cup \{\alpha\}$ is inconsistent. This can be seen as follows: There is a j such that $\alpha \equiv \alpha_j$. Since $\alpha_j \notin \mathfrak{M}^*$, a fortiori $\alpha_j \notin \mathfrak{M}_{j+1}$. Hence, case 1 of the definition of $\mathfrak{M}_{j+1}$ does not hold, i.e. $\mathfrak{M}_j \cup \{\alpha_j\}$ is inconsistent. But then $\mathfrak{M}^* \cup \{\alpha_j\}$ is certainly inconsistent, q.e.d.

3.4 Consequences of the maximal consistency of $\mathfrak{M}^*$

Lemma 1. If $\mathfrak{M}^* \vdash \alpha$, then $\alpha \in \mathfrak{M}^*$.

Proof. Let $\mathfrak{M}^* \vdash \alpha$. If $\alpha \notin \mathfrak{M}^*$, then $\mathfrak{M}^* \cup \{\alpha\}$ would be inconsistent, by the maximal consistency of $\mathfrak{M}^*$. In particular, we should have

$$\mathfrak{M}^* \cup \{\alpha\} \vdash \neg\alpha,$$

and hence, by Chap. IV, § 5.2 (5), also $\mathfrak{M}^* \vdash \neg\alpha$. But, together with $\mathfrak{M}^* \vdash \alpha$, this contradicts the consistency of $\mathfrak{M}^*$.

Lemma 2. If $\vdash \alpha$, then $\alpha \in \mathfrak{M}^*$.

P r o o f. This follows from Lemma 1, since $\mathfrak{M}^* \vdash \alpha$ follows from $\vdash \alpha$.

Lemma 3. If $\alpha_1,\ldots,\alpha_n \vdash \alpha$ and $\alpha_1,\ldots,\alpha_n \in \mathfrak{M}^*$, then $\alpha \in \mathfrak{M}^*$.

P r o o f. From $\alpha_1,\ldots,\alpha_n \vdash \alpha$ and $\alpha_1,\ldots,\alpha_n \in \mathfrak{M}^*$ it follows that $\mathfrak{M}^* \vdash \alpha$ and hence, by Lemma 1, that $\alpha \in \mathfrak{M}^*$.

Lemma 4. (a) $t = t \in \mathfrak{M}^*$.

(b) If $t_1 = t_2 \in \mathfrak{M}^*$, then $t_2 = t_1 \in \mathfrak{M}^*$.

(c) If $t_1 = t_2 \in \mathfrak{M}^*$ and $t_2 = t_3 \in \mathfrak{M}^*$, then $t_1 = t_3 \in \mathfrak{M}^*$.

(d) If $t_1 = t_1' \in \mathfrak{M}^*, \ldots, t_r = t_r' \in \mathfrak{M}^*$, then $P t_1 \ldots t_r \in \mathfrak{M}^*$ iff $P t_1' \ldots t_r' \in \mathfrak{M}^*$.

(e) If $t_1 = t_1' \in \mathfrak{M}^*, \ldots, t_r = t_r' \in \mathfrak{M}^*$, then $f t_1 \ldots t_r = f t_1' \ldots t_r' \in \mathfrak{M}^*$.

P r o o f. (a) follows from Lemma 2 because $\vdash t = t$.

Moreover, by using derived rules from Chap. IV, § 1.5 and by Lemma 3, we obtain

(b) because $t_1 = t_2 \vdash t_2 = t_1$	because of (ESy)
(c) because $t_1 = t_2,\ t_2 = t_3 \vdash t_1 = t_3$	because of (ET)
(d) because $t_1 = t_1', \ldots, t_r = t_r',\ P t_1 \ldots t_r \vdash P t_1' \ldots t_r'$	because of (ER)
and $t_1 = t_1', \ldots, t_r = t_r',\ P t_1' \ldots t_r' \vdash P t_1 \ldots t_r$	[using (b)]
(e) because $t_1 = t_1', \ldots, t_r = t_r' \vdash f t_1 \ldots t_r = f t_1' \ldots t_r'$	because of (ER').

Lemma 5. $\neg \alpha \in \mathfrak{M}^*$ iff $\alpha \notin \mathfrak{M}^*$.

P r o o f. (a) If $\neg \alpha \in \mathfrak{M}^*$ and $\alpha \in \mathfrak{M}^*$, then $\mathfrak{M}^*$ would be inconsistent.

(b) If $\neg \alpha \notin \mathfrak{M}^*$ and $\alpha \notin \mathfrak{M}^*$, then, by the maximal consistency of $\mathfrak{M}^*$, the sets $\mathfrak{M}^* \cup \{\neg \alpha\}$ and $\mathfrak{M}^* \cup \{\alpha\}$ would be inconsistent. But then $\mathfrak{M}^*$ itself would be inconsistent [Chap. IV, § 5.3 (21)].

Lemma 6. $(\alpha \wedge \beta) \in \mathfrak{M}^*$ iff $\alpha \in \mathfrak{M}^*$ and $\beta \in \mathfrak{M}^*$.

P r o o f. (a) Let $(\alpha \wedge \beta) \in \mathfrak{M}^*$. Then, by Lemma 3, $\alpha \in \mathfrak{M}^*$ and $\beta \in \mathfrak{M}^*$, since $(\alpha \wedge \beta) \vdash \alpha$ and $(\alpha \wedge \beta) \vdash \beta$.

(b) Let $\alpha \in \mathfrak{M}^*$ and $\beta \in \mathfrak{M}^*$. Then, since $\alpha, \beta \vdash (\alpha \wedge \beta)$ and by Lemma 3, $(\alpha \wedge \beta) \in \mathfrak{M}^*$.

§ 4. Completion of the proof of the completeness theorem

4.1 Survey. We have (cf. § 2) only to show that the set $\mathfrak{M}^*$ defined in § 3 is satisfiable. In order to do this, we define a domain of individuals ω (4.2) and an interpretation $\mathfrak{J}$

over ω (4.3). In 4.4 we show that this interpretation is a model of an arbitrary expression α if and only if α is an element of $\mathfrak{M}^*$[1]. From this it follows at once that $\mathfrak{J}$ is a model of $\mathfrak{M}^*$. Hence, $\mathfrak{M}^*$ is satisfiable, q.e.d.

<u>4.2 The domain of individuals ω.</u> We define a relation, depending on $\mathfrak{M}^*$, on the domain of the terms. This relation is an equivalence relation, and thus leads to a division of the domain of terms into classes. We lay down these classes of equivalent terms as the elements of ω.

<u>Definition.</u> $t \sim t'$ <u>is to mean that</u> $t = t' \in \mathfrak{M}^*$.

<u>$\sim$ is an equivalence relation</u>. In order to show this, we must prove that:

(1) $t = t \in \mathfrak{M}^*$.

(2) <u>If</u> $t_1 = t_2 \in \mathfrak{M}^*$, <u>then</u> $t_2 = t_1 \in \mathfrak{M}^*$.

(3) <u>If</u> $t_1 = t_2 \in \mathfrak{M}^*$ <u>and</u> $t_2 = t_3 \in \mathfrak{M}^*$, <u>then</u> $t_1 = t_3 \in \mathfrak{M}^*$.

However, we have already shown this in § 3, Lemma 4(a), (b) and (c). Moreover, by (d) and (e) of the same lemma, we have:

(4) If $t_1 \sim t_1', \ldots, t_r \sim t_r'$, then $Pt_1 \ldots t_r \in \mathfrak{M}^*$ iff $Pt_1' \ldots t_r' \in \mathfrak{M}^*$.

(5) If $t_1 \sim t_1', \ldots, t_r \sim t_r'$, then $ft_1 \ldots t_r \sim ft_1' \ldots t_r'$.

Let $\bar{t}$ be the equivalence class, with respect to the relation $\sim$, in which t lies.

<u>4.3 The interpretation $\mathfrak{J}$ over ω.</u> We must define $\mathfrak{J}(P)$ for every P and $\mathfrak{J}(f)$ for every f.

<u>Definition 1.</u> <u>Let P be an r-place predicate variable</u> $(r \geq 0)$. <u>Then let</u> $\mathfrak{J}(P)$ <u>be an r-place predicate such that</u>

(*) $\mathfrak{J}(P)$ <u>fits</u> $\bar{t}_1, \ldots, \bar{t}_r$ <u>if and only if</u> $Pt_1 \ldots t_r \in \mathfrak{M}^*$ - <u>for arbitrary elements</u> $\bar{t}_1, \ldots, \bar{t}_r$ <u>of</u> ω.

This definition makes use of the representatives $t_1, \ldots, t_r$ of the classes $\bar{t}_1, \ldots, \bar{t}_r$. However, by 4.2 (4), it is independent of the particular representatives chosen.

<u>Definition 2.</u> <u>Let f be an r-place function variable</u> $(r \geq 0)$. <u>Then let</u> $\mathfrak{J}(f)$ <u>be an r-place function such that</u>

---

[1] It would have sufficed to show that $\mathfrak{J}$ is a model of α if $\alpha \in \mathfrak{M}^*$. However, the proof of the above equivalence is technically simpler.

(**) $$\mathfrak{J}(f)(\bar{t}_1,\ldots,\bar{t}_r) = \overline{ft_1\ldots t_r}$$

<u>for arbitrary elements</u> $\bar{t}_1,\ldots,\bar{t}_r$ <u>of</u> ω.

As was the case with the last definition, this definition makes use of particular representatives of the classes $\bar{t}_1,\ldots,\bar{t}_r$. However, as 4.2 (5) shows, the definition does not depend on the choice of these representatives.

Remark. We assume that there are predicates $\mathfrak{J}(P)$ and functions $\mathfrak{J}(f)$ with the properties (*) and (**) respectively. These are assumptions which are generally accepted in mathematics. (*) determines which r-tuples of individuals the predicate $\mathfrak{J}(P)$ is to fit; and (**) determines the value of $\mathfrak{J}(f)$ for any given argument. From the extensionalist point of view, there is only one predicate which satisfies the condition for $\mathfrak{J}(P)$ given in (*) and only one function which fulfils the condition for $\mathfrak{J}(f)$ expressed in (**). However, for our purposes the mere existence of such predicates and functions is enough.

For the interpretation which we have just defined, we have:

(***) $$\mathfrak{J}(t) = \bar{t}.$$

This can easily be shown by induction on the structure of the terms:

(a) By (**), $\mathfrak{J}(x) = \bar{x}$, since x is a no-place function variable.

(b) We have

$$\begin{aligned} \mathfrak{J}(ft_1\ldots t_r) &= \mathfrak{J}(f)(\mathfrak{J}(t_1),\ldots,\mathfrak{J}(t_r)) && \text{(Chap. III, § 2.2)} \\ &= \mathfrak{J}(f)(\bar{t}_1,\ldots,\bar{t}_r) && \text{(induction hypothesis)} \\ &= \overline{ft_1\ldots t_r} \end{aligned}$$

<u>4.4 Satisfiability of $\mathfrak{M}^*$.</u> Finally, we prove the

<u>Theorem.</u> Mod $\mathfrak{J}\alpha$ <u>if and only if</u> $\alpha \in \mathfrak{M}^*$ [2].

Proof. It suffices to show that this theorem holds for an arbitrary expression α under the assumption that it holds for all expressions whose rank is less than the rank of α. We distinguish the following cases:

(a) α is a predicate expression, i.e. $\alpha \equiv Pt_1\ldots t_r$. Then we have

---

[2] Here and in the following, we shall not mention explicitly the fixed domain of individuals ω.

$\mathrm{Mod}\,\mathfrak{J}\,P\,t_1\ldots t_r$ iff $\mathfrak{J}(P)$ fits $\mathfrak{J}(t_1),\ldots,\mathfrak{J}(t_r)$

iff $\mathfrak{J}(P)$ fits $\bar{t}_1,\ldots,\bar{t}_r$ by (***)

iff $P\,t_1\ldots t_r \in \mathfrak{M}^*$ by (*)

(b) α is an equation, i.e. $\alpha \equiv t_1 = t_2$. We have

$\mathrm{Mod}\,\mathfrak{J}\,t_1 = t_2$ iff $\mathfrak{J}(t_1) = \mathfrak{J}(t_2)$

iff $\bar{t}_1 = \bar{t}_2$ by (***)

iff $t_1 \sim t_2$

iff $t_1 = t_2 \in \mathfrak{M}^*$

(c) α is a negation, i.e. $\alpha \equiv \neg\beta$. Then $R(\beta) < R(\alpha)$, and we have

$\mathrm{Mod}\,\mathfrak{J}\,\alpha$ iff $\mathrm{Mod}\,\mathfrak{J}\,\neg\beta$

iff not $\mathrm{Mod}\,\mathfrak{J}\,\beta$

iff not $\beta \in \mathfrak{M}^*$ (induction hypothesis)

iff $\neg\beta \in \mathfrak{M}^*$ (3.4, Lemma 5)

(d) α is a conjunction, i.e. $\alpha \equiv (\beta \wedge \gamma)$. Then $R(\beta) < R(\alpha)$, $R(\gamma) < R(\alpha)$ and

$\mathrm{Mod}\,\mathfrak{J}\,\alpha$ iff $\mathrm{Mod}\,\mathfrak{J}\,(\beta \wedge \gamma)$

iff $\mathrm{Mod}\,\mathfrak{J}\,\beta$ and $\mathrm{Mod}\,\mathfrak{J}\,\gamma$

iff $\beta \in \mathfrak{M}^*$ and $\gamma \in \mathfrak{M}^*$ (induction hypothesis)

iff $(\beta \wedge \gamma) \in \mathfrak{M}^*$ (3.4, Lemma 6)

iff $\alpha \in \mathfrak{M}^*$

(e) α is a generalisation.

(e_1) First of all, let $\mathrm{Mod}\,\mathfrak{J}\alpha$. There is a j such that $\alpha \equiv \alpha_j \equiv \bigwedge x\gamma_j$ (cf. 3.2). Then, if we choose y_1 as in 3.2, we have $\mathrm{Subst}\,\gamma_j\,x\,y_1\,\beta_j$. $\bar{y}_1$ is an element of ω, and, since $\mathrm{Mod}\,\mathfrak{J}\,\bigwedge x\gamma_j$, we have $\mathrm{Mod}\,\mathfrak{J}_x^{\bar{y}_1}\gamma_j$. If we note that $\bar{y}_1 = \mathfrak{J}(y_1)$, then the substitution theorem (Chap. III, §3.2) gives us $\mathrm{Mod}\,\mathfrak{J}\beta_j$. By Chap. III, §5.6, theorem 2, $R(\beta_j) = R(\gamma_j) < R(\alpha)$. Hence the induction hypothesis yields $\beta_j \in \mathfrak{M}^*$. Since $\mathfrak{M}^*$ is consistent, we therefore have $\neg\beta_j \notin \mathfrak{M}^*$. This shows that, in the definition (1), case 3 does not hold. Since α_j is a generalisation, case 1 is the only remaining possibility. But then α_j, i.e. α, is an element of $\mathfrak{M}_{j+1}$, and hence also of $\mathfrak{M}^*$.

(e_2) Finally, let $\alpha \in \mathfrak{M}^*$. α is a generalisation, i.e. $\alpha \equiv \bigwedge x\beta$. Now let $\bar{t}$ be an arbitrary element of ω. There is a j such that $t \equiv t_j$. There are infinitely many individual variables y with $y = t_j \in \mathfrak{S}_j$ (cf. 3.2). Hence, there is an individual variable y which does not occur in β such that $y = t_j \in \mathfrak{S}_j$. Then, by 3.2 (1), $y = t_j \in \mathfrak{M}_{j+1} \subset \mathfrak{M}^*$. It follows that $\mathfrak{I}(y) = \overline{y} = \overline{t_j} = \overline{t}$. Since y does not occur in β, there is, by Chap. II, § 5.4, Theorem 6, an expression δ with Subst $\beta\, x\, y\, \delta$. But then, by $(\mathrm{ExG}_{x,y})$, we can obtain $\vdash \bigwedge x \beta : \delta$, i.e. $\alpha \vdash \delta$. Now 3.4, Lemma 3 shows that $\delta \in \mathfrak{M}^*$. But $R(\delta) = R(\beta) < R(\alpha)$, and hence, by induction hypothesis, Mod $\mathfrak{I}\,\delta$. With the substitution theorem, this yields Mod $\mathfrak{I}_x^{\mathfrak{I}(y)}\beta$, i.e. Mod $\mathfrak{I}_x^{\overline{t}}\beta$. Since $\overline{t}$ was chosen to be an arbitrary element of ω, we thus have Mod $\mathfrak{I} \bigwedge x\beta$, i.e. Mod $\mathfrak{I}\alpha$, q.e.d.

Exercises. 1. Determine the number of elements in the Lindenbaum algebra of $\mathfrak{M}^*$ (cf. Chap. IV, § 5, Exercise 5).

2. $\mathfrak{M}^* \vdash \bigwedge x\, Px$ if and only if $\mathfrak{M}^* \vdash Pt$ for all t.

3. The assertion in Exercise 2 does not always hold if we omit *.

4. Each class $\overline{t}$ contains at least one individual variable.

§ 5. Consequences of the completeness theorem

The completeness theorem means (together with the theorem about the soundness of the rules of predicate logic) the equivalence of the relation of consequence to that of derivability, and also the equivalence of satisfiability and consistency. Thus, it is now possible to translate an important property of the syntactic notions of derivability and consistency into a property of the equivalent semantic notions of consequence and satisfiability. It is in this way that we obtain the so-called compactness theorems (5.1). In 5.2, we give a typical example of the application of the compactness theorem for satisfiability. We shall introduce a further example of such applications in Chap. VI, § 4.

The compactness theorems make use of the equivalence of the above-mentioned syntactic and semantic concepts, but they make no use of the way in which this equivalence has been proved. However, if we analyse the proof of the completeness theorem, we obtain a theorem about satisfiability, due to Skolem. We shall return to this in 5.3.

5.1 The compactness theorems[3]. In Chap. IV, § 1.2, we defined the relation $\mathfrak{M} \vdash \alpha$ by requiring that there should be finitely many elements $\alpha_1, \ldots, \alpha_n$ of $\mathfrak{M}$ with $\alpha_1, \ldots, \alpha_n \vdash \alpha$. Thus, trivially, we have the

[3] These theorems are sometimes also called finiteness theorems. The name compactness theorem is derived from topology in the following way: The Stone space is the topological space whose points σ are the maximal consistent sets of expressions and for which the sets $V_\alpha = \{\sigma \mid \alpha \notin \sigma\}$ form a basis of open sets. Now if $\mathfrak{M}$ is a set of expressions, then $\{V_\alpha \mid \alpha \in \mathfrak{M}\}$ is a covering for the space if and only if $\mathfrak{M}$ is inconsistent. Thus the "compactness theorem for consistency" is equivalent to the assertion that the Stone space is compact.

<u>Compactness theorem for derivability.</u> $\mathfrak{M} \vdash \alpha$ <u>if and only if there is a finite subset</u> $\mathfrak{J}$ <u>of</u> $\mathfrak{M}$ <u>such that</u> $\mathfrak{J} \vdash \alpha$.

If we replace the notion of derivability by the equivalent one of consequence, then we obtain the

<u>Compactness theorem for consequence.</u> $\mathfrak{M} \models \alpha$ <u>if and only if there is a finite subset</u> $\mathfrak{J}$ <u>of</u> $\mathfrak{M}$ <u>such that</u> $\mathfrak{J} \models \alpha$.

If $\mathfrak{M}$ is a consistent set, then, clearly, every finite subset $\mathfrak{J}$ of $\mathfrak{M}$ is also consistent [cf. Chap. IV, §5.3 (16)]. Conversely, if we assume that every finite subset $\mathfrak{J}$ of $\mathfrak{M}$ is consistent, then $\mathfrak{M}$ must also be consistent: For, if $\mathfrak{M}$ were inconsistent, $(\alpha \wedge \neg \alpha)$ would be derivable from $\mathfrak{M}$ and would hence also be derivable from some finite subset $\mathfrak{J}$ of $\mathfrak{M}$. But then $\mathfrak{J}$ would be inconsistent, contrary to hypothesis. Thus we have the

<u>Compactness theorem for consistency.</u> <u>A set</u> $\mathfrak{M}$ <u>is consistent if and only if every finite subset</u> $\mathfrak{J}$ <u>of</u> $\mathfrak{M}$ <u>is consistent.</u>

If we pass from consistency to the equivalent concept of satisfiability, then the above theorem gives us the

<u>Compactness theorem for satisfiability.</u> <u>A set</u> $\mathfrak{M}$ <u>is satisfiable if and only if every finite subset</u> $\mathfrak{J}$ <u>of</u> $\mathfrak{M}$ <u>is satisfiable</u>.

We can use the last theorem to show that an infinite axiom system $\mathfrak{M}$ is satisfiable. In order to do this, it suffices to prove that every finite subset $\mathfrak{J}$ of $\mathfrak{M}$ has a model. This method has become interesting for algebraic investigations. In the following, we discuss a simple example.

<u>5.2 Example. Non-archimedean ordered fields.</u> A field is defined as a model of the following axiom system:

$\bigwedge x \bigwedge y \bigwedge z (x + y) + z = x + (y + z)$	$\bigwedge x\, x + 0 = x$
$\bigwedge x \bigwedge y \bigwedge z (xy)z = x(yz)$	$\bigwedge x\, x + (-x) = 0$
$\bigwedge x \bigwedge y\, x + y = y + x$	$\bigwedge x\, x 1 = x$
$\bigwedge x \bigwedge y\, xy = yx$	$\bigwedge x (x \neq 0 \rightarrow x x^{-1} = 1)$
$\bigwedge x \bigwedge y \bigwedge z\, x(y + z) = xy + xz$	$0 \neq 1$

Here, in order to make the axioms easier to understand intuitively, we use the notations 0, 1, +, $\cdot$, $-$, $^{-1}$ which are usually used in mathematics (we write xy instead of $x \cdot y$). 0, 1 are to be understood as individual variables, $-$, $^{-1}$ as one-place and +, $\cdot$ as two-place function variables. If, instead, we used "f" for addition, our first axiom would read

$$\bigwedge x \bigwedge y \bigwedge z \;\; ffxyz = fxfyz.$$

It is not usual to make use of the negation function - and the reciprocation function $^{-1}$ in the formulation of the axioms for fields. However, from a logical point of view, the use of these functions is to be recommended, since by means of it we can dispense with the existential quantifier (or with the negator, if we represent $\bigvee$ by $\neg\bigwedge\neg$) in some axioms.

Note that 0^{-1} is also a term, although it is usual not to define the reciprocal of 0. In a field, the reciprocal of the zero element can be an arbitrary element of the field (for example, we could fix this by means of an additional axiom $0^{-1} = 0$). We use $\vee$, $\rightarrow$ and $\bigvee$ as abbreviations. Cf. Chap. II, § 1.5.

An ordered field is a field which, in addition, satisfies the following axioms (where $\leqslant$ is to be understood as a two-place predicate variable):

$$\bigwedge x \bigwedge y \bigwedge z\ (x \leqslant y \wedge y \leqslant z \rightarrow x \leqslant z)$$
$$\bigwedge x \bigwedge y\ (x \leqslant y \wedge y \leqslant x \rightarrow x = y)$$
$$\bigwedge x \bigwedge y\ (x \leqslant y \vee y \leqslant x)$$
$$\bigwedge x \bigwedge y \bigwedge z\ (x \leqslant y \rightarrow x + z \leqslant y + z)$$
$$\bigwedge x \bigwedge y \bigwedge z\ (x \leqslant y \wedge 0 \leqslant z \rightarrow xz \leqslant yz)$$

A non-archimedean ordered field is an ordered field for which, in addition, the following axioms (of which there are infinitely many) hold:

$$(*) \qquad \left\{ \begin{array}{r} 0 \leqslant u \\ 0 + 1 \leqslant u \\ (0 + 1) + 1 \leqslant u \\ ((0 + 1) + 1) + 1 \leqslant u \\ \cdots\cdots\cdots\cdots\cdots \end{array} \right.$$

Here, as well as the primitive notions 0, 1, -, $^{-1}$, +, $\cdot$ of the axioms for fields and the additional primitive notion $\leqslant$ of the axioms for ordered fields, we have a further primitive notion u. u is an individual variable which is different from 0 and 1. If we have a non-archimedean ordered field, then, in it, u is interpreted by an element of the field which is greater than or equal to the zero element, and also greater than or equal to every element of the field which can be obtained from the zero element by adding 1 finitely many times. Thus, this element of the field can be understood as an "infinitely large" element.

Let $\mathfrak{J}_o$ be the axiom system (consisting of 15 axioms) for ordered fields and $\mathfrak{J}_{on}$ the (infinite) axiom system for non-archimedean ordered fields. $\mathfrak{J}_o \subset \mathfrak{J}_{on}$.

We can ask whether there are any non-archimedean ordered fields (i.e. whether $\mathfrak{J}_{on}$ has a model). This is in fact the case. In elementary algebra it can be shown that, in the field formed by the rational functions in an unknown x over the field of the rational numbers, an ordering ($\leqslant$) can be introduced in such a way that this field becomes an ordered field in which the unknown x is an infinitely large element.

Here, we want to carry out the proof of the existence of a non-archimedean ordered field by means of the compactness theorem for satisfiability. In doing this, we assume only that the axiom system $\mathfrak{J}_o$ has a model (i.e. that there is an ordered field, e.g. the field of the rational numbers). Let $\mathfrak{J}$ be such a model.

It is sufficient to show that every finite subset of $\mathfrak{J}_{on}$ has a model. Let $\mathfrak{J}$ be such a finite subset. $\mathfrak{J}$ contains only finitely many axioms from (*). If $\mathfrak{J}$ does not contain any axioms from (*), then $\mathfrak{J} \subset \mathfrak{J}_o$. But then $\mathfrak{J}$ has a model, since, by hypothesis, $\mathfrak{J}_o$ is satisfiable. Thus, we may assume that $\mathfrak{J}$ contains at least one of the axioms from (*). Among all the axioms from (*) which occur in $\mathfrak{J}$, there is one which is the longest. Let this be $t_0 \leqslant u$. Then, if $t \leqslant u$ is any other one of the axioms from (*) which occur in $\mathfrak{J}$, t_0 can be obtained from t by adding 1 finitely many times.

We claim that $\mathfrak{J}_u^{\mathfrak{J}(t_0)}$ is a model of $\mathfrak{J}$ (where $\mathfrak{J}$ is one of the models of $\mathfrak{J}_0$ whose existence is asserted by our hypothesis). u does not occur in the axioms of $\mathfrak{J}_0$, and hence, by the coincidence theorem (Chap. III, § 3.1), $\mathfrak{J}_u^{\mathfrak{J}(t_0)}$ is a model of $\mathfrak{J}_0$ because $\mathfrak{J}$ is a model of $\mathfrak{J}_0$. Thus, we need only show that $\mathfrak{J}_u^{\mathfrak{J}(t_0)}$ is a model of those axioms of $\mathfrak{J}$ which belong to (*).

One such axiom is $t_0 \leqslant u$. We have

$$\begin{aligned} \text{Mod}\ \mathfrak{J}_u^{\mathfrak{J}(t_0)} t_0 \leqslant u \quad & \text{iff } \mathfrak{J}_u^{\mathfrak{J}(t_0)}(\leqslant) \text{ fits } \mathfrak{J}_u^{\mathfrak{J}(t_0)}(t_0),\ \mathfrak{J}_u^{\mathfrak{J}(t_0)}(u) \\ & \text{iff } \mathfrak{J}(\leqslant) \text{ fits } \mathfrak{J}(t_0),\ \mathfrak{J}(t_0) \\ & \text{iff Mod } \mathfrak{J} t_0 \leqslant t_0 . \end{aligned}$$

However, the last statement is true, since $t_0 \leqslant t_0$ is a consequence of $\mathfrak{J}_0$ and therefore $\mathfrak{J}$ is a model of $t_0 \leqslant t_0$.

Let us consider an element $t \leqslant u$ of $\mathfrak{J}$ which is different from $t_0 \leqslant u$ and which belongs to (*). We have

$$\begin{aligned} \text{Mod}\ \mathfrak{J}_u^{\mathfrak{J}(t_0)}\ t \leqslant u \quad & \text{iff } \mathfrak{J}_u^{\mathfrak{J}(t_0)}(\leqslant) \text{ fits } \mathfrak{J}_u^{\mathfrak{J}(t_0)}(t),\ \mathfrak{J}_u^{\mathfrak{J}(t_0)}(u) \\ & \text{iff } \mathfrak{J}(\leqslant) \text{ fits } \mathfrak{J}(t),\ \mathfrak{J}(t_0) \\ & \text{iff Mod } \mathfrak{J} t \leqslant t_0 . \end{aligned}$$

Now, for every term t, the expression $t \leqslant t + 1$ is a consequence of $\mathfrak{J}_0$. Since t_0 can be obtained from t by adding 1 finitely many times, it can be seen (bearing in mind the transitive laws for $\leqslant$ in $\mathfrak{J}_0$) that the expression $t \leqslant t_0$ is also a consequence of $\mathfrak{J}_0$. Hence, $\mathfrak{J}$ is a model of $t \leqslant t_0$, and thus $\mathfrak{J}_u^{\mathfrak{J}(t_0)}$ is a model of $t \leqslant u$.

5.3 The Löwenheim-Skolem theorem. We assume that the set of expressions $\mathfrak{M}$ is satisfiable. Then, trivially, $\mathfrak{M}$ is also consistent (Chap. IV, § 5.3). In the last section we proved the converse, namely that every consistent set is satisfiable. This brings us back to our initial assumption. Thus, apparently, we have gained nothing. However, if we inspect the proof of the satisfiability of a consistent set $\mathfrak{M}$ more closely, directing our attention particularly at the domain of individuals ω over which there is a model for $\mathfrak{M}$, we come to a remarkable conclusion. The domain of individuals ω which we considered in § 5 consisted of the equivalence classes of an equivalence relation over the set of all terms. It is possible to make a statement about the number of elements in ω: The set of all terms is denumerable; the set ω of the equivalence classes over a denumerable set may have a smaller number of elements, but it cannot have a larger number; hence, ω is either finite or denumerable (in other words, "at most denumerable"). - Summarising our argument, we have: A satisfiable set is consistent; a consistent set is satisfiable in a finite or denumerable domain of individuals; hence, a satisfiable set is always satisfiable over a finite or denumerable domain of individuals. Thus, we have the

Löwenheim-Skolem theorem. If a set $\mathfrak{M}$ of expressions has a model over any domain of individuals, then it has a model over a finite or denumerable domain.

Exercises. 1. In Chap. III, § 2 we noted that a domain of individuals ω uniquely determines the set of all predicates and functions over ω. We may ask ourselves whether the laws of logic still hold if we restrict ourselves to a proper subset of the set of predicates or of functions. In such a case, we would speak of a non-standard ontology, by contrast to the standard ontology, in which all predicates and functions are allowed. Examples of non-standard ontologies are those in which,

(a) for each n, the only n-place predicates which are allowed are those which fit either all n-tuples of individuals or no n-tuples of individuals. (Then, under the extensionalist point of view, there are only two n-place predicates for each n.)

(b) for each n, the only n-place predicates which are allowed are those which fit either finitely many n-tuples of individuals or all but a finite number of n-tuples of individuals.

For the sake of simplicity we assume that, in (a) and (b), all functions are allowed.

Show that the soundness of the predicate calculus (Chap. IV, § 3) holds for every non-standard ontology which, for each n, has at least one n-place predicate and at least one n-place function.

2. If we take the non-standard ontology in Exercise 1(a) as a basis, then the predicate calculus is sound but not complete. In order to prove this, show that, relative to this ontology,

$$\neg \bigwedge x\, Px \models \bigwedge x \neg Px$$

but that it is not true that

$$\neg \bigwedge x\, Px \vdash \bigwedge x \neg Px .$$

(The theorem of the soundness of the predicate calculus in relation to the standard ontology can be used to prove the latter assertion.)

The completeness proof given in this chapter breaks down for this non-standard ontology because not all the predicates $\mathfrak{J}(P)$ (cf. 4.3, Def. 1) exist. Show this for the case $\mathfrak{M} = \{Px, \neg Py\}$, where $y \not\equiv x$. (Show also that this set is consistent.) An analysis of the completeness proof shows that we only need to assume that, in the ontology we are taking as a basis, the predicates and functions defined in 4.3, Def. 1 and Def. 2 exist for every consistent set $\mathfrak{M}$. It has been shown that there are non-standard ontologies in which this requirement holds.

3. Prove the following extension of the Löwenheim-Skolem theorem: If a set $\mathfrak{M}$ of expressions which do not contain the equality sign possesses a model, then it has a model over a denumerable domain. Suggestion: Add to $\mathfrak{M}$ all expressions of the form $x = y$, where x, y are distinct individual variables.

4. Give an axiom system for fields of characteristic zero in the language of predicate logic, taking 0, 1, +, · as primitive notions. Show that every theorem of the theory of fields of characteristic zero (i.e. every specific statement - cf. Chap. I, § 3.9 - which follows from this system of axioms) also holds for all fields with a sufficiently high (prime) characteristic.

5. The fields of characteristic zero cannot be characterised by finitely many axioms within the framework of predicate logic.

6. Prove that: If a polynomial in several variables with whole-number coefficients is irreducible over every field of characteristic zero, then it is also irreducible over every field of sufficiently high (prime) characteristic.

7. If, for every n, there is a domain of individuals ω with at least n elements such that the set $\mathfrak{M}$ of expressions is satisfiable over ω, then $\mathfrak{M}$ is satisfiable over a denumerable domain of individuals.

8. A group is said to be orderable if a two-place relation $\leqslant$ can be defined in it which fulfils the order axioms (the first three axioms in the axiom system for ordered fields given in 5.2) and the axioms $\bigwedge x \bigwedge y \bigwedge z(x \leqslant y \rightarrow x + z \leqslant y + z \wedge z + x \leqslant z + y)$. Show that a denumerable group is orderable if every finitely generated subgroup is orderable.

VI. Peano's Axiom System

In modern mathematics, such strong preference is given to the axiomatic method that this is often regarded as a characteristic of mathematics in general. We do not want, here, to go into the question whether and to what extent this point of view is justified. We do, however, want to explain to what consequences it leads as far as what we could call the "objects" of a mathematical theory are concerned.

The axiomatic method consists in taking some axiom system $\mathfrak{A}$ as a basis and obtaining consequences from $\mathfrak{A}$. Such an axiom system $\mathfrak{A}$ consists of "mathematical statements", which can be built up as statement forms (cf. Chap. I, § 4.2) and which, in many cases, can be represented as expressions of the language of predicate logic. If the consequences are obtained by means of a calculus with the aid of rules (as is possible in predicate logic), then, as mathematicians, we are working entirely in a language with which we operate formally; in this case we speak of a formalised theory.

The concept of a formalised theory can be regarded as a development of the concept of an axiomatic theory. We should note that the assumption that the consequences can be obtained from the axioms formally by means of rules is not an essential part of the concept of an axiomatic theory. In fact the notion of consequence, as we constructed it, was at first defined not formally but semantically, with reference to concepts with a meaning such as those of a domain of individuals, a predicate and a function. The language of predicate logic has the property that this semantic notion of consequence is equivalent to a syntactic notion of derivability. However, there are languages with a semantic notion of consequence which cannot be replaced by an equivalent notion of derivability. An example of such a language is the language of second-order predicate logic, the essentials of which we shall introduce in § 1.

By what we have just said, an axiomatic theory which has a semantic notion of consequence is, basically, concerned with meaning and content. We can hold on to this fact if we want to find out what on earth such a theory is talking about. An axiom system $\mathfrak{A}$ determines which interpretations are models of $\mathfrak{A}$. The idea of saying that the "objects" of a mathematical theory based on a system of axioms $\mathfrak{A}$ are the set of the models of $\mathfrak{A}$ comes to mind very easily. For example, the "objects" of group theory are the models of the axiom system for group theory, i.e. the groups.

As far as the objects dealt with by them are concerned, we can detect two extreme tendencies in different mathematical theories: (1) There are theories like group theory and other modern mathematical theories which deal with very "many" objects; their wide applicability is due to this fact. (2) Other theories were built up with the intention of dealing with as few objects as possible, and in certain cases with only one object, so that such a theory "characterises" the given object. Geometry and arithmetic are of this nature.

In this chapter, we want to investigate the situation in arithmetic more thoroughly. In doing this, we shall find that:

(a) Peano's axiom system cannot be written down in the language of predicate logic; in order to write it down, we need the language of "second-order logic". We shall build up this language and its semantics in § 1.

(b) It is not possible, either in the language of predicate logic or in the language of second-order logic, to lay down any system of axioms which possesses only one model. On the contrary, if an interpretation is a model of a particular axiom system, then so is every interpretation which is isomorphic to it (§ 2). Thus, at the very best, an axiom system can determine an interpretation uniquely up to isomorphism.

(c) Peano's axiom system, written down in the language of second-order logic, is categorical (cf. Chap. I, § 4.3). In this sense, the natural numbers can be characterised in the language of second-order logic (§ 3).

(d) Although Peano's axiom system cannot be written down in the language of predicate logic, we can, in a fairly obvious way, replace it by an infinite axiom system which belongs to the language of predicate logic and which we could, at first sight, expect to characterise the natural numbers uniquely up to isomorphism. However, appearances are deceptive; for we can show that <u>every</u> axiom system in the language of predicate logic possesses, as well as this "natural interpretation", a model over an uncountable domain of individuals which is not isomorphic to the natural interpretation. Thus, in this sense, the natural numbers cannot be characterised in the language of predicate logic. As a corollary, we prove the incompleteness of second-order logic (§ 4).

§ 1. The language and semantics of second-order logic

Peano's axiom system (cf. Chap. I, § 3.7) is not formulated in the language of predicate logic; for the induction axiom makes a statement about <u>all</u> predicates. In order to do this, it is necessary to generalise a predicate variable; but, in the language of predicate logic, this is not allowed. If we make provision for this possibility, we obtain a language of what is called <u>second-order logic</u>.

The language of second-order logic is an extension of the language of predicate logic (first-order logic). It is this language (or, in fact, only a part of it; cf. the remark in 1.1) which we are going to present here. In doing this we shall, for the sake of brevity, limit ourselves to stating explicitly the points in which the language of second-order logic ($\mathfrak{L}_2$ for short) differs from the language of predicate logic ($\mathfrak{L}_1$ for short).

1.1 The expressions of $\mathfrak{L}_2$. We use the same symbols as in $\mathfrak{L}_1$. The notion of a term is taken over from $\mathfrak{L}_1$, and so is that of an atomic expression. In the formation of expressions, we want also to allow predicate variables to be generalised (and not only individual variables, as in $\mathfrak{L}_1$).

We could go further than this and also allow function variables to be generalised. Another possibility, which we shall not take up here, is the use of new symbols, which are second-order predicate variables (or function variables, as the case may be). If, for example, $\underline{P}$ were a one-place second-order predicate variable and P a normal one-place predicate variable, then we could use $\underline{P}\,P$ as an additional atomic expression (we could say that $\underline{P}\,P$ is obtained from Px "by raising the order"). We do not want to go into the question of the different possible "types" of such second-order variables. In $\mathfrak{L}_2$, these new second-order variables may not be generalised (otherwise we should have a language $\mathfrak{L}_3$).

We can achieve this by adding to the rules of the expression calculus (Chap. II, § 1.5) the following new rule:

Rule 5. We are allowed to pass from a row of symbols ζ to every row of symbols $\bigwedge P\zeta$, where P is an arbitrary predicate variable.

As variables for expressions of $\mathfrak{L}_2$ we use Greek letters, as we did for expressions of $\mathfrak{L}_1$.

1.2 Models and the relation of consequence. We take over from $\mathfrak{L}_1$ the notion of a domain of individuals ω and the notion of an interpretation $\mathfrak{J}$ (Chap. III, § 2). We also take over the definition of the relationship $\mathrm{Mod}_\omega \mathfrak{J}\alpha$: however, we have to supplement this to include the case that an expression begins with a generalised predicate variable P (i.e. with $\bigwedge P$). The definition is analogous to our earlier definition for the case that an expression begins with a generalised individual variable x (i.e. with $\bigwedge x$). If $\mathfrak{P}$ is a variable for predicates with the same number of places as P, then, in analogy to $\mathfrak{J}_x^{\mathfrak{r}}$, we introduce the notation $\mathfrak{J}_P^{\mathfrak{P}}$. $\mathfrak{J}_P^{\mathfrak{P}}$ is to be an interpretation over the same domain of individuals as $\mathfrak{J}$; it is to interpret all variables which are different from P just as $\mathfrak{J}$ does; and, finally, it is to be such that

$$\mathfrak{J}_P^{\mathfrak{P}}(P) = \mathfrak{P}.$$

Using this notation, we supplement the definition in Chap. III, § 2.3 by:

$$\mathrm{Mod}_\omega \mathfrak{J} \bigwedge P\alpha \quad \underline{\text{if and only if}} \quad \text{for every } \mathfrak{P}: \mathrm{Mod}_\omega \mathfrak{J}_P^{\mathfrak{P}} \alpha .$$

Thus, we have defined the notion of a model for arbitrary expressions α of $\mathfrak{L}_2$. With the aid of this notion of a model, we can define the notion of consequence and the notion of universal validity in the usual way (Chap. III, § 2.4).

1.3 Equality. An example of a universally valid expression is the expression

$$(*) \qquad x = y \leftrightarrow \bigwedge P(Px \leftrightarrow Py),$$

where P is a one-place predicate variable and x and y are distinct individual variables[1].

(*) expresses Leibniz's identitas indiscernibilium. (We could take (*) or the more general expression $t_1 = t_2 \leftrightarrow \bigwedge P(Pt_1 \leftrightarrow Pt_2)$ as a definition of equality.)

We must show that every interpretation $\mathfrak{J}$ is a model of (*). In order to do this, we have to prove (cf. also Chap. VII, § 1.3) that:

(a) If $\mathrm{Mod}_\omega \mathfrak{J} x = y$, then $\mathrm{Mod}_\omega \mathfrak{J} \bigwedge P(Px \leftrightarrow Py)$,

(b) If not $\mathrm{Mod}_\omega \mathfrak{J} x = y$, then not $\mathrm{Mod}_\omega \mathfrak{J} \bigwedge P(Px \leftrightarrow Py)$.

to (a): Let $\mathrm{Mod}_\omega \mathfrak{J} x = y$. Then $\mathfrak{J}(x) = \mathfrak{J}(y)$. If $\mathfrak{P}$ is an arbitrary one-place predicate, then we have to show that $\mathrm{Mod}_\omega \mathfrak{J}_P^{\mathfrak{P}}(Px \leftrightarrow Py)$, i.e. that $\mathrm{Mod}_\omega \mathfrak{J}_P^{\mathfrak{P}} Px$ if and only if $\mathrm{Mod}_\omega \mathfrak{J}_P^{\mathfrak{P}} Py$. But this follows immediately from the fact that

$$\mathfrak{J}_P^{\mathfrak{P}}(x) = \mathfrak{J}(x) = \mathfrak{J}(y) = \mathfrak{J}_P^{\mathfrak{P}}(y) .$$

to (b): Suppose that not $\mathrm{Mod}_\omega \mathfrak{J} x = y$, i.e. that $\mathfrak{J}(x) \neq \mathfrak{J}(y)$. Then there is a one-place predicate $\mathfrak{P}$ which fits $\mathfrak{J}(x)$ but does not fit $\mathfrak{J}(y)$. With this $\mathfrak{P}$, clearly $\mathrm{Mod}_\omega \mathfrak{J}_P^{\mathfrak{P}} Px$ but not $\mathrm{Mod}_\omega \mathfrak{J}_P^{\mathfrak{P}} Py$, hence not $\mathrm{Mod}_\omega \mathfrak{J}_P^{\mathfrak{P}}(Px \leftrightarrow Py)$, and therefore also not $\mathrm{Mod}_\omega \mathfrak{J} \bigwedge P(Px \leftrightarrow Py)$.

1.4 Rules of inference. In 4.5 we shall discover that there is no system of rules by means of which we can, as in ordinary predicate logic, derive all the consequences of an arbitrary set of expressions. Thus, every finite system of correct rules is necessarily incomplete. For this reason, we shall refrain here from giving rules of inference for second-order logic.

---

[1] Here, we take $(\alpha \leftrightarrow \beta)$ as an abbreviation (cf. Chap. II, § 1.5).

Exercises.

Preliminary. In the following, let P, Q, R and x, y, z be pairwise distinct. $\bigvee$, $\rightarrow$, $\leftrightarrow$ are to be regarded as abbreviations.

1. Find whether or not the following expressions are universally valid:

(a) $\bigwedge P \bigvee x\, Px$ (b) $\bigvee P \bigwedge x\, Px$

(c) $(\bigwedge P\, Px \rightarrow Qx)$ (d) $\bigwedge P\, (\bigwedge x\, Px \rightarrow Qy)$

(e) $\bigwedge P \bigwedge x \bigwedge y\, (x = y \rightarrow \neg(Px \wedge \neg Py))$ (f) $\bigwedge R \bigwedge x \bigvee P \bigwedge y\, (Rxy \leftrightarrow Px)$.

2. In 1.3 we showed that identity can be defined in second-order logic. Confirm that the following expressions are also possible definitions of $x = y$:

$$\bigwedge P\, (Px \rightarrow Py),\ \ \bigwedge R\, (\bigwedge x\, Rxx \rightarrow Rxy),\ \ \bigwedge R\, (Rxy \rightarrow Ryx),$$
$$\bigwedge R\, (Rxy \rightarrow \bigvee y\, Rxz),\ \ \bigwedge R\, (\bigwedge y\, Rxy \rightarrow Ryx)\ .$$

3. Show that the following expressions possess only models with finite domains of individuals:

(a) $\bigwedge R(\bigwedge x \bigwedge y \bigwedge z \bigwedge u(Rxy \wedge Rzu \rightarrow (x = z \leftrightarrow y = u)) \wedge \bigwedge x \bigvee yRxy \rightarrow \bigwedge y \bigvee xRxy)$

(b) $\bigwedge R(\bigwedge x \bigwedge y \bigwedge z(Rxy \wedge Ryz \rightarrow Rxz) \wedge \bigwedge x \neg Rxx \wedge \bigvee x \bigvee yRxy \rightarrow \bigvee x \bigwedge y \neg Rxy)$.

Show also that the given expressions possess models over every finite domain of individuals.

4. Show that the negations of the expressions given in Exercise 3 are satisfiable precisely over the infinite domains of individuals.

5. The compactness theorem does not hold in second-order logic. Hint: Exercise 3 and Chap. III, § 2, Exercise 5(c).

6. Extend second-order logic by adding a new quantifier $\mathrm{M}x$, meaning for infinitely many x. Find an expression α of second-order logic in which M does not occur, such that $\vDash \alpha \leftrightarrow \mathrm{M}xPx$.

§ 2. Isomorphic interpretations. Categoricity of axiom systems

2.1 Algebras. By an algebra we mean, in the simplest case, a finite sequence

(i) $$\mathfrak{A} = \langle \omega, \mathfrak{S}_1, \ldots, \mathfrak{S}_s \rangle ,$$

where ω is a (nonempty) domain of individuals and each $\mathfrak{S}_j$ is either a function over ω or a predicate over ω. ω is called the underlying set of $\mathfrak{A}$ and $\mathfrak{S}_1, \ldots, \mathfrak{S}_s$ are called the primitive notions of $\mathfrak{A}$. It is often convenient to use another notation for the algebra (i), by introducing the "indexing set" $I = \{1, \ldots, s\}$ and writing

(ii) $$\mathfrak{A} = \langle \omega, \{\langle i, \mathfrak{S}_i \rangle : i \in I\} \rangle .$$

More generally, we can, in (ii), admit arbitrary (and possibly infinite) indexing sets, and thus obtain the most general concept of an algebra.

Example: The algebra $\mathfrak{A}_0 = \langle \omega_0, \mathfrak{n}, \mathfrak{f} \rangle$, where ω_0 is the set of the natural numbers, $\mathfrak{n}$ is the zero element (i.e. a no-place function over ω_0) and $\mathfrak{f}$ the one-place successor function, which assigns to every number as its value the number which succeeds it. If we want to write this algebra in the form (ii), then we can choose the indexing set $\{1,2\}$. In connection with predicate logic, another possibility arises: If we want to symbolise statements about the given algebra $\mathfrak{A}_0$, we shall introduce an individual variable x as a symbol for $\mathfrak{n}$ and a one-place function variable f as a symbol for $\mathfrak{f}$. If we now choose an interpretation $\mathfrak{J}_0$ over ω_0 such that $\mathfrak{J}_0(x) = \mathfrak{n}$ and $\mathfrak{J}_0(f) = \mathfrak{f}$, then we can represent $\mathfrak{A}_0$ in the form $\langle \omega_0, \mathfrak{J}_0(x), \mathfrak{J}_0(f) \rangle$, where $\{x, f\}$ appears as the indexing set.

This example can be generalised: Every interpretation $\mathfrak{J}$ over a domain of individuals ω and every set $\mathfrak{V}$ of variables corresponds to a uniquely determined algebra

(iii) $$\mathfrak{A}_{\mathfrak{V}}(\mathfrak{J}) = \langle \omega, \{\langle v, \mathfrak{J}(v) \rangle : v \in \mathfrak{V}\} \rangle .$$

The primitive notions of this algebra are the $\mathfrak{J}(v)$ with $v \in \mathfrak{V}$. It is evident that

$$\mathfrak{A}_{\mathfrak{V}}(\mathfrak{J}) = \mathfrak{A}_{\mathfrak{V}'}(\mathfrak{J}') \quad \text{if and only if} \quad \mathfrak{V} = \mathfrak{V}' \text{ and } \mathfrak{J} \underset{\mathfrak{V}}{=} \mathfrak{J}'$$

(for the notation $\underset{\mathfrak{V}}{=}$ cf. Chap. III, § 3.1).

2.2 Isomorphisms of algebras. The concept of an isomorphism which is familiar to us from group theory and field theory can be generalised for arbitrary algebras. A 1-1 mapping Φ from ω onto ω' is called an isomorphism from the algebra (ii) onto the algebra

(ii)' $$\mathfrak{A}' = \langle \omega', \{\langle i, \mathfrak{G}_i' \rangle : i \in I\} \rangle$$

if the following conditions hold:

(a) for every $i \in I$, $\mathfrak{G}_i$ is a function or a predicate respectively if and only if $\mathfrak{G}_i'$ is a function or a predicate respectively,

(b) for every $i \in I$, $\mathfrak{G}_i$ and $\mathfrak{G}_i'$ have the same number of places,

(c) if $\mathfrak{G}_i$ is an r-place predicate over ω then, for all $\mathfrak{x}_1, \ldots, \mathfrak{x}_r \in \omega$,

$$\mathfrak{G}_i \text{ fits } \mathfrak{x}_1, \ldots, \mathfrak{x}_r \text{ if and only if } \mathfrak{G}_i' \text{ fits } \Phi(\mathfrak{x}_1), \ldots, \Phi(\mathfrak{x}_r) ,$$

(d) If $\mathfrak{G}_i$ is an r-place function over ω then, for all $\mathfrak{x}_1, \ldots, \mathfrak{x}_r \in \omega$,

$$\Phi(\mathfrak{G}_i(\mathfrak{x}_1, \ldots, \mathfrak{x}_r)) = \mathfrak{G}_i'(\Phi(\mathfrak{x}_1), \ldots, \Phi(\mathfrak{x}_r)) .$$[2]

---

[2] We could take the concept of an isomorphism even more generally, by allowing $\mathfrak{A}$ and $\mathfrak{A}'$ to have different indexing sets.

If $\mathfrak{S}_i$ is a no-place function, then condition (d) becomes

$$(d_0) \qquad \Phi(\mathfrak{S}_i) = \mathfrak{S}'_i .$$

Conditions (a) and (b) are independent of Φ. We say that $\mathfrak{A}$ and $\mathfrak{A}'$ are of the same type if these conditions hold. If $\mathfrak{J}$ and $\mathfrak{J}'$ are arbitrary interpretations and $\mathfrak{B}$ is any set of variables, then the algebras $\mathfrak{A}$ and $\mathfrak{A}'$ are clearly of the same type. Thus, Φ is an isomorphism from $\mathfrak{A}_{\mathfrak{B}}(\mathfrak{J})$ onto $\mathfrak{A}_{\mathfrak{B}}(\mathfrak{J}')$ if and only if Φ is a 1-1 mapping from ω onto ω' (where, naturally, ω is the domain of individuals which belongs to $\mathfrak{J}$ and ω' that which belongs to $\mathfrak{J}'$) which satisfies the following conditions:

(c') $\mathfrak{J}(P)$ fits $\mathfrak{x}_1, \ldots, \mathfrak{x}_r$ iff $\mathfrak{J}'(P)$ fits $\Phi(\mathfrak{x}_1), \ldots, \Phi(\mathfrak{x}_r)$
for every r-place $P \in \mathfrak{B}$ and all $\mathfrak{x}_1, \ldots, \mathfrak{x}_r \in \omega$,

(d') $\Phi(\mathfrak{J}(f)(\mathfrak{x}_1, \ldots, \mathfrak{x}_r)) = \mathfrak{J}'(f)(\Phi(\mathfrak{x}_1), \ldots, \Phi(\mathfrak{x}_r))$
for every r-place $f \in \mathfrak{B}$ and all $\mathfrak{x}_1, \ldots, \mathfrak{x}_r \in \omega$.

In particular, (d_0) becomes

(d'_0) $\Phi(\mathfrak{J}(x)) = \mathfrak{J}'(x)$ for every $x \in \mathfrak{B}$.

Two algebras $\mathfrak{A}$, $\mathfrak{A}'$ are said to be isomorphic if there is an isomorphism from $\mathfrak{A}$ onto $\mathfrak{A}'$. The relation of being isomorphic is reflexive, symmetric and transitive.

Lemma 1. Let Φ be a 1-1 mapping from the domain of individuals ω onto the domain of individuals ω'. Let $\mathfrak{J}$ be an arbitrary interpretation over ω. Then there is an interpretation $\mathfrak{J}'$ over ω' such that, for an arbitrary set of variables $\mathfrak{B}$, Φ is an isomorphism from the algebra $\mathfrak{A}_{\mathfrak{B}}(\mathfrak{J})$ onto the algebra $\mathfrak{A}_{\mathfrak{B}}(\mathfrak{J}')$.

Proof. We can take the conditions (c') and (d') as the definition of $\mathfrak{J}'$.

2.3 The model relationship in the case of isomorphic algebras $\mathfrak{A}_{\mathfrak{B}}(\mathfrak{J})$, $\mathfrak{A}_{\mathfrak{B}}(\mathfrak{J}')$

Lemma 2. Let Φ be an isomorphism from $\mathfrak{A}_{\mathfrak{B}}(\mathfrak{J})$ onto $\mathfrak{A}_{\mathfrak{B}}(\mathfrak{J}')$. Let $\mathfrak{B}(t = t) \subseteq \mathfrak{B}$. Then we have

$$(*) \qquad \Phi(\mathfrak{J}(t)) = \mathfrak{J}'(t) .$$

Proof by induction on the structure of t:
If t is an individual variable, then (*) is the same as 2.2 (d'_0).
If $t = ft_1 \ldots t_r$ is a compound term, then

$$\begin{aligned}\Phi(\mathfrak{J}(ft_1\ldots t_r)) &= \Phi(\mathfrak{J}(f)(\mathfrak{J}(t_1),\ldots,\mathfrak{J}(t_r))) && \text{(Chap. III, § 2.2)}\\ &= \mathfrak{J}'(f)(\Phi(\mathfrak{J}(t_1)),\ldots,\Phi(\mathfrak{J}(t_r))) && \text{(2.2 (d'))}\\ &= \mathfrak{J}'(f)(\mathfrak{J}'(t_1),\ldots,\mathfrak{J}'(t_r)) && \text{(induction hypothesis)}\\ &= \mathfrak{J}'(ft_1\ldots t_r)\,.\end{aligned}$$

Lemma 3. Let Φ be an isomorphism from $\mathfrak{A}_{\mathfrak{V}}(\mathfrak{J})$ onto $\mathfrak{A}_{\mathfrak{V}}(\mathfrak{J}')$. Let $\mathfrak{r}$ be an element of the domain of individuals ω which belongs to $\mathfrak{J}$. Then Φ is also an isomorphism from $\mathfrak{A}_{\mathfrak{V}\cup\{x\}}(\mathfrak{J}_x^{\mathfrak{r}})$ onto $\mathfrak{A}_{\mathfrak{V}\cup\{x\}}(\mathfrak{J}'^{\Phi(\mathfrak{r})}_x)$.

Proof. By hypothesis, we need only test for the variable x whether or not the conditions for isomorphism hold. By 2.2 (d_0') we must, in this case, show that $\Phi(\mathfrak{J}_x^{\mathfrak{r}}(x)) = \mathfrak{J}'^{\Phi(\mathfrak{r})}_x(x)$. But both sides are equal to $\Phi(\mathfrak{r})$.

Lemma 4. Let Φ be an isomorphism from $\mathfrak{A}_{\mathfrak{V}}(\mathfrak{J})$ onto $\mathfrak{A}_{\mathfrak{V}}(\mathfrak{J}')$. Let $\mathfrak{P}$ be a predicate over the domain of individuals ω which belongs to $\mathfrak{J}$. Let $\Phi(\mathfrak{P})$ be the predicate over the domain of individuals ω' (corresponding to $\mathfrak{J}'$) defined by: $\mathfrak{P}$ fits $\mathfrak{r}_1,\ldots,\mathfrak{r}_r$ iff $\Phi(\mathfrak{P})$ fits $\Phi(\mathfrak{r}_1),\ldots,\Phi(\mathfrak{r}_r)$ (for all $\mathfrak{r}_1,\ldots,\mathfrak{r}_r \in \omega$). Then Φ is also an isomorphism from $\mathfrak{A}_{\mathfrak{V}\cup\{P\}}(\mathfrak{J}_P^{\mathfrak{P}})$ onto $\mathfrak{A}_{\mathfrak{V}\cup\{P\}}(\mathfrak{J}'^{\Phi(\mathfrak{P})}_P)$, where P is an arbitrary r-place predicate variable.

Proof. By hypothesis, we need only test for the variable P whether or not the conditions for isomorphism hold. But, in this case, they correspond exactly to the definition of $\Phi(\mathfrak{P})$.

Isomorphism theorem. Let $\mathfrak{A}_{\mathfrak{V}}(\mathfrak{J})$ be isomorphic to $\mathfrak{A}_{\mathfrak{V}}(\mathfrak{J}')$. Let α be an expression (of the first or second order) such that $\mathfrak{V}(\alpha) \subseteq \mathfrak{V}$. Then

$$\text{Mod}\,\mathfrak{J}\,\alpha \quad \text{if and only if} \quad \text{Mod}\,\mathfrak{J}'\alpha\,.$$

(Briefly: Isomorphic algebras are either both a model or both not a model of a "specific" expression.)

Proof. We show by induction on the structure of α that the above equivalence holds for arbitrary $\mathfrak{V}$, $\mathfrak{J}$, $\mathfrak{J}'$, provided $\mathfrak{A}_{\mathfrak{V}}(\mathfrak{J})$ is isomorphic to $\mathfrak{A}_{\mathfrak{V}}(\mathfrak{J}')$.

1) Let $\alpha \equiv Pt_1\ldots t_r$ be a predicate expression such that $\mathfrak{V}(\alpha) \subseteq \mathfrak{V}$.

$$\begin{aligned}\text{Mod}\,\mathfrak{J}\,Pt_1\ldots t_r \quad &\text{iff} \quad \mathfrak{J}(P) \text{ fits } \mathfrak{J}(t_1),\ldots,\mathfrak{J}(t_r) && \\ &\text{iff} \quad \mathfrak{J}'(P) \text{ fits } \Phi(\mathfrak{J}(t_1)),\ldots,\Phi(\mathfrak{J}(t_r)) && \text{(2.2 (c'))}\\ &\text{iff} \quad \mathfrak{J}'(P) \text{ fits } \mathfrak{J}'(t_1),\ldots,\mathfrak{J}'(t_r) && \text{(Lemma 2)}\\ &\text{iff} \quad \text{Mod}\,\mathfrak{J}'\,Pt_1\ldots t_r\,.\end{aligned}$$

2) Let $\alpha \equiv t_1 = t_2$ be an equation such that $\mathfrak{B}(\alpha) \subseteq \mathfrak{B}$.

$$\begin{array}{lll} \mathrm{Mod}\,\mathfrak{J}\, t_1 = t_2 & \text{iff } \mathfrak{J}(t_1) = \mathfrak{J}(t_2) & \\ & \text{iff } \Phi(\mathfrak{J}(t_1)) = \Phi(\mathfrak{J}(t_2)) & (\Phi \text{ is 1-1}) \\ & \text{iff } \mathfrak{J}'(t_1) = \mathfrak{J}'(t_2) & (\text{Lemma 2}) \\ & \text{iff } \mathrm{Mod}\,\mathfrak{J}'\, t_1 = t_2 . & \end{array}$$

3) Let $\alpha \equiv \neg\beta$ be a negation such that $\mathfrak{B}(\alpha) \subseteq \mathfrak{B}$. Then also $\mathfrak{B}(\beta) \subseteq \mathfrak{B}$.

$$\begin{array}{lll} \mathrm{Mod}\,\mathfrak{J}\,\neg\beta & \text{iff not } \mathrm{Mod}\,\mathfrak{J}\beta & \\ & \text{iff not } \mathrm{Mod}\,\mathfrak{J}'\beta & (\text{induction hypothesis}) \\ & \text{iff } \mathrm{Mod}\,\mathfrak{J}'\neg\beta . & \end{array}$$

4) If $\alpha \equiv (\alpha_1 \wedge \alpha_2)$ is a conjunction such that $\mathfrak{B}(\alpha) \subseteq \mathfrak{B}$, then $\mathfrak{B}(\alpha_1) \subseteq \mathfrak{B}$ and $\mathfrak{B}(\alpha_2) \subseteq \mathfrak{B}$. We prove the required assertion as in 3).

5) Let $\alpha \equiv \bigwedge x\beta$ be a generalisation such that $\mathfrak{B}(\alpha) \subseteq \mathfrak{B}$. Then $\mathfrak{B}(\beta) \subseteq \mathfrak{B} \cup \{x\}$.

$$\begin{array}{lll} \mathrm{Mod}\,\mathfrak{J}\bigwedge x\beta & \text{iff (for all } \mathfrak{r} \in \omega)\ \mathrm{Mod}\,\mathfrak{J}^{\mathfrak{r}}_{x}\beta & \\ & \text{iff (for all } \mathfrak{r} \in \omega)\ \mathrm{Mod}\,\mathfrak{J}'^{\Phi(\mathfrak{r})}_{x}\beta & (\text{induction hypothesis, Lemma 3}) \\ & \text{iff (for all } \mathfrak{r}' \in \omega')\ \mathrm{Mod}\,\mathfrak{J}'^{\mathfrak{r}'}_{x}\beta & (\Phi(\omega) = \omega') \\ & \text{iff } \mathrm{Mod}\,\mathfrak{J}'\bigwedge x\beta . & \end{array}$$

6) If $\alpha \equiv \bigwedge P\beta$ (only in the case that α is a second-order expression) is a generalisation such that $\mathfrak{B}(\alpha) \subseteq \mathfrak{B}$, then $\mathfrak{B}(\beta) \subseteq \mathfrak{B} \cup \{P\}$. We prove the assertion as in 5), using Lemma 4. For the purposes of the proof, note that, as $\mathfrak{P}$ runs through all the r-place predicates over ω, $\Phi(\mathfrak{P})$ runs through all the r-place predicates over ω'.

2.4 Categorical axiom systems. A set $\mathfrak{S}$ of (first- or second-order) expressions may be regarded as an axiom system. The set of the primitive notions of $\mathfrak{S}$ is $\mathfrak{B}(\mathfrak{S}) = \bigcup_{\alpha \in \mathfrak{S}} \mathfrak{B}(\alpha)$. An algebra $\mathfrak{A}_{\mathfrak{B}}(\mathfrak{J})$ is said to be a model of $\mathfrak{S}$ if $\mathfrak{B}(\mathfrak{S}) \subseteq \mathfrak{B}$ and $\mathrm{Mod}\,\mathfrak{J}\mathfrak{S}$. By the isomorphism theorem which we have just proved we see that, if a given algebra is a model of an axiom system $\mathfrak{S}$, then so is every algebra which is isomorphic to it.

An axiom system $\mathfrak{S}$ is said to be categorical [c-categorical (where c is a finite or infinite cardinal number)] if there is an algebra $\mathfrak{A}$ which is a model of $\mathfrak{S}$ [and whose underlying set has cardinality c] and if any two algebras [whose underlying sets both have cardinality c] which are models of $\mathfrak{S}$ are isomorphic to each other. Thus, a categorical axiom system determines a model uniquely up to isomorphism, and a c-categorical axiom system together with the cardinal number c determines a model uniquely up to isomorphism. A categorical axiom system is c-categorical for some $c \geqslant 1$.

Exercises. 1. Show that the following axiom system is categorical (let $x \neq y$):

$$\bigwedge x Px, \quad \bigwedge x \bigwedge y \; x = y .$$

2. Show that the following axiom system is not categorical (let x, y, z be pairwise distinct):

$$\bigwedge x \; Rxx, \quad \bigwedge x \bigwedge y(Rxy \rightarrow Ryx), \quad \bigwedge x \bigwedge y \bigwedge z(Rxy \wedge Ryz \rightarrow Rxz),$$
$$\bigwedge x \bigwedge y \bigwedge z(x = y \vee y = z \vee x = z), \quad \bigvee x \bigvee y(x \neq y) .$$

Determine the number of non-isomorphic models.

3. Find out whether the axiom system

$$\bigwedge x \bigwedge y \bigwedge z(x = y \vee y = z \vee x = z), \quad \bigvee x \bigvee y(x \neq y \wedge Px \wedge \neg Py)$$

is categorical (let x, y, z be pairwise distinct).

4. Find the number of non-isomorphic models of the following axiom system (let $x_1, \ldots, x_4$ be pairwise distinct):

$$\bigwedge x_1 \bigwedge x_2 \bigwedge x_3 \bigwedge x_4 (x_1 = x_2 \vee x_1 = x_3 \vee x_1 = x_4 \vee x_2 = x_3 \vee x_2 = x_4 \vee x_3 = x_4),$$
$$\bigvee x_1 \bigvee x_2 (x_1 \neq x_2 \wedge Px_1 \wedge \neg Px_2) .$$

§ 3. The characterisability of the natural numbers in the language of second-order predicate logic

3.1 Survey. We take as a basis our thoughts in Chap. I, § 3.7 about Peano's system of axioms. This axiom system can be formulated in the language of second-order logic. The primitive notions of the axiom system are nought and successor of. We must represent these primitive notions by an individual variable and a one-place function variable. In order to fit in with the usual notation, we shall denote the individual variable which stands for nought by "0" and the one-place function variable which stands for successor of by " ' ". Moreover, as is customary, we shall write "x' " instead of " 'x", and so on.

We assume that x and y are individual variables which are different from each other and from 0. With the aid of a one-place predicate variable P, Peano's axiom system can be represented as follows in the language $\mathfrak{L}_2$ of second-order predicate logic (here, we take $\rightarrow$ as an abbreviation; cf. Chap. II, § 1.5) (see also Chap. I, § 3.7):

P1 $\bigwedge x\, x' \neq 0$

P2 $\bigwedge x \bigwedge y\, (x' = y' \rightarrow x = y)$

P3 $\bigwedge P\, (P0 \wedge \bigwedge x (Px \rightarrow Px') \rightarrow \bigwedge x\, Px)$.

Let $\mathfrak{S}_0$ be the set of these axioms.

Theorem. Peano's second-order axiom system $\mathfrak{S}_0$ is categorical. "The natural numbers can be characterised up to isomorphism in the language of second-order logic."

Proof. Let $\mathfrak{B}_0 = \mathfrak{B}(\mathfrak{S}_0) = \{0, '\}$. Let $\mathfrak{A}_{\mathfrak{B}_0}(\mathfrak{J}_1)$ and $\mathfrak{A}_{\mathfrak{B}_0}(\mathfrak{J}_2)$ be two models of $\mathfrak{S}_0$. We want to show that these algebras are isomorphic. Let ω_1 be the domain of individuals which belongs to $\mathfrak{J}_1$ and ω_2 that which belongs to $\mathfrak{J}_2$. For the sake of brevity, let us write:

$$\mathfrak{J}_1(0) = \mathfrak{n}_1, \qquad \mathfrak{J}_2(0) = \mathfrak{n}_2,$$

$$\mathfrak{J}_1(') = \mathfrak{f}_1, \qquad \mathfrak{J}_2(') = \mathfrak{f}_2 .$$

The fact that the two algebras are isomorphic follows from the

Theorem. There is a 1-1 mapping Φ from ω_1 onto ω_2 such that

(*) $\Phi(\mathfrak{n}_1) = \mathfrak{n}_2$

(**) $\Phi(\mathfrak{f}_1(\mathfrak{r})) = \mathfrak{f}_2(\Phi(\mathfrak{r}))$ for all $\mathfrak{r}$ in ω_1 .

This theorem follows from the following Lemma (which is formally weaker), which we shall prove in 3.5.

Lemma 1. There is a mapping Φ from ω_1 into ω_2 such that (*) and (**) hold.

From this Lemma we obtain the above theorem as follows: Since the assumptions about $\mathfrak{n}_1$, $\mathfrak{f}_1$ and $\mathfrak{n}_2$, $\mathfrak{f}_2$ are symmetrical, there also exists a mapping Ψ from ω_2 into ω_1 such that

(*') $\psi(\mathfrak{n}_2) = \mathfrak{n}_1$

(**') $\psi(\mathfrak{f}_2(\mathfrak{n})) = \mathfrak{f}_1(\psi(\mathfrak{n}))$ for all $\mathfrak{n}$ in ω_2 .

In 3.3 we show that

(***) $\psi(\Phi(\mathfrak{r})) = \mathfrak{r}$ for all $\mathfrak{r}$ in ω_1 .

By symmetry, we also have

(***') $\Phi(\psi(\mathfrak{n})) = \mathfrak{n}$ for all $\mathfrak{n}$ in ω_2 .

Now if $\Phi(\mathfrak{r}_1) = \Phi(\mathfrak{r}_2)$, then $\psi(\Phi(\mathfrak{r}_1)) = \psi(\Phi(\mathfrak{r}_2))$ and hence, by (***), $\mathfrak{r}_1 = \mathfrak{r}_2$. Thus, Φ is 1-1. In order to prove the above theorem we must also show that Φ is a mapping from ω_1 onto ω_2, i.e. that every element of ω_2 is an image under Φ of an element of ω_1. Let $\mathfrak{n}$ be an arbitrary element of ω_2. Then, by (***'), $\mathfrak{n} = \Phi(\psi(\mathfrak{n}))$, i.e. $\mathfrak{n}$ is the image under Φ of the element $\psi(\mathfrak{n})$ of ω_1.

3.2 Induction over the natural numbers. The method of proof by induction over the natural numbers rests on the fact that the interpretation $\mathfrak{J}$ of which we are thinking is a model of the induction axiom P3. For the following, let us take as a basis the interpretation $\mathfrak{J}_1$ (which is assumed to be a model of P3). If $\mathfrak{P}$ is an arbitrary one-place predicate over ω_1, then, by 1.2, the interpretation $\mathfrak{J}_{1P}^{\mathfrak{P}}$ is a model of $P0 \wedge \bigwedge x(Px \to Px') \to \bigwedge x Px$. This means:

If $\mathfrak{J}_{1P}^{\mathfrak{P}}$ is a model of P0 and of $\bigwedge x(Px \to Px')$, then $\mathfrak{J}_{1P}^{\mathfrak{P}}$ is also a model of $\bigwedge x Px$.

If we analyse this statement further by means of the definition of a model, we finally obtain:

If $\mathfrak{P}\mathfrak{n}_1$[3] and (if $\mathfrak{P}\mathfrak{r}$, then $\mathfrak{P}\mathfrak{f}_1(\mathfrak{r})$) for all $\mathfrak{r}$, then $\mathfrak{P}\mathfrak{r}$ for all $\mathfrak{r}$ in ω_1.

Let us assume that, for the property $\mathfrak{P}$, we could prove both the following assertions:

(1) $\mathfrak{P}\mathfrak{n}_1$.

(2) If $\mathfrak{P}\mathfrak{r}$, then $\mathfrak{P}\mathfrak{f}_1(\mathfrak{r})$ - for all $\mathfrak{r}$ in ω_1.

Then we should obtain

(3) $\mathfrak{P}\mathfrak{r}$ for all $\mathfrak{r}$ in ω_1.

The passage from (1) and (2) to (3) is called proof by induction over the natural numbers. This method of proof can be used in order to show that a property $\mathfrak{P}$ fits all the elements of ω_1.

3.3 Proof of (***). We choose the property $\mathfrak{P}$ which fits an element $\mathfrak{r}$ of ω_1 if and only if $\psi(\Phi(\mathfrak{r})) = \mathfrak{r}$. In order to prove (***) by induction over the natural numbers, it suffices to prove the statements (1) and (2) for the property $\mathfrak{P}$.

to (1): $\psi(\Phi(\mathfrak{n}_1)) = \psi(\mathfrak{n}_2)$ (by (*))

$= \mathfrak{n}_1$ (by (*'))

---

[3] This is to be an abbreviation for the statement that $\mathfrak{P}$ fits $\mathfrak{n}_1$.

to (2): Let $\psi(\Phi(\mathfrak{x})) = \mathfrak{x}$; then we have:

$$\begin{aligned} \psi(\Phi(\mathfrak{f}_1(\mathfrak{x}))) &= \psi(\mathfrak{f}_2(\Phi(\mathfrak{x}))) && \text{(by (**))} \\ &= \mathfrak{f}_1(\psi(\Phi(\mathfrak{x}))) && \text{(by (**'))} \\ &= \mathfrak{f}_1(\mathfrak{x}) && \text{(by hypothesis)} \end{aligned}$$

3.4 Peano relations. By way of preparation for the proof of Lemma 3.1 let us consider relations $\mathfrak{R}$ which can hold between elements of ω_1 and elements of ω_2 (i.e. whenever $\mathfrak{R}\mathfrak{x}\mathfrak{n}$, it is always the case that $\mathfrak{x} \in \omega_1$ and $\mathfrak{n} \in \omega_2$). We call such a relation a Peano relation if it fulfils the following two conditions:

(i) $$\mathfrak{R}\mathfrak{n}_1\mathfrak{n}_2 .$$

(ii) If $\mathfrak{R}\mathfrak{x}\mathfrak{n}$, then $\mathfrak{R}\mathfrak{f}_1(\mathfrak{x})\mathfrak{f}_2(\mathfrak{n})$ - for all $\mathfrak{x} \in \omega_1$ and $\mathfrak{n} \in \omega_2$.

Let $\mathfrak{R}_0$ be the intersection of all the Peano relations, i.e. let

$$\mathfrak{R}_0\mathfrak{x}\mathfrak{n} \quad \text{if and only if} \quad \mathfrak{R}\mathfrak{x}\mathfrak{n} \quad \text{for all Peano relations } \mathfrak{R} .$$

It follows from the definition of $\mathfrak{R}_0$ that:

(4) If $\mathfrak{R}_0\mathfrak{x}\mathfrak{n}$ and $\mathfrak{R}$ is a Peano relation, then $\mathfrak{R}\mathfrak{x}\mathfrak{n}$.

We assert:

(5) $\mathfrak{R}_0$ is a Peano relation.

Thus, $\mathfrak{R}_0$ is "the smallest Peano relation". In order to prove (5), we must show that $\mathfrak{R}_0$ fulfils the conditions (i) and (ii).

to (i): For every Peano relation $\mathfrak{R}$ we have $\mathfrak{R}\mathfrak{n}_1\mathfrak{n}_2$. Hence $\mathfrak{R}_0\mathfrak{n}_1\mathfrak{n}_2$.

to (ii): We assume that $\mathfrak{R}_0\mathfrak{x}\mathfrak{n}$ and have to show that $\mathfrak{R}_0\mathfrak{f}_1(\mathfrak{x})\mathfrak{f}_2(\mathfrak{n})$. Thus, we must prove that $\mathfrak{R}\mathfrak{f}_1(\mathfrak{x})\mathfrak{f}_2(\mathfrak{n})$ for an arbitrary Peano relation $\mathfrak{R}$. Let $\mathfrak{R}$ be an arbitrary Peano relation. Then $\mathfrak{R}\mathfrak{x}\mathfrak{n}$, because $\mathfrak{R}_0\mathfrak{x}\mathfrak{n}$ [see (4)], and hence $\mathfrak{R}\mathfrak{f}_1(\mathfrak{x})\mathfrak{f}_2(\mathfrak{n})$, since $\mathfrak{R}$ is a Peano relation.

The smallest Peano relation $\mathfrak{R}_0$ has the nature of a function, i.e.:

(6) For every element $\mathfrak{x}$ of ω_1 there is exactly one $\mathfrak{n}$ in ω_2 such that $\mathfrak{R}_0\mathfrak{x}\mathfrak{n}$.

We prove this assertion by induction over the natural numbers (3.2)[4]. In order to do this; we have to prove that:

---

[4] It is here and only here that we use the fact that $\mathfrak{J}_1$ is also a model of P1 and of P2.

(a) There is exactly one $\mathfrak{n}$ in $\mathfrak{w}_2$ such that $\mathfrak{R}_0\mathfrak{n}_1\mathfrak{n}$.

(b) If, for some $\mathfrak{x}$ in $\mathfrak{w}_1$, there is exactly one $\mathfrak{n}$ such that $\mathfrak{R}_0\mathfrak{x}\,\mathfrak{n}$, then there is exactly one $\mathfrak{z}$ in $\mathfrak{w}_2$ such that $\mathfrak{R}_0\mathfrak{f}_1(\mathfrak{x})\mathfrak{z}$.

to (a): Existence. $\mathfrak{R}_0\mathfrak{n}_1\mathfrak{n}_2$, since $\mathfrak{R}_0$ is a Peano relation.

Uniqueness. Suppose that $\mathfrak{R}_0\mathfrak{n}\,\mathfrak{v}$. We have to show that $\mathfrak{v} = \mathfrak{n}_2$. We prove this indirectly by assuming that $\mathfrak{v} \neq \mathfrak{n}_2$. Then we define a relation $\mathfrak{R}_0'$ by requiring that, for arbitrary $\mathfrak{x}$, $\mathfrak{n}$,

$\mathfrak{R}_0'\mathfrak{x}\,\mathfrak{n}$ if and only if $\mathfrak{R}_0\mathfrak{x}\,\mathfrak{n}$ but not also $\mathfrak{x} = \mathfrak{n}_1$ and $\mathfrak{n} = \mathfrak{v}$.

Then, certainly, not $\mathfrak{R}_0'\mathfrak{n}_1\mathfrak{v}$. $\mathfrak{R}_0'$ is a Peano relation: in order to show this, we must prove (i) and (ii) for it.

to (i): $\mathfrak{R}_0'\mathfrak{n}_1\mathfrak{n}_2$, since $\mathfrak{R}_0\mathfrak{n}_1\mathfrak{n}_2$ but not also $\mathfrak{n}_1 = \mathfrak{n}_1$ and $\mathfrak{n}_2 = \mathfrak{v}$ (because $\mathfrak{v} \neq \mathfrak{n}_2$).

to (ii): Let $\mathfrak{R}_0'\mathfrak{x}\,\mathfrak{n}$. We need to show that $\mathfrak{R}_0'\mathfrak{f}_1(\mathfrak{x})\mathfrak{f}_2(\mathfrak{n})$. We do this as follows: We have $\mathfrak{R}_0\mathfrak{x}\,\mathfrak{n}$ (since $\mathfrak{R}_0'\mathfrak{x}\,\mathfrak{n}$) and hence $\mathfrak{R}_0\mathfrak{f}_1(\mathfrak{x})\mathfrak{f}_2(\mathfrak{n})$ (since $\mathfrak{R}_0$ is a Peano relation). In order to prove that $\mathfrak{R}_0'\mathfrak{f}_1(\mathfrak{x})\mathfrak{f}_2(\mathfrak{n})$ we now have to show that $\mathfrak{f}_1(\mathfrak{x}) = \mathfrak{n}_1$ and $\mathfrak{f}_2(\mathfrak{n}) = \mathfrak{v}$ do not both hold. But $\mathfrak{f}_1(\mathfrak{x}) \neq \mathfrak{n}_1$, since $\mathfrak{I}_1$ is a model of P1.

Now, by applying (4) to $\mathfrak{R}_0$ and the Peano relation $\mathfrak{R}_0'$, we see that $\mathfrak{R}_0'\mathfrak{x}\,\mathfrak{n}$ whenever $\mathfrak{R}_0\mathfrak{x}\,\mathfrak{n}$. But this contradicts our assumption that $\mathfrak{R}_0\mathfrak{n}_1\mathfrak{v}$ but not $\mathfrak{R}_0'\mathfrak{n}_1\mathfrak{v}$.

to (b): We assume that for $\mathfrak{x}$ in $\mathfrak{w}_1$ there is exactly one $\mathfrak{n}$ such that $\mathfrak{R}_0\mathfrak{x}\,\mathfrak{n}$. We assert that there is then exactly one $\mathfrak{z}$ such that $\mathfrak{R}_0\mathfrak{f}_1(\mathfrak{x})\mathfrak{z}$.

Existence. $\mathfrak{R}_0\mathfrak{f}_1(\mathfrak{x})\mathfrak{f}_2(\mathfrak{n})$ because $\mathfrak{R}_0\mathfrak{x}\,\mathfrak{n}$, since $\mathfrak{R}_0$ is a Peano relation.

Uniqueness. We assume that, as well as $\mathfrak{R}_0\mathfrak{f}_1(\mathfrak{x})\mathfrak{f}_2(\mathfrak{n})$, also $\mathfrak{R}_0\mathfrak{f}_1(\mathfrak{x})\mathfrak{z}$ for a $\mathfrak{z}$ such that $\mathfrak{f}_2(\mathfrak{n}) \neq \mathfrak{z}$. We must refute this assumption. In order to do this, we define a relation $\mathfrak{R}_0'$ by requiring that, for arbitrary $\mathfrak{u}$, $\mathfrak{v}$:

$\mathfrak{R}_0'\mathfrak{u}\,\mathfrak{v}$ if and only if $\mathfrak{R}_0\mathfrak{u}\,\mathfrak{v}$ but not also $\mathfrak{f}_1(\mathfrak{x}) = \mathfrak{u}$ and $\mathfrak{z} = \mathfrak{v}$.

Then, certainly, not $\mathfrak{R}_0'\mathfrak{f}_1(\mathfrak{x})\mathfrak{z}$. $\mathfrak{R}_0'$ is a Peano relation; in order to show this, we must prove (i) and (ii) for it.

to (i): $\mathfrak{R}_0'\mathfrak{n}_1\mathfrak{n}_2$, since $\mathfrak{R}_0\mathfrak{n}_1\mathfrak{n}_2$ but not also $\mathfrak{f}_1(\mathfrak{x}) = \mathfrak{n}_1$ and $\mathfrak{z} = \mathfrak{n}_2$. For $\mathfrak{f}_1(\mathfrak{x}) \neq \mathfrak{n}_1$, since $\mathfrak{I}_1$ is a model of P1.

to (ii): Let $\mathfrak{R}_0'\mathfrak{u}\,\mathfrak{v}$. We need to show that $\mathfrak{R}_0'\mathfrak{f}_1(\mathfrak{u})\mathfrak{f}_2(\mathfrak{v})$. We do this as follows: We have $\mathfrak{R}_0\mathfrak{u}\,\mathfrak{v}$ (because $\mathfrak{R}_0'\mathfrak{u}\,\mathfrak{v}$) and hence $\mathfrak{R}_0\mathfrak{f}_1(\mathfrak{u})\mathfrak{f}_2(\mathfrak{v})$ (since $\mathfrak{R}_0$ is a Peano relation). In order to show that $\mathfrak{R}_0'\mathfrak{f}_1(\mathfrak{u})\mathfrak{f}_2(\mathfrak{v})$, we therefore have to show that $\mathfrak{f}_1(\mathfrak{x}) = \mathfrak{f}_1(\mathfrak{u})$ and $\mathfrak{z} = \mathfrak{f}_2(\mathfrak{v})$ do not both hold. If $\mathfrak{f}_1(\mathfrak{x}) = \mathfrak{f}_1(\mathfrak{u})$, then we should have $\mathfrak{x} = \mathfrak{u}$, since $\mathfrak{J}_1$ is a model of P2. Thus, we should have $\mathfrak{R}_0\mathfrak{x}\,\mathfrak{v}$. But then, by the assumption at the beginning of "to (b)", $\mathfrak{v} = \mathfrak{n}$. Now if, also, $\mathfrak{z} = \mathfrak{f}_2(\mathfrak{v})$, then we should have $\mathfrak{z} = \mathfrak{f}_2(\mathfrak{n})$, which would contradict the assumption about $\mathfrak{z}$.

If we now apply (4) to $\mathfrak{R}_0$ and the Peano relation $\mathfrak{R}_0'$, we see that $\mathfrak{R}_0'\mathfrak{f}_1(\mathfrak{x})\mathfrak{n}$ whenever $\mathfrak{R}_0\mathfrak{f}_1(\mathfrak{x})\mathfrak{n}$. But this contradicts our assumption that $\mathfrak{R}_0\mathfrak{f}_1(\mathfrak{x})\mathfrak{z}$ but not $\mathfrak{R}_0'\mathfrak{f}_1(\mathfrak{x})\mathfrak{z}$.

3.5 Proof of the Lemma in 3.1. We have just seen that the Peano relation $\mathfrak{R}_0$ has the nature of a function. Thus, for every $\mathfrak{x}$ in ω_1, there is exactly one $\mathfrak{n}$ in ω_2 such that $\mathfrak{R}_0\mathfrak{x}\,\mathfrak{n}$. Thus, there is a function Φ which is defined for every $\mathfrak{x}$ in ω_1 and which assigns to the element $\mathfrak{x}$ in ω_1 that $\mathfrak{n}$ in ω_2 for which $\mathfrak{R}_0\mathfrak{x}\,\mathfrak{n}$. By this definition, Φ is a mapping from ω_1 into ω_2. Moreover, by reason of the definition of Φ, we have the relation

$$(\substack{**\\**}) \qquad \mathfrak{R}_0\mathfrak{x}\,\Phi(\mathfrak{x}) .$$

In order to prove the Lemma, we must prove (*) and (**). (*) follows from

$\mathfrak{R}_0\mathfrak{n}_1\Phi(\mathfrak{n}_1)$	(by $(\substack{**\\**})$)
$\mathfrak{R}_0\mathfrak{n}_1\mathfrak{n}_2$	(since $\mathfrak{R}_0$ is a Peano relation)

and the fact that $\mathfrak{R}_0$ has the nature of a function. (**) follows from

$\mathfrak{R}_0\mathfrak{x}\,\Phi(\mathfrak{x})$	(by $(\substack{**\\**})$)
$\mathfrak{R}_0\mathfrak{f}_1(\mathfrak{x})\mathfrak{f}_2(\Phi(\mathfrak{x}))$	(since $\mathfrak{R}_0$ is a Peano relation)
$\mathfrak{R}_0\mathfrak{f}_1(\mathfrak{x})\Phi(\mathfrak{f}_1(\mathfrak{x}))$	(by $(\substack{**\\**})$)

and the fact that $\mathfrak{R}_0$ has the nature of a function.

Exercises. 1. Show that the following axiom system (with the primitive notions u, R) is categorical (let u, x, y, z and R, S be pairwise distinct):

$$\alpha, \ \bigwedge S(\alpha' \wedge \beta' \rightarrow \beta), \ \bigwedge x \bigwedge y(Rxy \vee x = y \vee Ryx),$$

where

$$\begin{aligned}\alpha \equiv\ & \bigwedge x \neg Rxx \\ & \wedge \bigwedge x \bigwedge y \bigwedge z(Rxy \wedge Ryz \rightarrow Rxz) \\ & \wedge \bigwedge x \bigvee y(Rxy \wedge \bigwedge z \neg(Rxz \wedge Rzy)) \\ & \wedge \bigwedge x(x = u \vee Rux) ,\end{aligned}$$

$$\beta \equiv \bigwedge x \bigwedge y(Rxy \rightarrow Sxy) ,$$

and α' is obtained from α by replacing R by S; β' is obtained from β by exchanging R and S.

Hint. First of all, check that a model of the above axiom system over the domain of the natural numbers is obtained by interpreting u as nought and R as the less-than relation. It is thus sufficient to show that every model is isomorphic to the given one. (The natural numbers could be characterised by the axiom system in this exercise instead of by Peano's axiom system.)

2. Express, in second-order language, that R is an order relation without a last element whose proper initial segments are finite (for this cf. also §1, Exercise 6). Show that the axiom system obtained in this way is categorical. Compare the models of this axiom system with the models of the axiom system in the previous exercise.

3. Find a categorical second-order axiom system for the whole numbers with the primitive notion $<$.

4. Let P be a one-place predicate variable. Find a second-order expression α with no free individual variables such that Mod $\mathfrak{J}\alpha$ if and only if $\mathfrak{J}(P)$ fits denumerably many elements.

5. Find a second-order expression α such that $\models_{\omega} \alpha$ if ω is uncountable and $\models_{\omega} \neg\alpha$ if ω is at most denumerable. (This shows that the Löwenheim-Skolem Theorem, Chap. V, §6.3, does not hold in second-order logic.)

6. Find a second-order expression α which characterises the fields of characteristic nought (cf. Chap. V, §5, Exercise 5).

§4. The non-characterisability of the natural numbers in the language of predicate logic

4.1 Arithmetic within second-order logic. In §3 we showed that Peano's axiom system $\mathfrak{S}_0$ is categorical, i.e. that any two models of $\mathfrak{S}_0$ are isomorphic. Since, by the theorem we proved in 2.3, we cannot distinguish between isomorphic algebras by means of the concept of a model, one model of $\mathfrak{S}_0$ is as good as another as far as arithmetic is concerned; all models of $\mathfrak{S}_0$ have an equal right to be regarded as natural numbers. For the following, we choose a fixed algebra $\mathfrak{A}_0 = \mathfrak{A}_{\mathfrak{B}_0}(\mathfrak{J}_0) = \langle \omega_0, \mathfrak{n}_0, \mathfrak{f}_0 \rangle$ which is a model of $\mathfrak{S}_0$. It is this to which we are referring when we speak of the (uniquely determined only up to isomorphism) natural model of $\mathfrak{S}_0$. In this connection, we shall talk of the elements of ω_0 as the natural numbers.

4.2 Arithmetic within predicate logic. In §1 we noted that the induction axiom, and thus Peano's axiom system, are not formulated in the language of predicate logic. However, many arithmetical statements can be formulated in the language of predicate logic. It is, therefore, natural to ask whether there is an axiom system, expressed in the language of predicate logic, of which all those statements of predicate logic which are valid for the natural numbers are consequences. The existence of such a system of axioms would carry with it the advantage that we could make use of the predicate calculus to derive the arithmetical statements which followed from it; for we have not given any derivation calculus for the language of second-order logic. This is not merely a temporary deficiency of this language; for we can show - in fact, as a corollary to the main theorem of this section (4.3) - that there is no adequate (i.e. both correct and complete) notion of derivation for the language of second-order logic.

We restrict ourselves at first to the arithmetical statements in the primitive notions which occur in Peano's axiom system. With regard to the addition of further primitive notions cf. 4.6 and the exercise.

We might at first think that any attempt to replace the induction axiom by axioms which can be written down in predicate logic must be doomed from the start, since the induction axiom says something about <u>all</u> properties. On the other hand it is also clear that the mathematician, in dealing with arithmetic, does not have to be interested in all properties; for the only properties which arise in arithmetic are those which can be formulated by means of the primitive notions of arithmetic, i.e. by means of the notions 0 and ' (and, of course, the use of logical means of expression). Any such property can be described by an expression α of predicate logic, which contains exactly one free variable x (apart from the primitive notions). Here we assume, for the sake of simplicity, that x does not occur bound in α. (For then we can operate with normal substitution, whereas otherwise we should have to make use of extended substitution.) We write $\alpha(t)$ for the expression which is obtained from α by substitution of t for x; thus $\alpha(x) = \alpha$. The property $\mathfrak{P}_\alpha$ over the domain ω_0 of the natural numbers (cf. 4.1) which is associated with such an α is defined as follows:

(1) $$\mathfrak{P}_\alpha \ \underline{\text{fits}}\ \mathfrak{x} \ \underline{\text{if and only if}}\ \text{Mod}\ \mathfrak{J}_{0\,x}^{\ \mathfrak{x}}\,\alpha .$$

The induction axiom P3 states, for the property $\mathfrak{P}_\alpha$:

(*) If $\mathfrak{P}_\alpha$ fits $\mathfrak{n}_0$ and if, whenever $\mathfrak{P}_\alpha$ fits some natural number $\mathfrak{x}$, it always also fits its successor $\mathfrak{f}_0(\mathfrak{x})$, then $\mathfrak{P}_\alpha$ fits every natural number.

But we can also obtain the statement (*) by using, instead of P3, the first-order axiom

P3_α: $$\alpha(0) \wedge \bigwedge x(\alpha(x) \rightarrow \alpha(x')) \rightarrow \bigwedge x\alpha .$$

(Note that P3_α contains only 0 and ' as free variables.) For then we have, first of all,

(**) $\text{Mod}\ \mathfrak{J}_0\text{P3}_\alpha$ <u>iff</u> if $\text{Mod}\ \mathfrak{J}_0\alpha(0)$ and $\text{Mod}\ \mathfrak{J}_0 \bigwedge x(\alpha(x) \rightarrow \alpha(x'))$, then $\text{Mod}\ \mathfrak{J}_0 \bigwedge x\alpha$.

But, by the substitution theorem (Chap. III, § 3.2),

(2) $$\text{Mod}\ \mathfrak{J}_{0\,x}^{\ \mathfrak{n}_0}\,\alpha \quad \text{if and only if} \quad \text{Mod}\ \mathfrak{J}_0\,\alpha(0)$$

and

(3) $$\text{Mod}\ \mathfrak{J}_{0\ x\ x}^{\ \mathfrak{x}\ \mathfrak{J}_{0\,x}^{\ \mathfrak{x}}(x')}\,\alpha \quad \text{if and only if} \quad \text{Mod}\ \mathfrak{J}_{0\,x}^{\ \mathfrak{x}}\,\alpha(x') .$$

But clearly

$$(4) \qquad \mathfrak{I}_0{}_{x\,x}^{\mathfrak{r}\;\mathfrak{I}_0{}_x^{\mathfrak{r}}(x')} = \mathfrak{I}_0{}_x^{\mathfrak{f}_0(x)} .$$

From (1),...,(4) we see that the left hand side of (**) is the same statement as (*).

These considerations show that, for every property $\mathfrak{P}_\alpha$ which can be formulated in arithmetic, we can replace the induction axiom P3 by the first-order axiom $P3_\alpha$. Thus we could expect that P3 could be replaced by the set of all these $P3_\alpha$'s in such a way that the resulting system of axioms $\mathfrak{S}_0'$ would have the same models as Peano's system of axioms $\mathfrak{S}_0$. Thus, we would have transformed Peano's axiom system into an equivalent system of axioms in the language of predicate logic. However, we find that $\mathfrak{S}_0'$ is not equivalent to $\mathfrak{S}_0$; $\mathfrak{S}_0'$ has more models than $\mathfrak{S}_0$.

In fact, it is not possible by any means whatever to replace $\mathfrak{S}_0$ by an axiom system $\mathfrak{S}$ in the language of predicate logic in such a way that $\mathfrak{S}_0$ and $\mathfrak{S}$ have the same models. This is an immediate consequence of a theorem of Skolem, to which we shall now turn our thoughts.

4.3 The main result of this section is the following

Theorem of Skolem. Let $\mathfrak{S}$ *be an arbitrary set of expressions of predicate logic such that* $\mathfrak{B}(\mathfrak{S}) = \mathfrak{B}_0 = \{0, '\}$. *Let* $\mathfrak{A}_0 = \mathfrak{A}_{\mathfrak{B}_0}(\mathfrak{I}_0) = \langle \omega_0, \mathfrak{n}_0, \mathfrak{f}_0 \rangle$ (cf. 4.1) *be a model of* $\mathfrak{S}$. *Then there is an algebra* $\mathfrak{A} = \mathfrak{A}_{\mathfrak{B}_0}(\mathfrak{I}) = \langle \omega_0, \mathfrak{n}, \mathfrak{f} \rangle$ *which is not isomorphic to* $\mathfrak{A}_0$ *and such that* $\mathfrak{A}$ *is a model of* $\mathfrak{S}$. *Thus* $\mathfrak{S}$ *is not* $\aleph_0$*-categorical* (where $\aleph_0$ is the cardinality of the countable set ω_0). "*The natural numbers cannot be characterised up to isomorphism* (even by giving the cardinality of the domain of individuals) *in the language of first-order logic.*"

As an immediate consequence, we obtain the following:

The natural numbers cannot be characterised by an axiom system within the framework of the means of expression provided by predicate logic (Skolem).

4.4 Proof of Skolem's theorem. In the following, let u be a fixed individual variable which is different from 0. We assign a term t_u to every $u \in \omega_0$ by means of the following inductive definition[5]:

[5] Here we make use of the fact that there is exactly one mapping t_u from the set of the natural numbers into the set of terms which satisfies both the conditions below.

$$t_{n_0} \equiv 0$$

$$t_{\mathfrak{f}_0(u)} \equiv t_u{}' .$$

It is easy to show by induction that

(***) $$\mathfrak{J}_0(t_u) = u$$

for every $u \in \omega_0$.

Let $\mathfrak{U}$ be the set of all the inequalities of the form $u \neq t_u$ with $u \in \omega_0$.

Thus, $\mathfrak{U} = \{u \neq 0, u \neq 0', u \neq 0'', \dots\}$.

Now we assert:

(a) There is no u such that Mod $\mathfrak{J}_0{}^u_u \mathfrak{U}$.

(b) There is an interpretation $\mathfrak{J}$ over ω_0 such that Mod $\mathfrak{J}\, \mathfrak{S} \cup \mathfrak{U}$.

Skolem's theorem follows from (a) and (b) as follows: $\mathfrak{A}_{\mathfrak{B}_0}(\mathfrak{J}_0)$ is not isomorphic to $\mathfrak{A}_{\mathfrak{B}_0}(\mathfrak{J})$. For, if it were, there would be an isomorphism Φ from $\mathfrak{A}_{\mathfrak{B}_0}(\mathfrak{J}_0)$ onto $\mathfrak{A}_{\mathfrak{B}_0}(\mathfrak{J})$. Then, by 2.3, Lemma 3, Φ would also be an isomorphism from $\mathfrak{A}_{\mathfrak{B}_0 \cup \{u\}}(\mathfrak{J}_0{}^u_u)$ onto $\mathfrak{A}_{\mathfrak{B}_0 \cup \{u\}}(\mathfrak{J}^{\Phi(u)}_u)$ for every $u \in \omega_0$. In particular, we could choose

$$u = \Phi^{-1}(\mathfrak{J}(u)) .$$

Then we should have $\mathfrak{J}^{\Phi(u)}_u = \mathfrak{J}$, and therefore $\mathfrak{A}_{\mathfrak{B}_0 \cup \{u\}}(\mathfrak{J}_0{}^u_u)$ and $\mathfrak{A}_{\mathfrak{B}_0 \cup \{u\}}(\mathfrak{J})$ would be isomorphic. Now by (b) we have Mod $\mathfrak{J}\, \mathfrak{U}$. Then, by the isomorphism theorem of 2.3, we should have Mod $\mathfrak{J}_0{}^u_u \mathfrak{U}$. But this contradicts (a).

It remains to prove (a) and (b).

Proof of (a): We consider the property $\mathfrak{P}$ over ω_0 which fits u if and only if not Mod $\mathfrak{J}_0{}^u_u \mathfrak{U}$. We have to prove that $\mathfrak{P}$ fits every $u \in \omega_0$. Because of P3 it suffices to show that:

(1) $\mathfrak{P}$ fits n_0, and

(2) If $\mathfrak{P}$ fits some u, then $\mathfrak{P}$ also fits $\mathfrak{f}_0(u)$.

to (1): We have to show that not Mod $\mathfrak{J}_0{}^{n_0}_u\,\mathfrak{U}$. In order to do this, we prove that not Mod $\mathfrak{J}_0{}^{n_0}_u\; u \neq t_{n_0}$, i.e. that Mod $\mathfrak{J}_0{}^{n_0}_u\; u = t_{n_0}$. In fact, $\mathfrak{J}_0{}^{n_0}_u(u) = n_0 = \mathfrak{J}_0{}^{n_0}_u(t_{n_0})$.

to (2): Suppose that $\mathfrak{P}$ fits $\mathfrak{u}$. Then there is an element $u \neq t_{\mathfrak{u}_1}$ of $\mathfrak{U}$ such that not Mod $\mathfrak{J}_0{}^{\mathfrak{u}}_u\; u \neq t_{\mathfrak{u}_1}$, i.e. such that Mod $\mathfrak{J}_0{}^{\mathfrak{u}}_u\; u = t_{\mathfrak{u}_1}$. Thus $\mathfrak{u} = \mathfrak{J}_0(t_{\mathfrak{u}_1}) = \mathfrak{u}_1$. It follows that $\mathfrak{f}_0(\mathfrak{u}) = \mathfrak{f}_0(\mathfrak{u}_1)$, and hence that Mod $\mathfrak{J}_0{}^{\mathfrak{f}_0(\mathfrak{u})}_u\; u = t_{\mathfrak{f}_0(\mathfrak{u}_1)}$, i.e. that not Mod $\mathfrak{J}_0{}^{\mathfrak{f}_0(\mathfrak{u})}_u\; u \neq t_{\mathfrak{f}_0(\mathfrak{u}_1)}$. This shows that not Mod $\mathfrak{J}_0{}^{\mathfrak{f}_0(\mathfrak{u})}_u\,\mathfrak{U}$. Thus, $\mathfrak{f}_0(\mathfrak{u})$ has the property $\mathfrak{P}$.

Proof of (b): Let $x_1, x_2, x_3, \ldots$ be pairwise distinct individual variables. Let $\alpha_2 \equiv x_2 \neq x_1$, $\alpha_n \equiv x_n \neq x_1 \wedge \ldots \wedge x_n \neq x_{n-1} \wedge \alpha_{n-1}$ for $n \geqslant 3$, $\beta_n \equiv \bigvee x_1 \ldots \bigvee x_n \alpha_n$ for $n \geqslant 2$. Let $\mathfrak{B}$ be the set of all the β_n $(n \geqslant 2)$. Let $\mathfrak{P}$ be an arbitrary finite subset of $\mathfrak{S} \cup \mathfrak{B} \cup \mathfrak{U}$. Then there is a $\mathfrak{v} \in \omega_0$ such that $\mathfrak{v}$ is different from all the $\mathfrak{u} \in \omega_0$ for which $u \neq t_{\mathfrak{u}} \in \mathfrak{P}$. For any such $\mathfrak{v}$, $\mathfrak{J}_0{}^{\mathfrak{v}}_u$ is (1) a model of $\mathfrak{S}$ (since Mod $\mathfrak{J}_0\,\mathfrak{S}$), (2) a model of $\mathfrak{B}$ (since ω_0 is infinite) and (3) a model of $\mathfrak{U} \cap \mathfrak{P}$ (by (***) and the choice of $\mathfrak{v}$). Thus, every such $\mathfrak{P}$ is satisfiable, and therefore (by the compactness theorem for satisfiability) so is $\mathfrak{S} \cup \mathfrak{B} \cup \mathfrak{U}$. By the Löwenheim-Skolem theorem and the choice of $\mathfrak{B}$, $\mathfrak{S} \cup \mathfrak{B} \cup \mathfrak{U}$ is satisfiable over a countable domain ω_0'. From this we can easily prove (cf. the Lemma at the end of 2.2) that $\mathfrak{S} \cup \mathfrak{B} \cup \mathfrak{U}$ is also satisfiable over ω_0, which at once gives us (b).

4.5 The incompleteness of second-order logic. If we analyse the proof in 4.4, we realise that the compactness theorem for satisfiability is an essential part of it. If the compactness theorem for satisfiability also held for second-order logic, then we could also use the proof in 4.4 to prove Skolem's theorem (4.3) for an arbitrary set $\mathfrak{S}$ of second-order expressions. However, this would contradict the categoricity of Peano's axiom system $\mathfrak{S}_0$. Therefore the compactness theorem for satisfiability does not hold for second-order logic.

In Chap. V, § 5.1 we proved the compactness theorem from the completeness theorem. Thus, in the case of second-order logic, we shall be able to infer incompleteness from the fact that the compactness theorem does not hold. More precisely, we have the

Incompleteness theorem. Let P be an arbitrary effective procedure with the property that, applied to an arbitrary set $\mathfrak{S}$ of expressions of second-order logic, it produces expressions of second-order logic which are consequences of $\mathfrak{S}$. Then P is incomplete in the sense that there is a set $\mathfrak{S}$ of second-order expressions and a second-order expression which follows from $\mathfrak{S}$ but which cannot be obtained from $\mathfrak{S}$ by means of the procedure[6].

We remark, without proof, that the incompleteness theorem also holds if we restrict ourselves to finite sets $\mathfrak{S}$ of expressions of second-order logic.

Proof. Let us consider any expression α which can be obtained from $\mathfrak{S}$ by means of the procedure P. It is a prerequisite of the effectiveness of P that, in obtaining α from $\mathfrak{S}$, we can refer only to finitely many elements of $\mathfrak{S}$.

Now let α be an arbitrary consequence of $\mathfrak{S}$. If the assertion of the incompleteness theorem were false, then α could be obtained by applying P to $\mathfrak{S}$. As we have just seen, α could then also be obtained by applying P to some finite subset $\mathfrak{S}_0$ of $\mathfrak{S}$. Then, by our definition of P, α would be a consequence of $\mathfrak{S}_0$. Thus we should have obtained the compactness theorem for the relation of consequence. But, as we have seen in Chap. V, §5.1, we could deduce from this the compactness theorem for satisfiability. But, as we have just shown, this does not hold for second-order logic.

4.6 Increasing the number of arithmetical primitive notions. In the language of second-order logic, addition and multiplication can (as was first shown by Dedekind) be defined by means of Peano's primitive notions 0, '. With the aid of the definitions, the theory of these can then be obtained as a consequence of Peano's axioms. This is true in particular of the following expressions:

$$\bigwedge x \; x + 0 = x$$
$$\bigwedge x \bigwedge y \; x + y' = (x + y)'$$
$$\bigwedge x \; x \cdot 0 = 0$$
$$\bigwedge x \bigwedge y \; x \cdot y' = x \cdot y + x ,$$

where + and · are to be taken as two-place function variables which are different from each other.

However, in the language of predicate logic we cannot give any definition of + and · in such a way that the above expressions are consequences of some arithmetically valid[7] expressions of predicate logic with the primitive notions 0 and '. Thus, if we adjoin the expressions we have mentioned to some set $\mathfrak{S}$ of expressions of predicate logic with the primitive notions 0 and ' which are arithmetically valid[7], then we obtain a substantially stronger theory.

This could lead us to suspect that it might be possible, by adding further primitive notions (e.g. +, ·) to Peano's primitive notions 0, ', to characterise the natural numbers in the language of predicate logic. However, this is not the case (cf. Exercise).

---

[6] For the sets of expressions of the usual predicate logic there is a procedure P_0 which consists in systematically forming all the possible derivations. By the application of P_0 to a set $\mathfrak{S}$ of expressions we mean the production of all expressions α for which there are expressions $\alpha_1, \ldots, \alpha_r \in \mathfrak{S}$ such that $\alpha_1 \ldots \alpha_r \alpha$ is derivable ($r \geqslant 0$). Then the procedure P_0 produces only such expressions α as are consequences of $\mathfrak{S}$, and it produces all such expressions.

[7] I.e. which are valid in $\mathfrak{J}_0$ (4.1).

<u>Exercise.</u> Let $\mathfrak{S}$ be an arbitrary set of expressions of predicate logic whose primitive notions include 0 and '. Let $\mathfrak{A}_{\mathfrak{B}_0}(\mathfrak{K}) = \langle \omega_0, \mathfrak{n}_0, \mathfrak{f}_0 \rangle$ be an algebra which is a model of $\mathfrak{S}$ and satisfies $\mathfrak{K}(0) = \mathfrak{J}_0(0)$, $\mathfrak{K}(') = \mathfrak{J}_0(')$. Then there is an algebra $\mathfrak{A}_{\mathfrak{B}_0}(\mathfrak{J}) = \langle \omega_0, \mathfrak{n}, \mathfrak{f} \rangle$ which is a model of $\mathfrak{S}$ and is not isomorphic to $\mathfrak{A}_{\mathfrak{B}_0}(\mathfrak{K})$. (Suggestion: Show that the proof of Skolem's Theorem (4.4) can be carried through virtually unchanged.)

VII. Extensions of the Language, Normal Forms

§ 1. Extensions of the language of predicate logic

1.1 Statement of the problem. In Chap. II, we built up the language of predicate logic on the basis of the junctors $\wedge$ and $\neg$ and the quantifier $\bigwedge$. Other junctors, such as $\vee$, $\rightarrow$, $\leftrightarrow$ (Chap. I, § 6.2), and the quantifier $\bigvee$ (Chap. I, § 7.5) can be defined on this basis (see Chap. II, § 1.5; cf. also Chap. I, § 6, Exercise 5).

Thus the connectives we have just mentioned can, in principle, be dispensed with. However, they are still worth using, if only for reasons of economy, particularly when we want to apply logic, e.g. in symbolisations of everyday statements. This leads to an extension of the language of predicate logic. In this section we want to consider a language which is formed by taking the connectives $\vee$, $\rightarrow$, $\bigvee$ in addition to the connectives $\wedge$, $\neg$, $\bigwedge$. The semantics of the extended language can be built up analogously to those of the previous language. The system of rules of predicate logic is extended by additional rules, which are concerned with the new connectives. It is easy to see that these rules are sound; the completeness of the extended set of rules can be proved by reference to the completeness of predicate logic in $\wedge$, $\neg$, $\bigwedge$.

The method we use here can also be applied to extensions which are obtained by adding connectives other than those which we have considered here.

1.2 The language of the extended predicate logic (cf. Chap. II). In addition to the symbols in Chap. II, § 1.2 we shall use

(4') Two further junctors: The disjunctor and the implicator.

(5') One further quantifier: The existential quantifier.

The assumptions about the symbols, set down in Chap. II, § 1.2, are to be altered in the obvious way.

As names we use:

"$\vee$" for the disjunctor,
"$\rightarrow$" for the implicator,
"$\bigvee$" for the existential quantifier.

In the following, when we speak of "rows of symbols", we shall also allow the symbols $\vee$, $\rightarrow$, $\bigvee$ to be used in constructing them.

The extended expressions, in the extended predicate logic, are to correspond to the expressions in predicate logic. The extended expression calculus is defined by adding to the rules of the expression calculus (Chap. II, § 1.5) the following three rules:

Rule 3' : We are allowed to pass from two (not necessarily different) rows of symbols ζ_1, ζ_2 to the row of symbols $(\zeta_1 \vee \zeta_2)$.

Rule 3'': We are allowed to pass from two (not necessarily different) rows of symbols ζ_1, ζ_2 to the row of symbols $(\zeta_1 \rightarrow \zeta_2)$.

Rule 4' : We are allowed to pass from a row of symbols ζ to any row of symbols $\bigvee x \zeta$, where x is an arbitrary individual variable.

Definition. Those rows of symbols which can be derived by the rules of the extended expression calculus (and only such rows) are called extended expressions. The expressions we have considered up till now will also be called ordinary expressions in order to distinguish them from the extended expressions. Every ordinary expression is an extended expression.

In this section we shall use "ρ", "σ", "τ", possibly with indices, as variables for extended expressions. From § 2 onwards we shall use "α", "β", "γ" as before.

The extended expressions can be classified in the obvious way (cf. Chap. II, § 1.5) as atomic expressions[1], negations, conjunctions, disjunctions, implications, generalisations or particularisations.

The elementary questions of decidability for the extended expressions can be treated as in Chap. II, § 2. The same holds for what was said in Chap. II, § 3 about proofs and definitions by induction; here, of course, the new connectives $\vee$, $\rightarrow$, $\bigvee$ must be taken into account.

Free occurrence of a variable and substitution can be defined for extended expressions analogously to the definitions in Chap. II, § 4 and § 5, by treating $\vee$ and $\rightarrow$ in the same way as $\wedge$ and $\bigvee$ in the same way as $\bigwedge$.

---

[1] These are the same whether we are talking about ordinary expressions or extended ones.

1.3 The semantics of the extended predicate logic. We must extend the definition of a model by adding regulations for disjunctions, implications and particularisations. To this end we define:

(2') $\mathrm{Mod}_{\omega}\mathfrak{J}(\sigma \vee \tau)$ if and only if $\mathrm{Mod}_{\omega}\mathfrak{J}\sigma$ or $\mathrm{Mod}_{\omega}\mathfrak{J}\tau$

(2'') $\mathrm{Mod}_{\omega}\mathfrak{J}(\sigma \rightarrow \tau)$ if and only if (if $\mathrm{Mod}_{\omega}\mathfrak{J}\sigma$, then $\mathrm{Mod}_{\omega}\mathfrak{J}\tau$)

(3') $\mathrm{Mod}_{\omega}\mathfrak{J} \bigvee x\sigma$ if and only if $\mathrm{Mod}_{\omega}\mathfrak{J}^{\mathfrak{r}}_{x}\sigma$ for at least one element $\mathfrak{r}$ of ω.

The definability of $\vee$, $\rightarrow$, $\bigvee$ by means of $\wedge$, $\neg$, $\bigwedge$ is expressed in

Theorem 1.

(a) $\mathrm{Mod}_{\omega}\mathfrak{J}(\sigma \vee \tau)$ if and only if $\mathrm{Mod}_{\omega}\mathfrak{J}\neg(\neg\sigma \wedge \neg\tau)$.

(b) $\mathrm{Mod}_{\omega}\mathfrak{J}(\sigma \rightarrow \tau)$ if and only if $\mathrm{Mod}_{\omega}\mathfrak{J}\neg(\sigma \wedge \neg\tau)$.

(c) $\mathrm{Mod}_{\omega}\mathfrak{J} \bigvee x\sigma$ if and only if $\mathrm{Mod}_{\omega}\mathfrak{J}\neg\bigwedge x\neg\sigma$.

Proof.

$\mathrm{Mod}_{\omega}\mathfrak{J}(\sigma \vee \tau)$ iff $\mathrm{Mod}_{\omega}\mathfrak{J}\sigma$ or $\mathrm{Mod}_{\omega}\mathfrak{J}\tau$ (2')

iff not (not $\mathrm{Mod}_{\omega}\mathfrak{J}\sigma$ and not $\mathrm{Mod}_{\omega}\mathfrak{J}\tau$) (Chap. I, § 6.2 (1))

iff not ($\mathrm{Mod}_{\omega}\mathfrak{J}\neg\sigma$ and $\mathrm{Mod}_{\omega}\mathfrak{J}\neg\tau$)

iff not $\mathrm{Mod}_{\omega}\mathfrak{J}(\neg\sigma \wedge \neg\tau)$

iff $\mathrm{Mod}_{\omega}\mathfrak{J}\neg(\neg\sigma \wedge \neg\tau)$.

Thus, we have proved (a). The same method can be used to prove (b) with the aid of Chap. I, § 6.2 (3) and (c) with the aid of Chap. I, § 7.5 (***).

The relation of consequence for extended expressions and sets of expressions can be defined with the aid of the definition of a model just as it was for ordinary expressions and sets of expressions. The same holds for the satisfiability of a set of extended expressions.

The fact, stated in Theorem 1, that $\vee$, $\rightarrow$, $\bigvee$ can be expressed in terms of $\wedge$, $\neg$, $\bigwedge$ suggests the idea of assigning to every extended expression σ, by means of an effective procedure, an ordinary expression σ^* such that $\sigma \mathrel{=\!\!\mid\!\!\models} \sigma^*$. To this end we define inductively:

$$\begin{aligned}
\sigma^* &\equiv \sigma \text{ if } \sigma \text{ is an atomic extended expression,}\\
[\neg\sigma]^* &\equiv \neg\sigma^*,\\
(\sigma\wedge\tau)^* &\equiv (\sigma^*\wedge\tau^*),\\
(\sigma\vee\tau)^* &\equiv \neg(\neg\sigma^*\wedge\neg\tau^*),\\
(\sigma\rightarrow\tau)^* &\equiv \neg(\sigma^*\wedge\neg\tau^*),\\
[\bigwedge x\sigma]^* &\equiv \bigwedge x\sigma^*,\\
[\bigvee x\sigma]^* &\equiv \neg\bigwedge x\neg\sigma^*.
\end{aligned}$$

It is easy to see by induction that σ^* is always an ordinary expression and that

$$\mathrm{Mod}_\omega \mathfrak{J}\sigma \text{ if and only if } \mathrm{Mod}_\omega \mathfrak{J}\sigma^*$$

always holds.

It follows from the last statement that

Theorem 2. $\sigma \mathrel{=\!\!\mid\!\models} \sigma^*$.

The coincidence theorem (Chap. III, § 3.1) and the substitution theorem (Chap. III, §3.2) clearly also hold for extended expressions.

1.4 An extended predicate calculus. We now extend the application of the rules of the predicate calculus (Chap. IV, § 2.3) by allowing them to be used for arbitrary extended expressions and sequents built up from arbitrary extended expressions. We also add further rules, which are concerned with the new connectives $\vee$, $\rightarrow$, $\bigvee$ and which we shall present straight away in the symbolism which we used and explained in Chap. IV, § 2.3:

Additional rules of the extended predicate calculus

(D) $\dfrac{\Sigma_1\sigma : \rho \qquad \Sigma_2\tau : \rho}{\Sigma_{12}(\sigma\vee\tau) : \rho}$ (D') $\dfrac{\Sigma : \sigma}{\Sigma : (\sigma\vee\tau)}$ (D'') $\dfrac{\Sigma : \tau}{\Sigma : (\sigma\vee\tau)}$

(I) $\dfrac{\Sigma_1\neg\sigma : \rho \qquad \Sigma_2\tau : \rho}{\Sigma_{12}(\sigma\rightarrow\tau) : \rho}$ (I') $\dfrac{\Sigma\sigma : \tau}{\Sigma : (\sigma\rightarrow\tau)}$

(P_x) $\dfrac{\Sigma : \sigma}{\Sigma : \bigvee x\sigma}$ (P_x') $\dfrac{\Sigma\sigma : \tau}{\Sigma\bigvee x\sigma : \tau}$ if x does not occur free in τ or Σ.

Note that all the additional rules are rules for introducing one of the connectives $\vee$, $\rightarrow$, $\bigvee$ in the antecedent or the succedent. (D') and (D'') are also called thinning rules. Note the analogy between (D) and (I), between (D) and (P_x'), between (D') and (D'') respectively and (P_x).

The concept of a derivation is taken over from ordinary predicate logic. $\vdash' \sigma_1 \ldots \sigma_n \sigma$ or $\sigma_1, \ldots, \sigma_n \vdash' \sigma$ is to mean that the sequent $\sigma_1 \ldots \sigma_n \sigma$ is derivable in the extended predicate calculus.

For ordinary expressions we have, trivially:

(a) If $\alpha_1, \ldots, \alpha_n \vdash \alpha$, then $\alpha_1, \ldots, \alpha_n \vdash' \alpha$,

for, in order to derive $\alpha_1 \ldots \alpha_n \alpha$ in the extended predicate calculus, we can use the same derivation as in the ordinary predicate calculus. But, conversely, it is also true that

(b) If $\alpha_1, \ldots, \alpha_n \vdash' \alpha$, then $\alpha_1, \ldots, \alpha_n \vdash \alpha$.

This is not as trivial as (a), for in the extended predicate calculus we have more rules at our disposal than in the ordinary predicate calculus.

We could prove the truth of (b) purely by investigating the two calculi. However, we prefer to obtain (b) by making a detour via the soundness of the extended predicate calculus and the completeness of the ordinary predicate calculus.

For an arbitrary set $\mathfrak{S}$ of extended expressions and for an arbitrary extended expression σ we define the relation $\mathfrak{S} \vdash' \sigma$ (analogously to $\mathfrak{S} \vdash \alpha$) by requiring that there should be finitely many elements $\sigma_1, \ldots, \sigma_n$ in $\mathfrak{S}$ such that $\sigma_1, \ldots, \sigma_n \vdash' \sigma$.

As for the ordinary predicate calculus, it is easy to show that all the rules of the extended predicate calculus produce sound sequents when applied to sound sequents. This gives us the

Theorem of the soundness of the extended predicate calculus. If $\mathfrak{S} \vdash' \sigma$, then $\mathfrak{S} \models \sigma$.
Now we can prove (b) at once: Let $\alpha_1, \ldots, \alpha_n \vdash' \alpha$. Thus we have $\{\alpha_1, \ldots, \alpha_n\} \vdash' \alpha$, and hence $\{\alpha_1, \ldots, \alpha_n\} \models \alpha$. By the completeness of the ordinary predicate calculus this gives us $\{\alpha_1, \ldots, \alpha_n\} \vdash \alpha$, and hence $\alpha_1, \ldots, \alpha_n \vdash \alpha$.

Because of (a) and (b) we can write "$\vdash$" instead of "$\vdash'$", where in the case of extended expressions we mean the extended relation of derivation and in the case of ordinary expressions we mean either the ordinary or the extended relation of derivation.

Every one of the derived rules which we introduced in Chap. IV, § 4 is also a derived rule in the extended predicate calculus, since, in justifying it, we made use only of the defining rules of ordinary predicate logic, and since these rules are also defining rules of the extended predicate calculus. In analogy to Theorem 1, we also have

<u>Theorem 3.</u>

(a) $$(\sigma \vee \tau) \dashv\vdash \neg(\neg\sigma \wedge \neg\tau)$$

(b) $$(\sigma \rightarrow \tau) \dashv\vdash \neg(\sigma \wedge \neg\tau)$$

(c) $$\bigvee x\sigma \dashv\vdash \neg\bigwedge x \neg\sigma$$

By way of <u>proof</u> we give the following derivations:

<u>to</u> (a):

$(\neg\sigma \wedge \neg\tau) : (\neg\sigma \wedge \neg\tau)$	(A)	$\sigma : \sigma$	(A)
$(\neg\sigma \wedge \neg\tau) : \neg\sigma$	(C')	$\sigma : (\sigma \vee \tau)$	(D')
$(\neg\sigma \wedge \neg\tau) : \neg\tau$	(C'')	$\neg(\sigma \vee \tau) : \neg\sigma$	(CaPo)
$\sigma : \neg(\neg\sigma \wedge \neg\tau)$	(CaPo')	$\tau : \tau$	(A)
$\tau : \neg(\neg\sigma \wedge \neg\tau)$	(CaPo')	$\tau : (\sigma \vee \tau)$	(D'')
$(\sigma \vee \tau) : \neg(\neg\sigma \wedge \neg\tau)$	(D)	$\neg(\sigma \vee \tau) : \neg\tau$	(CaPo)
		$\neg(\sigma \vee \tau) : (\neg\sigma \wedge \neg\tau)$	(C)
		$\neg(\neg\sigma \wedge \neg\tau) : (\sigma \vee \tau)$	(CaPo'')

<u>to</u> (b):

$(\sigma \wedge \neg\tau) : (\sigma \wedge \neg\tau)$	(A)	$\sigma : \sigma$	(A)
$(\sigma \wedge \neg\tau) : \sigma$	(C')	$\neg\tau : \neg\tau$	(A)
$(\sigma \wedge \neg\tau) : \neg\tau$	(C'')	$\sigma\, \neg\tau : (\sigma \wedge \neg\tau)$	(C)
$\neg\sigma : \neg(\sigma \wedge \neg\tau)$	(CaPo)	$\neg(\sigma \wedge \neg\tau)\sigma : \tau$	(CaPo'')
$\tau : \neg(\sigma \wedge \neg\tau)$	(CaPo')	$\neg(\sigma \wedge \neg\tau) : (\sigma \rightarrow \tau)$	(I')
$(\sigma \rightarrow \tau) : \neg(\sigma \wedge \neg\tau)$	(I)		

<u>to</u> (c):

$\bigwedge x \neg\sigma : \bigwedge x \neg\sigma$	(A)	$\sigma : \sigma$	(A)
$\bigwedge x \neg\sigma : \neg\sigma$	(G)	$\sigma : \bigvee x\sigma$	(P_x)
$\sigma : \neg\bigwedge x \neg\sigma$	(CaPo')	$\neg\bigvee x\sigma : \neg\sigma$	(KP_1)
$\bigvee\sigma : \neg\bigwedge x \neg\sigma$	(P'_x)	$\neg\bigvee x\sigma : \bigwedge x \neg\sigma$	(G_x)
		$\neg\bigwedge x \neg\sigma : \bigvee x\sigma$	(CaPo'')

As a consequence of Theorem 3 and with regard to the definition of σ^* (cf. 1.3) it is easy to prove

<u>Theorem 4.</u> $\sigma \dashv\vdash \sigma^*$.

<u>1.5 The completeness of the extended predicate calculus.</u> We prove the

<u>Completeness theorem.</u> If $\mathfrak{S} \vDash \sigma$, <u>then</u> $\mathfrak{S} \vdash \sigma$.

Proof. Let $\mathfrak{S}^*$ be the set of the τ^* such that $\tau \in \mathfrak{S}$. Then, by Theorem 2, $\mathfrak{S}^* \models \sigma^*$. Since these are ordinary expressions, the completeness theorem for the ordinary predicate calculus then gives us $\mathfrak{S}^* \vdash \sigma^*$. It follows from Theorem 4 that $\mathfrak{S} \vdash \sigma$, q.e.d.

§ 2. Derived rules and derivability relations with the connectives $\vee$, $\rightarrow$, $\bigvee$

2.1 Derived rules for $\vee$, $\rightarrow$

Notation	Name	Rule	Justification	
(IIn)	Introduction of the implicator	$\beta : (\alpha \rightarrow \beta)$	$\beta\alpha : \beta$	(AEx)
			$\beta : (\alpha \rightarrow \beta)$	(I')
(XQ'')	"Ex contradictione	$\neg\alpha : (\alpha \rightarrow \beta)$	$\neg\alpha\alpha : \beta$	(XQ)
	quodlibet (3)"		$\neg\alpha : (\alpha \rightarrow \beta)$	(I')
(RD)	Removal of the disjunctor	$\Sigma_1\alpha : \gamma$	$\Sigma_1\alpha : \gamma$	
	in the succedent	$\Sigma_2\beta : \gamma$	$\Sigma_2\beta : \gamma$	
		$\underline{\Sigma_3 : (\alpha \vee \beta)}$	$\Sigma_3 : (\alpha \vee \beta)$	
		$\Sigma_{123} : \gamma$	$\Sigma_{12}(\alpha \vee \beta) : \gamma$	(D)
			$\Sigma_{123} : \gamma$	(CuRu)
(RI)	Removal of the implicator	$\underline{\Sigma : (\alpha \rightarrow \beta)}$	$\Sigma : (\alpha \rightarrow \beta)$	
		$\Sigma\alpha : \beta$	$\alpha\neg\alpha : \beta$	(XQ)
			$\beta : \beta$	(A)
			$\alpha(\alpha \rightarrow \beta) : \beta$	(I)
			$\Sigma\alpha : \beta$	(CuRu)
(MPn)	Modus Ponens	$\Sigma_1 : \alpha$	$\Sigma_1 : \alpha$	
		$\underline{\Sigma_2 : (\alpha \rightarrow \beta)}$	$\Sigma_2 : (\alpha \rightarrow \beta)$	
		$\Sigma_{12} : \beta$	$\Sigma_2\alpha : \beta$	(RI)
			$\Sigma_{12} : \beta$	(CuRu)
(ChRu)	Chain rule	$\Sigma_1 : (\alpha \rightarrow \beta)$	$\Sigma_1 : (\alpha \rightarrow \beta)$	
		$\underline{\Sigma_2 : (\beta \rightarrow \gamma)}$	$\Sigma_2 : (\beta \rightarrow \gamma)$	
		$\Sigma_{12} : (\alpha \rightarrow \gamma)$	$\Sigma_1\alpha : \beta$	(RI)
			$\Sigma_2\beta : \gamma$	(RI)
			$\Sigma_{12}\alpha : \gamma$	(CuRu)
			$\Sigma_{12} : (\alpha \rightarrow \gamma)$	(I')

2.2 Derivability relations

Name	Rule	Justification	
Self-implication	$\vdash (\alpha \rightarrow \alpha)$	$\alpha : \alpha$	(A)
		$: (\alpha \rightarrow \alpha)$	(I')
Commutative law	$(\alpha \vee \beta) \dashv\vdash (\beta \vee \alpha)$	$\alpha : \alpha$	(A)
		$\alpha : (\beta \vee \alpha)$	(D'')
		$\beta : \beta$	(A)
		$\beta : (\beta \vee \alpha)$	(D')
		$(\alpha \vee \beta) : (\beta \vee \alpha)$	(D)
Associative law	$(\alpha \vee (\beta \vee \gamma)) \dashv\vdash ((\alpha \vee \beta) \vee \gamma)$	$\alpha : \alpha$	(A)
		$\alpha : (\alpha \vee \beta)$	(D')
		$\alpha : ((\alpha \vee \beta) \vee \gamma)$	(D')
		$\beta : \beta$	(A)
		$\beta : (\alpha \vee \beta)$	(D'')
		$\beta : ((\alpha \vee \beta) \vee \gamma)$	(D')
		$\gamma : \gamma$	(A)
		$\gamma : ((\alpha \vee \beta) \vee \gamma)$	(D'')
		$(\beta \vee \gamma) : ((\alpha \vee \beta) \vee \gamma)$	(D)
		$(\alpha \vee (\beta \vee \gamma)) : ((\alpha \vee \beta) \vee \gamma)$	(D)

(The second derivability relation is proved in a similar way.)

Law of excluded middle	$\vdash (\alpha \vee \neg \alpha)$	$\alpha : \alpha$	(A)
		$\alpha : (\alpha \vee \neg \alpha)$	(D')
		$\neg \alpha : \neg \alpha$	(A)
		$\neg \alpha : (\alpha \vee \neg \alpha)$	(D'')
		$: (\alpha \vee \neg \alpha)$	(R)

First distributive law $(\alpha \wedge (\beta \vee \gamma)) \dashv\vdash ((\alpha \wedge \beta) \vee (\alpha \wedge \gamma))$

Justification:

$\alpha : \alpha$	(A)	$(\alpha \wedge \beta) : (\alpha \wedge \beta)$	(A)
$\beta : \beta$	(A)	$(\alpha \wedge \beta) : \alpha$	(C')
$\gamma : \gamma$	(A)	$(\alpha \wedge \beta) : \beta$	(C'')
$\alpha\beta : (\alpha \wedge \beta)$	(C)	$(\alpha \wedge \beta) : (\beta \vee \gamma)$	(D')
$\alpha\beta : ((\alpha \wedge \beta) \vee (\alpha \wedge \gamma))$	(D')	$(\alpha \wedge \beta) : (\alpha \wedge (\beta \vee \gamma))$	(C)
$\alpha\gamma : (\alpha \wedge \gamma)$	(C)	$(\alpha \wedge \gamma) : (\alpha \wedge \gamma)$	(A)
$\alpha\gamma : ((\alpha \wedge \beta) \vee (\alpha \wedge \gamma))$	(D'')	$(\alpha \wedge \gamma) : \alpha$	(C')

$\alpha(\beta \vee \gamma) : ((\alpha \wedge \beta) \vee (\alpha \wedge \gamma))$	(D)	$(\alpha \wedge \gamma) : \gamma$	(C'')
$(\alpha \wedge (\beta \vee \gamma)) : ((\alpha \wedge \beta) \vee (\alpha \wedge \gamma))$	(AnU)	$(\alpha \wedge \gamma) : (\beta \vee \gamma)$	(D'')
		$(\alpha \wedge \gamma) : (\alpha \wedge (\beta \vee \gamma))$	(C)
		$((\alpha \wedge \beta) \vee (\alpha \wedge \gamma)) : (\alpha \wedge (\beta \vee \gamma))$	(D)

Second distributive law $((\alpha \vee \beta) \wedge (\alpha \vee \gamma)) \dashv\vdash (\alpha \vee (\beta \wedge \gamma))$

Justification:

$\alpha : \alpha$	(A)	$\alpha : \alpha$	(A)
$\alpha : (\alpha \vee (\beta \wedge \gamma))$	(D')	$\alpha : (\alpha \vee \beta)$	(D')
$\beta : \beta$	(A)	$\alpha : (\alpha \vee \gamma)$	(D')
$\gamma : \gamma$	(A)	$\alpha : ((\alpha \vee \beta) \wedge (\alpha \vee \gamma))$	(C)
$\beta\gamma : (\beta \wedge \gamma)$	(C)	$\beta : \beta$	(A)
$\beta\gamma : (\alpha \vee (\beta \wedge \gamma))$	(D'')	$\beta : (\alpha \vee \beta)$	(D'')
$(\alpha \vee \beta)\gamma : (\alpha \vee (\beta \wedge \gamma))$	(D)	$\gamma : \gamma$	(A)
$(\alpha \vee \beta)(\alpha \vee \gamma) : (\alpha \vee (\beta \wedge \gamma))$	(D)	$\gamma : (\alpha \vee \gamma)$	(D'')
$((\alpha \vee \beta) \wedge (\alpha \vee \gamma)) : (\alpha \vee (\beta \wedge \gamma))$	(AnU)	$(\beta \wedge \gamma) : ((\alpha \vee \beta) \wedge (\alpha \vee \gamma))$	(UU)
		$(\alpha \vee (\beta \wedge \gamma)) : ((\alpha \vee \beta) \wedge (\alpha \vee \gamma))$	(D)

Implication and disjunction $(\alpha \rightarrow \beta) \dashv\vdash (\neg\alpha \vee \beta)$

Justification:

$\neg\alpha : \neg\alpha$	(A)	$\neg\alpha\alpha : \beta$	(XQ)
$\neg\alpha : (\neg\alpha \vee \beta)$	(D')	$\neg\alpha : (\alpha \rightarrow \beta)$	(I')
$\beta : \beta$	(A)	$\beta\alpha : \beta$	(AEx)
$\beta : (\neg\alpha \vee \beta)$	(D'')	$\beta : (\alpha \rightarrow \beta)$	(I')
$(\alpha \rightarrow \beta) : (\neg\alpha \vee \beta)$	(I)	$(\neg\alpha \vee \beta) : (\alpha \rightarrow \beta)$	(D)

Disjunction and implication $(\alpha \vee \beta) \dashv\vdash (\neg\alpha \rightarrow \beta)$

Justification:

$\alpha\neg\alpha : \beta$	(XQ)	$\neg\neg\alpha : \alpha$	(NN')
$\alpha : (\neg\alpha \rightarrow \beta)$	(I')	$\neg\neg\alpha : (\alpha \vee \beta)$	(D')
$\beta\neg\alpha : \beta$	(AEx)	$\beta : \beta$	(A)
$\beta : (\neg\alpha \rightarrow \beta)$	(I')	$\beta : (\alpha \vee \beta)$	(D'')
$(\alpha \vee \beta) : (\neg\alpha \rightarrow \beta)$	(D)	$(\neg\alpha \rightarrow \beta) : (\alpha \vee \beta)$	(I)

Removal of the disjunctor in the antecedent $\neg\alpha(\alpha \vee \beta) \vdash \beta$

$\neg\alpha\alpha : \beta$	(XQ)
$\beta : \beta$	(A)
$\neg\alpha(\alpha \vee \beta) : \beta$	(D)

Particularisation and generalisation $\bigvee x\alpha \dashv\vdash \neg\bigwedge x\neg\alpha$

Justification: See 1.4, Theorem 3 (c).

Generalisation and particularisation $\bigwedge x\alpha \dashv\vdash \neg\bigvee x \neg\alpha$

Justification:

$\bigwedge x\alpha : \bigwedge x\alpha$	(A)	$\neg\alpha : \neg\alpha$	(A)
$\bigwedge x\alpha : \alpha$	(G)	$\neg\alpha : \bigvee x\neg\alpha$	(P_x)
$\neg\alpha : \neg\bigwedge x\alpha$	(CaPo)	$\neg\bigvee x\neg\alpha : \alpha$	(CaPo'')
$\bigvee x\neg\alpha : \neg\bigwedge x\alpha$	(P'_x)	$\neg\bigvee x\neg\alpha : \bigwedge x\alpha$	(G_x)
$\bigwedge x\alpha : \neg\bigvee x\neg\alpha$	(CaPo')		

Negation of the universal quantifier $\neg\bigwedge x\alpha \dashv\vdash \bigvee x\neg\alpha$

Justification:

$\neg\bigvee x\neg\alpha : \bigwedge x\alpha$	(see above)	$\bigwedge x\alpha : \neg\bigvee x\neg\alpha$	(see above)
$\neg\bigwedge x\alpha : \bigvee x\neg\alpha$	(CaPo'')	$\bigvee x\neg\alpha : \neg\bigwedge x\alpha$	(CaPo')

Negation of the existential quantifier $\neg\bigvee x\alpha \dashv\vdash \bigwedge x\neg\alpha$

Justification:

$\neg\bigwedge x\neg\alpha : \bigvee x\alpha$	(see above)	$\bigvee x\alpha : \neg\bigwedge x\neg\alpha$	(see above)
$\neg\bigvee x\alpha : \bigwedge x\neg\alpha$	(CaPo'')	$\bigwedge x\neg\alpha : \neg\bigvee x\alpha$	(CaPo')

2.3 Derived rules for $\bigvee$

Notation	Name	Rule	Justification	
(PP_x)	Simultaneous particularisation over x	$\dfrac{\Sigma\alpha : \beta}{\Sigma\bigvee x\alpha : \bigvee x\beta}$ if x does not occur free in Σ	$\Sigma\alpha : \beta$	
			$\Sigma\alpha : \bigvee x\beta$	(P_x)
			$\Sigma\bigvee x\alpha : \bigvee x\beta$	(P'_x)
$(ExP_{x,t})$	Extended particularisation rule	$\dfrac{\Sigma : \beta}{\Sigma : \bigvee x\alpha}$ if Subst $\alpha\, x\, t\, \beta$	$\Sigma : \beta$	
			$\bigwedge x\neg\alpha : \bigwedge x\neg\alpha$	(A)
			$\bigwedge x\neg\alpha : \neg\beta$	$(ExG_{x,t})$
			$\Sigma\bigwedge x\neg\alpha : \neg\bigwedge x\neg\alpha$	(X)
			$\Sigma : \neg\bigwedge x\neg\alpha$	(SeDe)
			$\neg\bigwedge x\neg\alpha : \bigvee x\alpha$	(see 2.2)
			$\Sigma : \bigvee x\alpha$	(CuRu)
$(SP_{x,y})$	Substitution and particularisation rule	$\dfrac{\Sigma\beta : \gamma}{\Sigma\bigvee x\alpha : \gamma}$ if Subst $\alpha\, x\, y\, \beta$, Subst $\beta\, x\, y\, \alpha$, y not free in $\Sigma\gamma$	$\Sigma\beta : \gamma$	
			$\Sigma\neg\gamma : \neg\beta$	(CaPo)
			$\Sigma\neg\gamma : \bigwedge x\neg\alpha$	$(SG_{x,y})$
			$\Sigma\neg\bigwedge x\neg\alpha : \gamma$	(CaPo'')
			$\bigvee x\alpha : \neg\bigwedge x\neg\alpha$	(see 2.2)
			$\Sigma\bigvee x\alpha : \gamma$	(CuRu)

Notation	Name	Rule	Justification	
$(\mathrm{ReP}_{x,y})$	Renaming of bound variables in particularisations	$\bigvee x\alpha : \bigvee y\beta$ if Subst α x y β and Subst β x y α	$\beta : \beta$	(A)
			$\beta : \bigvee y\beta$	(P_y)
			$\bigvee x\alpha : \bigvee y\beta$	$(\mathrm{SP}_{x,y})$

§ 3. Further derivability relations connected with quantification

3.1 Summary. The following derivability relations hold:

(1) $\bigwedge x \neg\alpha \dashv\vdash \neg\bigvee x\alpha$ $\qquad$ $\bigvee x \neg\alpha \dashv\vdash \neg\bigwedge x\alpha$

(2) $\bigwedge x(\alpha \wedge \beta) \dashv\vdash (\bigwedge x\alpha \wedge \bigwedge x\beta)$ $\qquad$ $\bigvee x(\alpha \wedge \beta) \vdash (\bigvee x\alpha \wedge \bigvee x\beta)$

(3) $\bigwedge x(\alpha \vee \beta) \dashv (\bigwedge x\alpha \vee \bigwedge x\beta)$ $\qquad$ $\bigvee x(\alpha \vee \beta) \dashv\vdash (\bigvee x\alpha \vee \bigvee x\beta)$

(4) $\bigwedge x(\alpha \rightarrow \beta) \vdash (\bigwedge x\alpha \rightarrow \bigwedge x\beta)$ $\qquad$ $\bigvee x(\alpha \rightarrow \beta) \dashv (\bigvee x\alpha \rightarrow \bigvee x\beta)$

Note the missing derivability relations in (2), (3), (4). cf. Exercise 1 concerning their nonexistence.

Under the condition that x does not occur free in α, we have

(5) $\bigwedge x\alpha \dashv\vdash \alpha$ $\qquad$ $\bigvee x\alpha \dashv\vdash \alpha$

(6) $\bigwedge x(\alpha \wedge \beta) \dashv\vdash (\alpha \wedge \bigwedge x\beta)$ $\qquad$ $\bigvee x(\alpha \wedge \beta) \dashv\vdash (\alpha \wedge \bigvee x\beta)$

(7) $\bigwedge x(\alpha \vee \beta) \dashv\vdash (\alpha \vee \bigwedge x\beta)$ $\qquad$ $\bigvee x(\alpha \vee \beta) \dashv\vdash (\alpha \vee \bigvee x\beta)$

(8) $\bigwedge x(\alpha \rightarrow \beta) \dashv\vdash (\alpha \rightarrow \bigwedge x\beta)$ $\qquad$ $\bigvee x(\alpha \rightarrow \beta) \dashv\vdash (\alpha \rightarrow \bigvee x\beta)$

But for arbitrary α, β we have only:

(5') $\bigwedge x\alpha \vdash \alpha$ $\qquad$ $\bigvee x\alpha \dashv \alpha$

(6') $\bigwedge x(\alpha \wedge \beta) \vdash (\alpha \wedge \bigwedge x\beta)$

(7') $\qquad$ $\bigvee x(\alpha \vee \beta) \dashv (\alpha \vee \bigvee x\beta)$

(8') $\qquad$ $\bigvee x(\alpha \rightarrow \beta) \dashv (\alpha \rightarrow \bigvee x\beta)$

Cf. Exercise 2 concerning the nonexistence of the missing derivability relations.

Under the condition that x does not occur free in β, we also have:

(9) $\bigwedge x(\alpha \rightarrow \beta) \dashv\vdash (\bigvee x\alpha \rightarrow \beta)$ $\qquad$ $\bigwedge x(\alpha \rightarrow \beta) \dashv\vdash (\bigvee x\alpha \rightarrow \beta)$

Here, note the transformation of $\bigwedge$ into $\bigvee$ and vice versa. We also have

(10) $\bigwedge x\alpha \vdash \bigvee x\alpha$

The following derivability relations concern the iteration of quantifiers:

(11) $\bigwedge x \bigwedge y \alpha \dashv\vdash \bigwedge y \bigwedge x \alpha$ $\quad\quad$ $\bigvee x \bigvee y \alpha \dashv\vdash \bigvee y \bigvee x \alpha$

(12) $\bigvee x \bigwedge y \alpha \vdash \bigwedge y \bigvee x \alpha$

<u>If</u> Subst $\alpha y x \beta$, <u>then</u>

(13) $\bigwedge x \bigwedge y \alpha \vdash \bigwedge x \beta$ $\quad\quad$ $\bigvee x \bigvee y \alpha \dashv \bigvee x \beta$

<u>3.2 Proof of the derivability relations (1) to (13)</u>

<u>to</u> (1): This was shown in 2.2.

<u>to</u> (2):

$\bigwedge x(\alpha \wedge \beta) : \bigwedge x(\alpha \wedge \beta)$	(A)	$(\bigwedge x\alpha \wedge \bigwedge x\beta) : (\bigwedge x\alpha \wedge \bigwedge \alpha\beta)$	(A)
$\bigwedge x(\alpha \wedge \beta) : (\alpha \wedge \beta)$	(G)	$(\bigwedge x\alpha \wedge \bigwedge x\beta) : \bigwedge x\alpha$	(C')
$\bigwedge x(\alpha \wedge \beta) : \alpha$	(C')	$(\bigwedge x\alpha \wedge \bigwedge x\beta) : \alpha$	(G)
$\bigwedge x(\alpha \wedge \beta) : \bigwedge x\alpha$	(G_x)	$(\bigwedge x\alpha \wedge \bigwedge x\beta) : \bigwedge x\beta$	(C'')
$\bigwedge x(\alpha \wedge \beta) : \beta$	(C'')	$(\bigwedge x\alpha \wedge \bigwedge x\beta) : \beta$	(G)
$\bigwedge x(\alpha \wedge \beta) : \bigwedge x\beta$	(G_x)	$(\bigwedge x\alpha \wedge \bigwedge x\beta) : (\alpha \wedge \beta)$	(C)
$\bigwedge x(\alpha \wedge \beta) : (\bigwedge x\alpha \wedge \bigwedge x\beta)$	(C)	$(\bigwedge x\alpha \wedge \bigwedge \alpha\beta) : \bigwedge x(\alpha \wedge \beta)$	(G_x)

$(\alpha \wedge \beta) : (\alpha \wedge \beta)$	(A)
$(\alpha \wedge \beta) : \alpha$	(C')
$\bigvee x(\alpha \wedge \beta) : \bigvee x\alpha$	(PP_x)
$(\alpha \wedge \beta) : \beta$	(C'')
$\bigvee x(\alpha \wedge \beta) : \bigvee x\beta$	(PP_x)
$\bigvee x(\alpha \wedge \beta) : (\bigvee x\alpha \wedge \bigvee x\beta)$	(C)

<u>to</u> (3):

$\alpha : \alpha$	(A)
$\alpha : (\alpha \vee \beta)$	(D')
$\bigwedge x\alpha : \bigwedge x(\alpha \vee \beta)$	(GG_x)
$\beta : \beta$	(A)
$\beta : (\alpha \vee \beta)$	(D'')
$\bigwedge x\beta : \bigwedge x(\alpha \vee \beta)$	(GG_x)
$(\bigwedge x\alpha \vee \bigwedge x\beta : \bigwedge x(\alpha \vee \beta)$	(D)

$\alpha : \alpha$	(A)	$\alpha : \alpha$	(A)
$\alpha : \bigvee x\alpha$	(P_x)	$\alpha : (\alpha \vee \beta)$	(D')
$\alpha : (\bigvee x\alpha \vee \bigvee x\beta)$	(D')	$\bigvee x\alpha : \bigvee x(\alpha \vee \beta)$	(PP_x)
$\beta : \beta$	(A)	$\beta : \beta$	(A)
$\beta : \bigvee x\beta)$	(P_x)	$\beta : (\alpha \vee \beta)$	(D'')
$\beta : (\bigvee x\alpha \vee \bigvee x\beta)$	(D'')	$\bigvee x\beta : \bigvee x(\alpha \vee \beta)$	(PP_x)
$(\alpha \vee \beta) : (\bigvee x\alpha \vee \bigvee x\beta)$	(D)	$(\bigvee x\alpha \vee \bigvee x\beta) : \bigvee x(\alpha \vee \beta)$	(D)
$\bigwedge x(\alpha \vee \beta) : (\bigvee x\alpha \vee \bigvee x\beta)$	(P'_x)		

<u>to</u> (4):

$\bigwedge x(\alpha \to \beta) : \bigwedge x(\alpha \to \beta)$	(A)
$\bigwedge x(\alpha \to \beta) : (\alpha \to \beta)$	(G)
$\bigwedge x(\alpha \to \beta)\alpha : \beta$	(RI)
$\bigwedge x(\alpha \to \beta) \bigwedge x\alpha : \bigwedge x\beta$	(GG_x)
$\bigwedge x(\alpha \to \beta) : (\bigwedge x\alpha \to \bigwedge x\beta)$	(I')

$\beta\alpha : \beta$	(AEx)
$\beta : (\alpha \to \beta)$	(I')
$\bigvee x\beta : \bigvee x(\alpha \to \beta)$	(PP_x)
$\neg\alpha\alpha : \beta$	(XQ)
$\neg\alpha : (\alpha \to \beta)$	(I')
$\neg(\alpha \to \beta) : \alpha$	(CaPo'')
$\neg(\alpha \to \beta) : \bigvee x\alpha$	(P_x)
$\neg\bigvee x\alpha : (\alpha \to \beta)$	(CaPo'')
$\neg\bigvee x\alpha : \bigvee x(\alpha \to \beta)$	(P_x)
$(\bigvee x\alpha \to \bigvee x\beta) : \bigvee x(\alpha \to \beta)$	(I)

In the following we shall, at various points, use the assumption that x does not occur free in α (in (5),...,(8)) or in β (in (9)). These points will be marked on the right with a "*".

<u>to</u> (5) and (5'):

$\bigwedge x\alpha : \bigwedge x\alpha$	(A)
$\bigwedge x\alpha : \alpha$	(G)

$\alpha : \alpha$	(A)
$\alpha : \bigwedge x\alpha$	(G_x)*

$\alpha : \alpha$	(A)
$\alpha : \bigvee x\alpha$	(P_x)

$\alpha : \alpha$	(A)
$\bigvee x\alpha : \alpha$	(P'_x)*

<u>to</u> (6) and (6'):

$\bigwedge x(\alpha \wedge \beta) : \bigwedge x(\alpha \wedge \beta)$	(A)
$\bigwedge x(\alpha \wedge \beta) : (\alpha \wedge \beta)$	(G)
$\bigwedge x(\alpha \wedge \beta) : \alpha$	(C')
$\bigwedge x(\alpha \wedge \beta) : \beta$	(C'')
$\bigwedge x(\alpha \wedge \beta) : \bigwedge x\beta$	(G_x)
$\bigwedge x(\alpha \wedge \beta) : (\alpha \wedge \bigwedge x\beta)$	(C)

$(\alpha \wedge \bigwedge x\beta) : (\alpha \wedge \bigwedge \alpha\beta)$	(A)
$(\alpha \wedge \bigwedge x\beta) : \alpha$	(C')
$(\alpha \wedge \bigwedge x\beta) : \bigwedge x\beta$	(C'')
$(\alpha \wedge \bigwedge x\beta) : \beta$	(G)
$(\alpha \wedge \bigwedge x\beta) : (\alpha \wedge \beta)$	(C)
$(\alpha \wedge \bigwedge x\beta) : \bigwedge x(\alpha \wedge \beta)$	(G_x)*

$(\alpha \wedge \beta) : (\alpha \wedge \beta)$	(A)
$(\alpha \wedge \beta) : \alpha$	(C')
$(\alpha \wedge \beta) : \beta$	(C'')
$(\alpha \wedge \beta) : \bigvee x\beta$	(P_x)
$(\alpha \wedge \beta) : (\alpha \wedge \bigvee x\beta)$	(C)
$\bigvee x(\alpha \wedge \beta) : (\alpha \wedge \bigvee z\beta)$	(P'_x)*

$\alpha : \alpha$	(A)
$\beta : \beta$	(A)
$\alpha\beta : (\alpha \wedge \beta)$	(C)
$\alpha \bigvee x\beta : \bigvee x(\alpha \wedge \beta)$	(PP_x)*
$(\alpha \wedge \bigvee x\beta) : \bigvee x(\alpha \wedge \beta)$	(AnU)

to (7) and (7'):

$\alpha : \alpha$	(A)	$\alpha : \alpha$	(A)
$\alpha : (\alpha \vee \bigwedge x\beta)$	(D')	$\alpha : (\alpha \vee \beta)$	(D')
$\neg\,\alpha(\alpha \vee \beta) : \beta$	(§ 2.2)	$\alpha : \bigwedge x(\alpha \vee \beta)$	(G_x)*
$\neg\,\alpha \bigwedge x(\alpha \vee \beta) : \bigwedge x\beta$	(GG_x)*	$\beta : \beta$	(A)
$\neg\,\alpha \bigwedge x(\alpha \vee \beta) : (\alpha \vee \bigwedge x\beta)$	(D'')	$\beta : (\alpha \vee \beta)$	(D'')
$\bigwedge x(\alpha \vee \beta) : (\alpha \vee \bigwedge x\beta)$	(R)	$\bigwedge x\beta : \bigwedge x(\alpha \vee \beta)$	(GG_x)
		$(\alpha \vee \bigwedge x\beta) : \bigwedge x(\alpha \vee \beta)$	(D)

$\alpha : \alpha$	(A)	$\alpha : \alpha$	(A)
$\alpha : (\alpha \vee \bigvee x\beta)$	(D')	$\alpha : (\alpha \vee \beta)$	(D')
$\beta : \bigvee x\beta$	((5))	$\alpha : \bigvee x(\alpha \vee \beta)$	(P_x)
$\beta : (\alpha \vee \bigvee x\beta)$	(D'')	$\beta : \beta$	(A)
$(\alpha \vee \beta) : (\alpha \vee \bigvee x\beta)$	(D)	$\beta : (\alpha \vee \beta)$	(D'')
$\bigvee x(\alpha \vee \beta) : (\alpha \vee \bigvee x\beta)$	(P'_x)*	$\bigvee x\beta : \bigvee x(\alpha \vee \beta)$	(PP_x)
		$(\alpha \vee \bigvee x\beta) : \bigvee x(\alpha \vee \beta)$	(D)

to (8) and (8'):

$(\alpha \to \beta) : (\alpha \to \beta)$	(A)	$(\alpha \to \bigwedge x\beta) : (\alpha \to \bigwedge x\beta)$	(A)
$(\alpha \to \beta)\alpha : \beta$	(RI)	$\alpha(\alpha \to \bigwedge x\beta) : \bigwedge x\beta$	(RI)
$\bigwedge x(\alpha \to \beta)\alpha : \bigwedge x\beta$	(GG_x)*	$\alpha(\alpha \to \bigwedge x\beta) : \beta$	(G)
$\bigwedge x(\alpha \to \beta) : (\alpha \to \bigwedge x\beta)$	(I')	$(\alpha \to \bigwedge x\beta) : (\alpha \to \beta)$	(I')
		$(\alpha \to \bigwedge x\beta) : \bigwedge x(\alpha \to \beta)$	(G_x)*

$(\alpha \to \beta) : (\alpha \to \beta)$	(A)	$(\alpha \to \bigvee x\beta) : (\alpha \to \bigvee x\beta)$	(A)
$(\alpha \to \beta)\alpha : \beta$	(RI)	$\alpha(\alpha \to \bigvee x\beta) : \bigvee x\beta$	(RI)
$\bigvee x(\alpha \to \beta)\alpha : \bigvee x\beta$	(PP_x)*	$\beta : (\alpha \to \beta)$	(IIn)
$\bigvee x(\alpha \to \beta) : (\alpha \to \bigvee x\beta)$	(I')	$\bigvee x\beta : \bigvee x(\alpha \to \beta)$	(PP_x)
		$(\alpha \to \bigvee x\beta)\alpha : \bigvee x(\alpha \to \beta)$	(CuRu)
		$\neg\,\alpha : (\alpha \to \beta)$	(XQ'')
		$\neg\,\alpha : \bigvee x(\alpha \to \beta)$	(P_x)
		$(\alpha \to \bigvee x\beta) : \bigvee x(\alpha \to \beta)$	(R)

to (9):

$\bigwedge x(\alpha \to \beta) : \bigwedge x(\alpha \to \beta)$	(A)	$(\bigvee x\alpha \to \beta) : (\bigvee x\alpha \to \beta)$	(A)
$\bigwedge x(\alpha \to \beta) : (\alpha \to \beta)$	(G)	$(\bigvee x\alpha \to \beta) \bigvee x\alpha : \beta$	(RI)
$\bigwedge x(\alpha \to \beta)\alpha : \beta$	(RI)	$\alpha : \bigvee x\alpha$	((5))
$\bigwedge x(\alpha \to \beta) \bigvee x\alpha : \beta$	(P'_x)*	$(\bigvee x\alpha \to \beta)\alpha : \beta$	(CuRu)
$\bigwedge x(\alpha \to \beta) : (\bigvee x\alpha \to \beta)$	(I')	$(\bigvee x\alpha \to \beta) : (\alpha \to \beta)$	(I')
		$(\bigvee x\alpha \to \beta) : \bigwedge x(\alpha \to \beta)$	(G_x)*

$(\alpha \to \beta) : (\alpha \to \beta)$	(A)
$\alpha(\alpha \to \beta) : \beta$	(RI)
$\bigwedge x\alpha : \alpha$	((5))
$(\alpha \to \beta) \bigwedge x\alpha : \beta$	(CuRu)
$(\alpha \to \beta) : (\bigwedge x\alpha \to \beta)$	(I')
$\bigvee x(\alpha \to \beta) : (\bigwedge x\alpha \to \beta)$	(P'_x)*

$(\bigwedge x\alpha \to \beta) : (\bigwedge x\alpha \to \beta)$	(A)
$(\bigwedge x\alpha \to \beta) \bigwedge x\alpha : \beta$	(RI)
$\beta : (\alpha \to \beta)$	(IIn)
$((\bigwedge x\alpha \to \beta) \bigwedge x\alpha : (\alpha \to \beta)$	(CuRu)
$(\bigwedge x\alpha \to \beta) \bigwedge x\alpha : \bigvee x(\alpha \to \beta)$	(P_x)
$\neg \alpha : (\alpha \to \beta)$	(XQ'')
$\bigvee x \neg \alpha : \bigvee x(\alpha \to \beta)$	(PP_x)
$\neg \bigwedge x\alpha : \bigvee x \neg \alpha$	((I))
$\neg \bigwedge x\alpha : \bigvee x(\alpha \to \beta)$	(CuRu)
$(\bigwedge x\alpha \to \beta) : \bigvee x(\alpha \to \beta)$	(R)

<u>to</u> (10):

$\bigwedge x\alpha : \bigwedge x\alpha$	(A)
$\bigwedge x\alpha : \alpha$	(G)
$\bigwedge x\alpha : \bigvee x\alpha$	(P_x)

<u>to</u> (11): By symmetry, we need only show one direction of the assertion in each case.

$\bigwedge y\alpha : \bigwedge y\alpha$	(A)
$\bigwedge y\alpha : \alpha$	(G)
$\bigwedge x \bigwedge y\alpha : \bigwedge x\alpha$	(GG_x)
$\bigwedge x \bigwedge y\alpha : \bigwedge y \bigwedge x\alpha$	(G_y)

$\alpha : \alpha$	(A)
$\alpha : \bigvee x\alpha$	(P_x)
$\bigvee y\alpha : \bigvee y \bigvee x\alpha$	(PP_y)
$\bigvee x \bigvee y\alpha : \bigvee y \bigvee x\alpha$	(P'_x)

<u>to</u> (12):

$\bigwedge y\alpha : \bigwedge y\alpha$	(A)
$\bigwedge y\alpha : \alpha$	(G)
$\bigvee x \bigwedge y\alpha : \bigvee x\alpha$	(PP_x)
$\bigvee x \bigwedge y\alpha : \bigwedge y \bigvee x\alpha$	(G_y)

<u>to</u> (13):

$\bigwedge y\alpha : \bigwedge y\alpha$	(A)
$\bigwedge y\alpha : \beta$	$(ExG_{y,x})$
$\bigwedge x \bigwedge y\alpha : \bigwedge x\beta$	(GG_x)

$\beta : \beta$	(A)
$\beta : \bigvee y\alpha$	$(ExP_{y,x})$
$\bigvee x\beta : \bigvee x \bigvee y\alpha$	(PP_x)

<u>Exercises. 1.</u> Find semantic examples which show that the derivability relations which do not appear in 3.1(1),..., (4) do not hold.

<u>2.</u> Find semantic examples which show that the derivability relations which do not appear in 3.1(5'),..., (8') (compare with (5),..., (8)) do not hold.

§4. Conjunctive and disjunctive normal forms

4.1 Statement of the problem. Here we limit ourselves to expressions which are formed by using only the connectives $\neg$, $\wedge$, $\vee$, i.e. without the use of the connectives $\rightarrow$, $\bigwedge$, $\bigvee$. For any such expression α, we can find effectively two expressions $\alpha_\wedge$, and $\alpha_\vee$ such that

(*) $\alpha_\wedge \dashv\vDash \alpha$ and $\alpha_\vee \dashv\vDash \alpha$.

(**) $\alpha_\wedge$ is an iterated conjunction of iterated disjunctions of atomic or negated atomic expressions. $\alpha_\vee$ is an iterated disjunction of iterated conjunctions of atomic or negated atomic expressions.

For example, if

$$\alpha \equiv \neg(\neg P \wedge Q) \wedge (P \vee R),$$

where P, Q, R are no-place predicate variables, then we can put

$$\alpha_\wedge \equiv (P \vee \neg Q) \wedge (P \vee R),$$

$$\alpha_\vee \equiv P \vee (\neg Q \wedge R).$$

(In $\alpha_\vee$, P is to be understood as a conjunction with only one member.) $\alpha_\wedge$ and $\alpha_\vee$ are not uniquely determined by the requirements (*) and (**); however, they can be uniquely determined by means of additional conditions.

$\alpha_\wedge$ is called a conjunctive normal form and $\alpha_\vee$ a disjunctive normal form of α.

In the following, we shall only show that every expression α (in $\neg$, $\wedge$, $\vee$) possesses a conjunctive normal form. We can use the same procedure to show that α has a disjunctive normal form. We can also pass directly from a conjunctive normal form to a disjunctive one (see Exercise 2).

4.2 Iterated conjunctions and disjunctions. We define by induction:

$$(1) \qquad \bigwedge_{i=1}^{1} \alpha_i \equiv \alpha_1$$

$$\bigwedge_{i=1}^{n+1} \alpha_i \equiv \left(\bigwedge_{i=1}^{n} \alpha_i \wedge \alpha_{n+1}\right).$$

Then we have:

$$(2) \quad \left(\bigwedge_{i=1}^{n} \alpha_i \wedge \bigwedge_{j=1}^{m} \beta_j\right) \dashv\vDash \bigwedge_{k=1}^{n+m} \gamma_k, \text{ where } \gamma_k \equiv \begin{cases} \alpha_k & \text{for } k = 1,\dots,n, \\ \beta_{k-n} & \text{for } k = n+1,\dots,n+m. \end{cases}$$

Here we are dealing with a generalisation of the associative law $(\alpha \wedge \beta) \wedge \gamma \dashv\vDash \alpha \wedge (\beta \wedge \gamma)$.

Correspondingly, we can generalise the distributive laws $\alpha \vee (\beta \wedge \gamma) \dashv\models (\alpha \vee \beta) \wedge (\alpha \vee \gamma)$ and $(\alpha \wedge \beta) \vee \gamma \dashv\models (\alpha \vee \gamma) \wedge (\beta \vee \gamma)$ to

(3) $$\left(\bigwedge_{i=1}^{n} \alpha_i \vee \bigwedge_{j=1}^{m} \beta_j\right) \dashv\models \bigwedge_{k=1}^{nm} \gamma_k,$$

where $\gamma_{i+n(j-1)} \equiv (\alpha_i \vee \beta_j)$ for $i = 1,\ldots,n;\ j = 1,\ldots,m$.

$\bigvee_{i=1}^{n} \alpha_i$ can be defined in a way corresponding to (1). It is easy to obtain the following statements, which are dual to (2), (3):

(2') $$\left(\bigvee_{i=1}^{n} \alpha_i \vee \bigvee_{j=1}^{m} \beta_i\right) \dashv\models \bigvee_{k=1}^{n+m} \gamma_k, \text{ where } \gamma_k \equiv \begin{cases} \alpha_k & \text{for } k = 1,\ldots,n, \\ \beta_{k-n} & \text{for } k = n+1,\ldots,n+m. \end{cases}$$

(3') $$\left(\bigvee_{i=1}^{n} \alpha_i \wedge \bigvee_{j=1}^{m} \beta_j\right) \dashv\models \bigvee_{k=1}^{nm} \gamma_k,$$

where $\gamma_{i+n(j-1)} \equiv (\alpha_i \wedge \beta_j)$ for $i = 1,\ldots,n;\ j = 1,\ldots,m$.

4.3 Producing conjunctive normal forms. We show by induction on the structure of α that we can effectively assign to α a conjunctive normal form $\alpha_\wedge$ which satisfies the conditions (*), (**).

For technical reasons, we shall prove by induction on the structure of α that we can effectively assign to α and $\neg\alpha$ conjunctive normal forms $\alpha_\wedge$ and $[\neg\alpha]_\wedge$ such that (*) and (**) hold.

(a) Let α be an atomic expression. Then, clearly, we can put:

$$\alpha_\wedge \equiv \alpha, \quad [\neg\alpha]_\wedge \equiv \alpha^2$$

(b) We prove the assertion for a negation $\neg\alpha$ under the assumption that it has already been proved for α. Thus, by our assumption, we can find conjunctive normal forms $\alpha_\wedge$ and $[\neg\alpha]_\wedge$ such that $\alpha \dashv\models \alpha_\wedge$ and $\neg\alpha \dashv\models [\neg\alpha]_\wedge$. We have to show that there are also such normal forms for $\neg\alpha$ and $\neg\neg\alpha$. We need only consider $\neg\neg\alpha$. But $\neg\neg\alpha \dashv\models \alpha$, so that we can take $\alpha_\wedge$ as the conjunctive normal form of $\neg\neg\alpha$.

(c) We prove the assertion for a conjunction $(\alpha \wedge \beta)$ under the assumption that it has already been proved for α and β. Thus, by our assumption, we can find conjunctive normal forms $\alpha_\wedge$, $[\neg\alpha]_\wedge$, $\beta_\wedge$, $[\neg\beta]_\wedge$ such that $\alpha \dashv\models \alpha_\wedge$, $\neg\alpha \dashv\models [\neg\alpha]_\wedge$, $\beta \dashv\models \beta_\wedge$, $\neg\beta \dashv\models [\neg\beta]_\wedge$. It follows that $(\alpha \wedge \beta) \dashv\models (\alpha_\wedge \wedge \beta_\wedge)$. If we now transform $(\alpha_\wedge \wedge \beta_\wedge)$ by means of 4.2(2), it can be seen that the result is a conjunctive normal form γ. Put $(\alpha \wedge \beta)_\wedge \equiv \gamma$. We also have

---

[2] These conjunctive normal forms are one-member conjunctions of one-member disjunctions

$$\neg(\alpha \wedge \beta) \;\dashv\models\; (\neg\alpha \vee \neg\beta).$$

Now $(\neg\alpha \vee \neg\beta) \dashv\models ([\neg\alpha]_\wedge \vee [\neg\beta]_\wedge)$. If we transform the right-hand side by means of 4.2(3), also using 4.2(2'), then we obtain a conjunctive normal form δ. Put $[\neg(\alpha \wedge \beta)]_\wedge \equiv \delta$.

(d) Finally, we prove the assertion for a disjunction $(\alpha \vee \beta)$ under the assumption that it has already been proved for α and β. Thus, by our assumption, we can find conjunctive normal forms $\alpha_\wedge$, $[\neg\alpha]_\wedge$, $\beta_\wedge$, $[\neg\beta]_\wedge$ such that $\alpha \dashv\models [\neg\alpha]_\wedge$, $\beta \dashv\models \beta_\wedge$, $\neg\beta \dashv\models [\neg\beta]_\wedge$. It follows that $(\alpha \vee \beta) \dashv\models (\alpha_\wedge \vee \beta_\wedge)$. If we transform the right-hand side by 4.2(3), also using 4.2(2'), then we obtain a conjunctive normal form γ. Put $(\alpha \vee \beta)_\wedge \equiv \gamma$. Moreover, we have $\neg(\alpha \vee \beta) \dashv\models (\neg\alpha \wedge \neg\beta)$. But

$$(\neg\alpha \wedge \neg\beta) \;\dashv\models\; ([\neg\alpha]_\wedge \wedge [\neg\beta]_\wedge).$$

The right-hand side can be transformed by 4.2(2) into a conjunctive normal form δ. Put $[\neg(\alpha \vee \beta)]_\wedge \equiv \delta$.

Exercises. 1. Put the following expressions into conjunctive and disjunctive normal form:

(a) $((P_1 \wedge P_2) \vee (P_3 \vee \neg P_4))$

(b) $(\neg(P_1 \vee (P_2 \wedge \neg P_4)) \vee \neg\neg(P_3 \wedge \neg(\neg P_1 \vee P_4)))$.

2. Show how to change a conjunctive normal form directly into a disjunctive normal form.

3. By an $\leftrightarrow$- expression we mean a row of symbols which is built up from proposition variables and the two-place connective $\leftrightarrow$ in the usual way (cf. Chap. I, §6.2). By an $\leftrightarrow$- normal form we mean an $\leftrightarrow$- expression which has one of the following forms:

$$P_1,\ (P_1 \leftrightarrow P_2),\ ((P_1 \leftrightarrow P_2) \leftrightarrow P_3),\ (((P_1 \leftrightarrow P_2) \leftrightarrow P_3) \leftrightarrow P_4), \ldots$$

Show that every $\leftrightarrow$- expression is semantically equivalent to an $\leftrightarrow$- normal form.

4. By an $\neg, \leftrightarrow$- expression we mean a row of symbols which is built up from proposition variables, the negator and the two-place connective $\leftrightarrow$ in the usual way. By $\neg, \leftrightarrow$- normal forms we mean all expressions which have one of the following forms:

$$T_1,\ (T_1 \leftrightarrow T_2),\ ((T_1 \leftrightarrow T_2) \leftrightarrow T_3), \ldots,$$

where the T_j are negated or non-negated proposition variables. Show that every $\neg, \to$- expression is semantically equivalent to a $\neg, \to$- normal form.

§5. Prenex normal forms

5.1 Prefixes. In this section we consider expressions which are formed using only the connectives $\neg$, $\wedge$, $\bigwedge$, $\bigvee$, i.e. without the use of the connectives $\vee$, $\to$.

By a prefix we mean the empty row of symbols or a row of symbols which can be obtained by concatenation of rows of symbols of the form $\bigwedge x$ or $\bigvee x$, where the individual variables after the quantifiers may be different from each other. Examples of prefixes are $\bigwedge x_1$, $\bigwedge x_2$, $\bigvee x_1$, $\bigwedge x_1 \bigvee x_1$, $\bigwedge x_1 \bigwedge x_2 \bigvee x_3$. We shall use P, sometimes with indices, as variables for prefixes.

If α is an arbitrary expression ($\neg$, $\wedge$, $\bigwedge$, $\bigvee$) and P is a prefix, then $P\alpha$ is also an expression.

The inverse prefix P^{-1} of P is obtained by substituting $\bigwedge$ for $\bigvee$ and $\bigvee$ for $\bigwedge$ everywhere in P. Thus, for example, $[\bigwedge x_1 \bigwedge x_2 \bigvee x_3]^{-1} \equiv \bigvee x_1 \bigvee x_2 \bigwedge x_3$. If P is empty, then P^{-1} is also to be empty.

5.2 Prenex normal forms. An expression of the form $P\alpha$ is called a prenex normal form if α does not contain any quantifiers. If $P\alpha$ is a prenex normal form, then P and α are uniquely determined. P is called the prefix of $P\alpha$ and α its kernel.

β is called a prenex normal form of α if β is a prenex normal form and $\alpha \dashv\vDash \beta$. In 5.3 we prove the

Theorem. Every expression α has a prenex normal form which can be found effectively.

5.3 Proof. We prove the theorem by induction on the structure of α.

(a) Let α be atomic. Then α does not contain any quantifiers. Thus, α is itself a prenex normal form (with an empty prefix). Then α is a prenex normal form of α.

(b) Let α be the negation of β and, by induction hypothesis, let β possess a prenex normal form $P\beta_0$. Making use of the relations

$$\neg \bigwedge x\gamma \dashv\vDash \bigvee x \neg \gamma \quad \text{and} \quad \neg \bigvee x\gamma \dashv\vDash \bigwedge x \neg \gamma$$

we see that

$$\neg P\beta_0 \dashv\vDash P^{-1} \neg \beta_0.$$

$P^{-1} \neg \beta_0$ is a prenex normal form. Hence, since $\alpha \dashv\vDash \neg P\beta_0$, $P^{-1} \neg \beta_0$ is a prenex normal form of α.

(c) Let α be the conjunction of α_1 and α_2. By induction hypothesis, let α_1 and α_2 possess prenex normal forms $P_1\alpha_{10}$ and $P_2\alpha_{20}$. Then we have

$$(*) \qquad \alpha \dashv\vDash (P_1\alpha_{10} \wedge P_2\alpha_{20}).$$

We now need to find a prenex normal form of the expression on the right.

In 5.4 and 5.5 we shall prove the following two lemmas:

Lemma 1. Let P be an arbitrary prefix, α an arbitrary expression and $\mathfrak{J}$ a finite set of individual variables. Then we can find a prefix P' and an expression α' such that

(1) $P\alpha \dashv\vDash P'\alpha'$,

(2) the same individual variables occur free in $P\alpha$ as in $P'\alpha'$,

(3) P' does not contain any of the variables in $\mathfrak{J}$.

Lemma 2. Suppose that none of the variables in P_1 occur free in α_{20} and none of the variables in P_2 occur free in α_{10}. Then

$$(P_1\alpha_{10} \wedge P_2\alpha_{20}) \dashv\vDash P_1P_2(\alpha_{10} \wedge \alpha_{20}).$$

Now it is possible, with the aid of these two lemmas, to find a prenex normal form of the right-hand side of (*). Let $\mathfrak{J}_2$ be the set of those individual variables which occur free in α_{20}. By Lemma 1 we can, for P_1, α_{10} and $\mathfrak{J}_2$, find a P_1' and an α_{10}' with the properties given in the lemma. Let $\mathfrak{J}_1$ be the set of those individual variables which occur free in $P_1'\alpha_{10}'$. Again by Lemma 1, there are, for P_2, α_{20} and $\mathfrak{J}_1$, a P_2' and an α_{20}' with the properties given in the lemma.

Because

$$P_1\alpha_{10} \dashv\vDash P_1'\alpha_{10}' \quad \text{and} \quad P_2\alpha_{20} \dashv\vDash P_2'\alpha_{20}'$$

we have

(**) $$(P_1\alpha_{10} \wedge P_2\alpha_{20}) \dashv\vDash (P_1'\alpha_{10}' \wedge P_2'\alpha_{20}').$$

By construction, none of the individual variables in P_1' occur free in α_{20}' and none of the individual variables in P_2' occur free in α_{10}'[3]. Thus, by Lemma 2, we have:

(***) $$(P_1'\alpha_{10}' \wedge P_2'\alpha_{20}') \dashv\vDash P_1'P_2'(\alpha_{10}' \wedge \alpha_{20}').$$

The assertion then follows from (*), (**) and (***).

(d) Let α be a generalisation or a particularisation of β. By induction hypothesis, let β possess a prenex normal form $P\beta_0$. Now if $\alpha \equiv \bigwedge x\beta$ or $\alpha \equiv \bigvee x\beta$, then clearly $\bigwedge xP\beta_0$ or $\bigvee xP\beta_0$ respectively is a prenex normal form of α.

---

[3] We can see this as follows: (1) let z occur in P_1' and occur free in α_{20}'. Then either z occurs free in $P_2'\alpha_{20}'$ or z occurs in P_2'. In the first case, z also occurs free in $P_2\alpha_{20}$ and therefore also occurs free in α_{20}. But then z cannot occur in P_1', by the choice of $\mathfrak{J}_2$. In the second case, by the choice of $\mathfrak{J}_1$, z does not occur in $P_1'\alpha_{10}'$, and hence it does not occur in P_1'. Thus we see that there is no such z. (2) z cannot occur in P_2' and occur free in α_{10}', since this would contradict the choice of $\mathfrak{J}_1$.

5.4 Proof of Lemma 1. Let P be empty. Then $P' \equiv P$ and $\alpha' \equiv \alpha$ fulfils the requirements. Now it is sufficient to show that: If the statement of Lemma 1 holds for P and α, then it also holds for QxP and α, where Q is either $\bigwedge$ or $\bigvee$ and x is an arbitrary individual variable.

By hypothesis, there is a P' and an α' such that the statements (1), (2), (3) hold.

From (1) we have

(0') $$QxP\alpha \dashv\models QxP'\alpha'.$$

Clearly, the same variables occur free in $QxP\alpha$ as in $QxP'\alpha'$.

Now we distinguish two cases:

Case 1. x does not occur free in $P\alpha$, and therefore also not in $P'\alpha'$. Then

$$QxP\alpha \dashv\models P'\alpha'.$$

Hence, P' and α' fulfil the requirements.

Case 2. x occurs free in $P\alpha$ and therefore also in $P'\alpha'$. Let y be an individual variable which is not in $\mathfrak{J}$ and does not occur in $P'\alpha'$. By Chap. II, § 5.4, Theorem 6, there is an expression α'' such that

(i) $$\text{Subst}\,\alpha' xy\alpha'' \quad \text{and} \quad \text{Subst}\,\alpha'' yx\alpha'.$$

Now we want to show that, for every P_1 in which neither x or y occurs, the statements

(ii) $$\text{Subst}\,P_1\alpha'\,xy\,P_1\alpha'' \quad \text{and} \quad \text{Subst}\,P_1\alpha''\,yx\,P_1\alpha'$$

hold. (ii) holds, as we have just seen, for an empty P_1. Thus it suffices to show that (ii) holds for QzP_1 if (ii) holds for P_1, where z is any variable which is different from x and y.

x occurs free in α', and therefore also in $QzP_1\alpha'$. Since z is different from y, by the definition of substitution (cf. 1.2 and also Chap. II, § 5.3) we have

$$\text{Subst}\,QzP_1\alpha'\,xy\,QzP_1\alpha''.$$

y occurs free in α'' and therefore also in $QzP_1\alpha''$. Since z is different from x we have, by the definition of substitution, $\text{Subst}\,QzP_1\alpha''\,yx\,QzP_1\alpha'$

Thus, (ii) has been proved for QzP_1.

Since x occurs free in $P'\alpha'$ and y does not occur in $P'\alpha'$, we can apply (ii) to the prefix P'. Thus we have

$$\text{Subst}\, P'\alpha'\, x\, y\, P'\alpha'' \quad \text{and} \quad \text{Subst}\, P'\alpha''\, y\, x\, P'\alpha'.$$

Now, using the derived rule ($\text{ReG}_{x,y}$), we obtain

$$\bigwedge x P'\alpha' \dashv\vDash \bigwedge y P'\alpha''$$

and, using the derived rule ($\text{ReP}_{x,y}$), we have

$$\bigvee x P'\alpha' \dashv\vDash \bigvee y P'\alpha''.$$

Thus we have, in general,

(1') $$QxP'\alpha' \dashv\vDash QyP'\alpha''.$$

If z occurs free in $QxP'\alpha'$, then z also occurs free in α' and $z \neq x$. Hence, by (i), z also occurs free in α''. z also occurs free in $QyP'\alpha''$, since z is not bound by the quantifiers in P' (because z occurs free in $QxP'\alpha'$) and $z \neq y$ (because z occurs in $P'\alpha'$ whereas y does not).

Conversely: If z occurs free in $QyP'\alpha''$, then z occurs free in α'' and $z \neq y$. Hence, by (i), z also occurs free in α'. z also occurs free in $QxP'\alpha'$, since z is not bound by the quantifiers in P' (because z occurs free in $QyP'\alpha''$) and $z \neq x$ (z occurs in α'', but x does not; for $\text{Subst}\,\alpha' xy\alpha''$ and $x \neq y$, since x occurs in $P'\alpha'$ whereas y does not). Thus we have:

(2') The same variables occur free in $QxP'\alpha'$ as in $QyP'\alpha''$.

Since y is not in $\mathfrak{J}$, (3) gives us:

(3') QyP' does not contain any of the variables in $\mathfrak{J}$.

(0'), (1'), (2'), (3') show that the statements of Lemma 1 also hold for QxP and α. Thus we have proved Lemma 1.

<u>5.5 Proof of Lemma 2.</u> It is sufficient to show that:

If none of the variables in P occur free in β, then

(4) $$(P\alpha \wedge \beta) \dashv\vDash P(\alpha \wedge \beta).$$

From this we obtain, first of all: If none of the variables in P occur free in α, then

(4') $$(\alpha \wedge P\beta) \dashv\vDash P(\alpha \wedge \beta).$$

For we have

$(\alpha \wedge P\beta) \dashv\models (P\beta \wedge \alpha)$

$(P\beta \wedge \alpha) \dashv\models P(\beta \wedge \alpha)$ by (4)

$(\beta \wedge \alpha) \dashv\models (\alpha \wedge \beta)$

$P(\beta \wedge \alpha) \dashv\models P(\alpha \wedge \beta)$ (by repeated use of simultaneous generalisations and particularisations)

Now Lemma 2 can be proved from (4) and (4') as follows:

$(P_1\alpha_{10} \wedge P_2\alpha_{20}) \dashv\models P_1(\alpha_{10} \wedge P_2\alpha_{20})$ by (4)

$(\alpha_{10} \wedge P_2\alpha_{20}) \dashv\models P_2(\alpha_{10} \wedge \alpha_{20})$ by (4')

$P_1(\alpha_{10} \wedge P_2\alpha_{20}) \dashv\models P_1P_2(\alpha_{10} \wedge \alpha_{20})$ (by repeated use of simultaneous generalisations and particularisations)

We now prove (4). (4) is trivial if P is empty. Thus it is sufficient to show, under the assumption that (4) holds, that

(4a) $$(QxP\alpha \wedge \beta) \dashv\models QxP(\alpha \wedge \beta)$$

if neither x nor any of the variables in P occurs free in β. This can be seen as follows:

$(QxP\alpha \wedge \beta) \dashv\models Qx(P\alpha \wedge \beta)$ [3.1(6)]

$Qx(P\alpha \wedge \beta) \dashv\models QxP(\alpha \wedge \beta)$ [simultaneous generalisation or particularisation of (4)].

Exercise. Put the following expressions into prenex normal form ($\rightarrow$ and $\leftrightarrow$ are to be regarded as abbreviations; let $x \not\equiv y$):

(a) $(Px \wedge \bigwedge x\, Px)$
(b) $(Px \rightarrow \bigwedge x\, Px)$
(c) $(\bigwedge x\, Px \wedge Px)$
(d) $(\bigwedge x\, Px \rightarrow Px)$
(e) $(\bigwedge x\, Px \leftrightarrow Px)$
(f) $(\bigvee x\, Px \rightarrow (\bigvee y(\bigwedge x\, Rxy \rightarrow \bigwedge y\, Rxy) \rightarrow \bigwedge x(Px \leftrightarrow \bigvee y\, Py)))$.

VIII. The Theorems of A. Robinson, Craig and Beth

In this chapter we shall prove two important new results about first-order predicate logic: Craig's interpolation lemma (1957) and Beth's definability theorem (1953). The interpolation lemma states that, for any expressions α, γ such that $\alpha \models \gamma$, there is an "interpolating" expression β such that any variable which occurs free in β occurs free in both α and γ and such that $\alpha \models \beta$ and $\beta \models \gamma$. The definability theorem states that an "implicit definition" within the framework of the language of ordinary predicate logic can always be changed into an "explicit definition". More precisely, the following holds (in this introduction we shall restrict ourselves to the case of a one-place predicate variable P): Let α be an expression which "implicity defines" P in the sense that, in every interpretation $\mathfrak{J}$ which satisfies α, the predicate $\mathfrak{J}(P)$ is uniquely determined by the $\mathfrak{J}$-images of the other variables. (This can also be formulated as follows: If P' is a one-place predicate variable which does not occur in α and which is different from P, and if α' if obtained from α by replacing P by P' everywhere in α, then $\alpha \wedge \alpha' \models \bigwedge x(Px \leftrightarrow P'x)$.)

Then P can be "explicitly defined by means of an expression β", i.e. there is an expression β and a variable x which does not occur free in α such that $\alpha \models \bigwedge x(Px \leftrightarrow \beta)$.

In § 5 we prove a satisfiablity theorem due to A. Robinson (1956). From this we deduce Craig's interpolation lemma and finally, in § 6, Beth's definability theorem. In the proof of A. Robinson's satisfiability theorem we use the concept of an elementary embedding (§ 3), which was introduced by Tarski and Vaught (1957).

§ 1. Embeddings of algebras, subalgebras, chains of algebras

1.1 Notation. In this chapter we shall consider algebras of the sort $\mathfrak{A}_{\mathfrak{B}}(\mathfrak{J})$, which were introduced in Chap. VI, § 2.1. Here, $\mathfrak{B}$ is a set of variables (of all sorts). The underlying set which belongs to $\mathfrak{A}_{\mathfrak{B}}(\mathfrak{J})$ is identical with the domain of individuals belonging to $\mathfrak{J}$; we shall denote it by ω. Correspondingly, ω', ω_j etc. are the underlying sets of the algebras $\mathfrak{A}_{\mathfrak{B}}(\mathfrak{J}')$, $\mathfrak{A}_{\mathfrak{B}}(\mathfrak{J}_j)$ etc.

1.2 Embeddings. In Chap. VI, §2.2, we introduced the concept of an isomorphism from $\mathfrak{A}_{\mathfrak{B}}(\mathfrak{I})$ into $\mathfrak{A}_{\mathfrak{B}}(\mathfrak{I}')$. An isomorphism Φ is a 1-1 mapping from ω onto ω' with the properties (c') and (d') which we named earlier. If we omit the requirement that Φ should be a mapping from ω onto ω' and require instead only that Φ should be a mapping from ω into ω', then we call Φ an "embedding". Thus we have

Definition 1. A 1-1 mapping Φ from ω into ω' is called an embedding of $\mathfrak{A}_{\mathfrak{B}}(\mathfrak{I})$ into $\mathfrak{A}_{\mathfrak{B}}(\mathfrak{I}')$ if

(1) $\mathfrak{I}(P)$ fits $\mathfrak{r}_1, \ldots, \mathfrak{r}_r$ if and only if $\mathfrak{I}'(P)$ fits $\Phi(\mathfrak{r}_1), \ldots, \Phi(\mathfrak{r}_r)$ for all $P \in \mathfrak{B}$ and all $\mathfrak{r}_1, \ldots, \mathfrak{r}_r \in \omega$,

(2) $\Phi(\mathfrak{I}(f)(\mathfrak{r}_1, \ldots, \mathfrak{r}_r)) = \mathfrak{I}'(f)(\Phi(\mathfrak{r}_1), \ldots, \Phi(\mathfrak{r}_r))$ for all $f \in \mathfrak{B}$ and all $\mathfrak{r}_1, \ldots, \mathfrak{r}_r \in \omega$.

For a no-place function variable x, (2) becomes

(2_0) $\Phi(\mathfrak{I}(x)) = \mathfrak{I}'(x)$ for all $x \in \mathfrak{B}$.

Definition 2. $\mathfrak{A}_{\mathfrak{B}}(\mathfrak{I})$ is said to be embeddable into $\mathfrak{A}_{\mathfrak{B}}(\mathfrak{I}')$ if there is an embedding Φ of $\mathfrak{A}_{\mathfrak{B}}(\mathfrak{I})$ into $\mathfrak{A}_{\mathfrak{B}}(\mathfrak{I}')$.

Remarks.

(a) The relation of embeddability is reflexive and transitive.

(b) Every isomorphism is an embedding. Thus, if $\mathfrak{A}_{\mathfrak{B}}(\mathfrak{I})$ is isomorphic to $\mathfrak{A}_{\mathfrak{B}}(\mathfrak{I}')$ then $\mathfrak{A}_{\mathfrak{B}}(\mathfrak{I})$ is embeddable into $\mathfrak{A}_{\mathfrak{B}}(\mathfrak{I}')$.

(c) If Φ is an embedding of $\mathfrak{A}_{\mathfrak{B}}(\mathfrak{I})$ into $\mathfrak{A}_{\mathfrak{B}}(\mathfrak{I}')$ then, for arbitrary $x_1, \ldots, x_n$ and $\mathfrak{r}_1, \ldots \mathfrak{r}_n \in \omega$ and an arbitrary term t such that $\mathfrak{B}(t=t) \subset \mathfrak{B} \cup \{x_1, \ldots, x_n\}$,

$$\Phi(\mathfrak{I}_{x_1 \ldots x_n}^{\mathfrak{r}_1 \ldots \mathfrak{r}_n}(t)) = \mathfrak{I}'^{\Phi(\mathfrak{r}_1) \ldots \Phi(\mathfrak{r}_n)}_{x_1 \ \ldots \ x_n}(t) .$$

Proof of (c) by induction on the structure of t: If $t \equiv x$ and $x \equiv x_m$, where m is maximal, then, in the above, both sides are equal to $\Phi(\mathfrak{r}_m)$; if, on the other hand, $x \not\equiv x_m$ for every m, then we need to show that $\Phi(\mathfrak{I}(x)) = \mathfrak{I}'(x)$, which follows from ($2_0$) since $x \in \mathfrak{B}$. We prove the assertion for a compound term by means of (2).

1.3 Subalgebras. If the identity mapping Φ is an embedding of $\mathfrak{A}_{\mathfrak{B}}(\mathfrak{I})$ into $\mathfrak{A}_{\mathfrak{B}}(\mathfrak{I}')$, then we call $\mathfrak{A}_{\mathfrak{B}}(\mathfrak{I})$ a subalgebra of $\mathfrak{A}_{\mathfrak{B}}(\mathfrak{I}')$. Thus we have

Definition 3. $\mathfrak{A}_{\mathfrak{B}}(\mathfrak{I})$ is said to be a subalgebra of $\mathfrak{A}_{\mathfrak{B}}(\mathfrak{I}')$ - or $\mathfrak{A}_{\mathfrak{B}}(\mathfrak{I}')$ an extension of $\mathfrak{A}_{\mathfrak{B}}(\mathfrak{I})$ - if $\omega \subset \omega'$ and

(1') $\mathfrak{I}(P)$ fits $\mathfrak{r}_1,\ldots,\mathfrak{r}_r$ if and only if $\mathfrak{I}'(P)$ fits $\mathfrak{r}_1,\ldots,\mathfrak{r}_r$ for all $P\in\mathfrak{B}$ and all $\mathfrak{r}_1,\ldots,\mathfrak{r}_r\in\omega$,

(2') $\mathfrak{I}(f)(\mathfrak{r}_1,\ldots,\mathfrak{r}_r) = \mathfrak{I}'(f)(\mathfrak{r}_1,\ldots,\mathfrak{r}_r)$ for all $f\in\mathfrak{B}$ and all $\mathfrak{r}_1,\ldots,\mathfrak{r}_r\in\omega$.

Remarks.

(d) If $\mathfrak{A}_{\mathfrak{B}}(\mathfrak{I})$ is a subalgebra of $\mathfrak{A}_{\mathfrak{B}}(\mathfrak{I}')$, then $\mathfrak{A}_{\mathfrak{B}}(\mathfrak{I})$ is embeddable into $\mathfrak{A}_{\mathfrak{B}}(\mathfrak{I}')$.

(e) The relation "is a subalgebra of" is reflexive and transitive.

(f) If $\mathfrak{A}_{\mathfrak{B}}(\mathfrak{I})$ is a subalgebra of $\mathfrak{A}_{\mathfrak{B}}(\mathfrak{I}')$ then, for arbitrary $x_1,\ldots,x_n$ and $\mathfrak{r}_1,\ldots,\mathfrak{r}_n\in\omega$ and an arbitrary term t such that $\mathfrak{B}(t=t)\subset\mathfrak{B}\cup\{x_1,\ldots,x_n\}$,

$$\mathfrak{I}^{\mathfrak{r}_1\ldots\mathfrak{r}_n}_{x_1\ldots x_n}(t) = \mathfrak{I}'^{\mathfrak{r}_1\ldots\mathfrak{r}_n}_{x_1\ldots x_n}(t) .$$

1.4 The union of a chain of algebras. If, for each $j = 0, 1, 2, \ldots$, $\mathfrak{A}_{\mathfrak{B}}(\mathfrak{I}_j)$ is a subalgebra of $\mathfrak{A}_{\mathfrak{B}}(\mathfrak{I}_{j+1})$, then we say that the $\mathfrak{A}_{\mathfrak{B}}(\mathfrak{I}_j)$ form a chain of algebras. We have the

Theorem. For every chain $\mathfrak{A}_{\mathfrak{B}}(\mathfrak{I}_j)$ of algebras there is an algebra $\mathfrak{A}_{\mathfrak{B}}(\mathfrak{I})$ such that (1) $\omega = \bigcup\omega_j$ and (2) $\mathfrak{A}_{\mathfrak{B}}(\mathfrak{I}_j)$ is a subalgebra of $\mathfrak{A}_{\mathfrak{B}}(\mathfrak{I})$ for each j.

Proof. The algebra $\mathfrak{A}_{\mathfrak{B}}(\mathfrak{I})$ which we shall now construct is also called the union of the chain of algebras $\mathfrak{A}_{\mathfrak{B}}(\mathfrak{I}_j)$. The domain of individuals of $\mathfrak{I}$ is determined by the requirement (1). Clearly, it suffices to define $\mathfrak{I}$ for the elements of $\mathfrak{B}$.

First of all, let P be an r-place predicate variable in $\mathfrak{B}$. Let $\mathfrak{r}_1,\ldots,\mathfrak{r}_r$ be elements of ω. Then there is a smallest index j_0 such that each of $\mathfrak{r}_1,\ldots,\mathfrak{r}_r\in\omega_{j_0}$. We define $\mathfrak{I}(P)$ as follows:

(*) $\mathfrak{I}(P)$ fits $\mathfrak{r}_1,\ldots,\mathfrak{r}_r$ iff $\mathfrak{I}_{j_0}(P)$ fits $\mathfrak{r}_1,\ldots,\mathfrak{r}_r$.

Finally, let f be an r-place function variable in $\mathfrak{B}$. Let $\mathfrak{r}_1,\ldots,\mathfrak{r}_r$ be elements of ω. Defining j_0 as above, we put

(**) $\mathfrak{I}(f)(\mathfrak{r}_1,\ldots,\mathfrak{r}_r) = \mathfrak{I}_{j_0}(f)(\mathfrak{r}_1,\ldots,\mathfrak{r}_r)$.

Since the $\mathfrak{A}_{\mathfrak{B}}(\mathfrak{I}_j)$ form a chain, (*) and (**) hold, not only for j_0, but also for every $j\geq j_0$. Hence it is easy to show that, for each j, $\mathfrak{A}_{\mathfrak{B}}(\mathfrak{I}_j)$ is a subalgebra of $\mathfrak{A}_{\mathfrak{B}}(\mathfrak{I})$.

Exercises. 1. Let Q be the set of rational numbers and R the set of real numbers. Moreover, let $\mathfrak{B} = \{P\}$, where P is a one-place predicate variable. Let $\mathfrak{I}_1$ have Q as its domain of individuals, and let $\mathfrak{I}_1(P)$ fit $\mathfrak{r}$ if and only if $\mathfrak{r}$ is the square of a rational number. Let $\mathfrak{I}_2$ have R as its domain of individuals, and let $\mathfrak{I}_2$ fit $\mathfrak{r}$ if and only if $\mathfrak{r}$ is the square of a real number. Is $\mathfrak{A}_{\mathfrak{B}}(\mathfrak{I}_1)$ a subalgebra of $\mathfrak{A}_{\mathfrak{B}}(\mathfrak{I}_2)$?

2. There is an infinite algebra with no proper subalgebras.
3. Every nondenumerable algebra has infinitely many subalgebras.
4. Let I be a nonempty index set. For each $i \in I$, let $\mathfrak{A}_{\mathfrak{B}}(\mathfrak{I}_i)$ be an algebra. For each pair $i, j \in I$, let there be an index $k \in I$ such that $\mathfrak{A}_{\mathfrak{B}}(\mathfrak{I}_i)$ and $\mathfrak{A}_{\mathfrak{B}}(\mathfrak{I}_j)$ are subalgebras of $\mathfrak{A}_{\mathfrak{B}}(\mathfrak{I}_k)$. Show that there is an algebra $\mathfrak{A}_{\mathfrak{B}}(\mathfrak{I})$ such that all the $\mathfrak{A}_{\mathfrak{B}}(\mathfrak{I}_i)$ are subalgebras of $\mathfrak{A}_{\mathfrak{B}}(\mathfrak{I})$.

§ 2. Theories

2.1 Theories. The theory of an algebra.

Definition. By $\Theta_{\mathfrak{B}}(\mathfrak{I})$, the theory determined by $\mathfrak{B}$ and $\mathfrak{I}$, we mean the set of expressions α such that $\mathfrak{B}(\alpha) \subset \mathfrak{B}$ and $\mathrm{Mod}\,\mathfrak{I}\alpha$. Thus we have

$$\Theta_{\mathfrak{B}}(\mathfrak{I}) = \{\alpha \mid \mathfrak{B}(\alpha) \subset \mathfrak{B} \text{ and } \mathrm{Mod}\,\mathfrak{I}\alpha\} .$$

An algebra $\mathfrak{A}_{\mathfrak{B}}(\mathfrak{I})$ determines $\mathfrak{B}$, and it also determines $\mathfrak{I}$ for the elements of $\mathfrak{B}$. But, in order to determine $\Theta_{\mathfrak{B}}(\mathfrak{I})$, we need only know the values of $\mathfrak{I}$ for the elements of $\mathfrak{B}$. Thus, an algebra $\mathfrak{A}_{\mathfrak{B}}(\mathfrak{I})$ determines the theory $\Theta_{\mathfrak{B}}(\mathfrak{I})$ uniquely. For this reason, we can also call $\Theta_{\mathfrak{B}}(\mathfrak{I})$ the theory of the algebra $\mathfrak{A}_{\mathfrak{B}}(\mathfrak{I})$.

2.2 Different algebras with the same theory. In general, different algebras have different theories. This is shown by

Example 1. Let $\mathfrak{B}$ be empty, $\omega = \{1\}$, $\omega' = \{1, 2\}$, $\mathfrak{I}$ and $\mathfrak{I}'$ be arbitrary interpretations over ω and ω' respectively, $\alpha \equiv \bigwedge x \bigwedge y\; x = y$ (where $x \not\equiv y$). Then $\alpha \in \Theta_{\mathfrak{B}}(\mathfrak{I})$, $\alpha \notin \Theta_{\mathfrak{B}}(\mathfrak{I}')$, $\neg\alpha \notin \Theta_{\mathfrak{B}}(\mathfrak{I})$, $\neg\alpha \in \Theta_{\mathfrak{B}}(\mathfrak{I}')$.

However, in certain cases, different algebras have the same theory. For a start, the isomorphism theorem of Chap. VI, § 2.3 has as an immediate corollary the following

Theorem. Isomorphic algebras have the same theory.

But non-isomorphic algebras can also have the same theory. This is illustrated by

Example 2. Let $\mathfrak{B}$ be empty, ω denumerable, $\omega \subset \omega'$, ω' not denumberable, and let $\mathfrak{I}$, $\mathfrak{I}'$ be arbitrary interpretations over ω, ω' respectively. Then $\mathfrak{A}_{\mathfrak{B}}(\mathfrak{I})$ and $\mathfrak{A}_{\mathfrak{B}}(\mathfrak{I}')$ are not isomorphic, since there is no 1-1 mapping from ω onto ω'. However, $\Theta_{\mathfrak{B}}(\mathfrak{I}) = \Theta_{\mathfrak{B}}(\mathfrak{I}')$: (a) If $\alpha \in \Theta_{\mathfrak{B}}(\mathfrak{I}')$, then α has no free variables and $\mathrm{Mod}\,\mathfrak{I}'\alpha$. By the Löwenheim-Skolem theorem, there is an interpretation $\mathfrak{I}''$ over a denumerable domain (which, by Lemma 1 of Chap. VI, § 2.2, we can take to be the domain ω) such that

$\mathrm{Mod}\,\mathfrak{J}''\alpha$. It follows from the coincidence theorem that $\mathrm{Mod}\,\mathfrak{J}\alpha$. (b) If $\alpha \notin \Theta_{\mathfrak{B}}(\mathfrak{J}')$, then either not $\mathfrak{B}(\alpha) \subset \mathfrak{B}$, in which case $\alpha \notin \Theta_{\mathfrak{B}}(\mathfrak{J})$, or $\mathfrak{B}(\alpha) \subset \mathfrak{B}$ and $\neg\alpha \in \Theta_{\mathfrak{B}}(\mathfrak{J}')$, in which case it follows as in (a) that $\neg\alpha \in \Theta_{\mathfrak{B}}(\mathfrak{J})$, and hence that $\alpha \notin \Theta_{\mathfrak{B}}(\mathfrak{J})$.

In both example 1 and example 2, $\mathfrak{A}_{\mathfrak{B}}(\mathfrak{J})$ is a subalgebra of $\mathfrak{A}_{\mathfrak{B}}(\mathfrak{J}')$. In the following section, we shall introduce a stronger version of the general concept of an embedding and a corresponding version of the notion of a subalgebra, so that, if $\mathfrak{A}_{\mathfrak{B}}(\mathfrak{J})$ is a subalgebra of $\mathfrak{A}_{\mathfrak{B}}(\mathfrak{J}')$ in the stronger sense, then the two algebras have the same theory.

2.3 Some lemmas.

Lemma 1. *If* $\mathrm{Mod}\,\mathfrak{J}'\,\Theta_{\mathfrak{B}}(\mathfrak{J})$, *then* $\Theta_{\mathfrak{B}}(\mathfrak{J}) = \Theta_{\mathfrak{B}}(\mathfrak{J}')$.

Proof. (a) Let $\alpha \in \Theta_{\mathfrak{B}}(\mathfrak{J})$. Then $\mathfrak{B}(\alpha) \subset \mathfrak{B}$ and $\mathrm{Mod}\,\mathfrak{J}'\alpha$, and hence $\alpha \in \Theta_{\mathfrak{B}}(\mathfrak{J}')$. (b) Let $\alpha \notin \Theta_{\mathfrak{B}}(\mathfrak{J})$. If not $\mathfrak{B}(\alpha) \subset \mathfrak{B}$, then $\alpha \notin \Theta_{\mathfrak{B}}(\mathfrak{J}')$. If, on the other hand, $\mathfrak{B}(\alpha) \subset \mathfrak{B}$, then $\neg\alpha \in \Theta_{\mathfrak{B}}(\mathfrak{J})$, $\mathrm{Mod}\,\mathfrak{J}'\neg\alpha$, not $\mathrm{Mod}\,\mathfrak{J}'\alpha$, $\alpha \notin \Theta_{\mathfrak{B}}(\mathfrak{J}')$.

Lemma 2. *If* $\Theta_{\mathfrak{B}}(\mathfrak{J}) \models \alpha$, *then there is a* $\delta \in \Theta_{\mathfrak{B}}(\mathfrak{J})$ *such that* $\delta \models \alpha$.

Proof. By the compactness theorem for the relation of consequence, there are finitely many elements $\delta_1, \ldots, \delta_r \in \Theta_{\mathfrak{B}}(\mathfrak{J})$ such that $\{\delta_1, \ldots, \delta_r\} \models \alpha$. We may assume that $r \geq 1$, since $\Theta_{\mathfrak{B}}(\mathfrak{J})$ is nonempty (e.g. $\bigwedge x\, x{=}x \in \Theta_{\mathfrak{B}}(\mathfrak{J})$). Then $\delta \equiv \delta_1 \wedge \ldots \wedge \delta_r$ has the required properties. We can use the same sort of method to prove

Lemma 3. *If* $\Theta_{\mathfrak{B}_1}(\mathfrak{J}_1) \cup \Theta_{\mathfrak{B}_2}(\mathfrak{J}_2)$ *is not satisfiable, then there are elements* $\delta_i \in \Theta_{\mathfrak{B}_i}(\mathfrak{J}_i)$ $(i = 1, 2)$ *such that* $\{\delta_1, \delta_2\}$ *is not satisfiable*.

Exercises. 1. Let $\mathfrak{J}_1$ be an interpretation over $\underline{Q}$ and $\mathfrak{J}_2$ an interpretation over $\underline{R}$ (cf. §1, Exercise 1). Let $\mathfrak{B}$ be empty. Show that: $\mathfrak{A}_{\mathfrak{B}}(\mathfrak{J}_1)$ is not isomorphic to $\mathfrak{A}_{\mathfrak{B}}(\mathfrak{J}_2)$, (b) $\Theta_{\mathfrak{B}}(\mathfrak{J}_1) = \Theta_{\mathfrak{B}}(\mathfrak{J}_2)$ (use the Löwenheim-Skolem theorem).

2. Let $\mathfrak{B}(\mathfrak{M}) \subset \mathfrak{B}$. Show that the following two assertions are equivalent: (a) $\alpha \in \mathfrak{M}$ or $\neg\alpha \in \mathfrak{M}$ for each α such that $\mathfrak{B}(\alpha) \subset \mathfrak{B}$. (b) $\mathfrak{M}$ is the set of all expressions α such that $\mathfrak{B}(\alpha) \subset \mathfrak{B}$ or there is an $\mathfrak{J}$ such that $\mathfrak{M} = \Theta_{\mathfrak{B}}(\mathfrak{J})$.

3. The structures $\mathfrak{A}_{\mathfrak{B}}(\mathfrak{J})$ and $\mathfrak{A}_{\mathfrak{B}}(\mathfrak{J}')$ are said to be *elementarily equivalent* if $\Theta_{\mathfrak{B}}(\mathfrak{J}) = \Theta_{\mathfrak{B}}(\mathfrak{J}')$. Show that elementarily equivalent structures with finite underlying sets are isomorphic.

4. Let $\mathfrak{A}_{\mathfrak{B}}(\mathfrak{J})$ and $\mathfrak{A}_{\mathfrak{B}}(\mathfrak{J}')$ be structures with finite underlying sets. Let $\mathrm{Mod}\,\mathfrak{J}\alpha$ if and only if $\mathrm{Mod}\,\mathfrak{J}'\alpha$ for every α such that $\mathfrak{B}(\alpha) \subset \mathfrak{B}$ which is a prenex normal form with no existential quantifiers. Show, using Exercise 3, that $\mathfrak{A}_{\mathfrak{B}}(\mathfrak{J})$ is isomorphic to $\mathfrak{A}_{\mathfrak{B}}(\mathfrak{J}')$.

§ 3. Elementary embeddings of algebras. Elementary subalgebras. Elementary chains of algebras

3.1 Elementary embeddings.

Definition 1. A 1-1 mapping Φ from ω into ω' is said to be an elementary embedding of $\mathfrak{a}_{\mathfrak{B}}(\mathfrak{J})$ into $\mathfrak{a}_{\mathfrak{B}}(\mathfrak{J}')$ if

$$(3) \qquad \Theta_{\mathfrak{B}\cup\{x_1,\dots,x_n\}}\left(\mathfrak{J}\frac{\mathfrak{r}_1\dots\mathfrak{r}_n}{x_1\dots x_n}\right) = \Theta_{\mathfrak{B}\cup\{x_1,\dots,x_n\}}\left(\mathfrak{J}'\frac{\Phi(\mathfrak{r}_1)\dots\Phi(\mathfrak{r}_n)}{x_1\ \dots\ x_n}\right)$$

for arbitrary $n \geqslant 0$, arbitrary sequences $x_1,\dots,x_n$ of individual variables and sequences $\mathfrak{r}_1,\dots,\mathfrak{r}_n$ of elements of ω.

Definition 2. $\mathfrak{a}_{\mathfrak{B}}(\mathfrak{J})$ is said to be elementarily embeddable into $\mathfrak{a}_{\mathfrak{B}}(\mathfrak{J}')$ if there is an elementary embedding from $\mathfrak{a}_{\mathfrak{B}}(\mathfrak{J})$ into $\mathfrak{a}_{\mathfrak{B}}(\mathfrak{J}')$.

Remark.

(a) The relation of elementary embeddability is reflexive and transitive.

Theorem 1. If Φ is an elementary embedding of $\mathfrak{a}_{\mathfrak{B}}(\mathfrak{J})$ into $\mathfrak{a}_{\mathfrak{B}}(\mathfrak{J}')$, then Φ is an embedding of $\mathfrak{a}_{\mathfrak{B}}(\mathfrak{J})$ into $\mathfrak{a}_{\mathfrak{B}}(\mathfrak{J}')$.

Proof. We must show that Φ satisfies the conditions (1) and (2).

To (1): Let $P \in \mathfrak{B}$ be an r-place predicate variable and $\mathfrak{r}_1,\dots,\mathfrak{r}_r \in \omega$. We choose r pairwise distinct individual variables $x_1,\dots,x_r$. Then we have

$$\begin{aligned}\mathfrak{J}(P) \text{ fits } \mathfrak{r}_1,\dots,\mathfrak{r}_r \quad &\text{iff } \operatorname{Mod}\mathfrak{J}\frac{\mathfrak{r}_1\dots\mathfrak{r}_r}{x_1\dots x_r}Px_1\dots x_r\\ &\text{iff } \operatorname{Mod}\mathfrak{J}'\frac{\Phi(\mathfrak{r}_1)\dots\Phi(\mathfrak{r}_r)}{x_1\ \dots\ x_r}Px_1\dots x_r \qquad \text{by (3)}\\ &\text{iff } \mathfrak{J}'(P) \text{ fits } \Phi(\mathfrak{r}_1),\dots,\Phi(\mathfrak{r}_r).\end{aligned}$$

To (2): Let $f \in \mathfrak{B}$ be an r-place function variable and $\mathfrak{r}_1,\dots,\mathfrak{r}_r \in \omega$. We choose $r+1$ pairwise distinct individual variables $x_1,\dots,x_r$, x. Then we have

$$\begin{aligned}\mathfrak{J}(f)(\mathfrak{r}_1,\dots,\mathfrak{r}_r) = \mathfrak{r} \quad &\text{iff } \operatorname{Mod}\mathfrak{J}\frac{\mathfrak{r}_1\dots\mathfrak{r}_r\mathfrak{r}}{x_1\dots x_r x}fx_1\dots x_r = x\\ &\text{iff } \operatorname{Mod}\mathfrak{J}'\frac{\Phi(\mathfrak{r}_1)\dots\Phi(\mathfrak{r}_r)\Phi(\mathfrak{r})}{x_1\ \dots\ x_r\ \ x}fx_1\dots x_r = x \qquad \text{(by (3))}\\ &\text{iff } \mathfrak{J}'(f)(\Phi(\mathfrak{r}_1),\dots,\Phi(\mathfrak{r}_r)) = \Phi(\mathfrak{r}) .\end{aligned}$$

It follows that $\Phi(\mathfrak{J}(f)(\mathfrak{r}_1,\dots,\mathfrak{r}_r)) = \mathfrak{J}'(f)(\Phi(\mathfrak{r}_1),\dots,\Phi(\mathfrak{r}_r))$.

Theorem 2. *If* $\mathfrak{a}_{\mathfrak{B}}(\mathfrak{J})$ *is elementarily embeddable into* $\mathfrak{a}_{\mathfrak{B}}(\mathfrak{J}')$, *then* $\Theta_{\mathfrak{B}}(\mathfrak{J}) = \Theta_{\mathfrak{B}}(\mathfrak{J}')$.

Proof. This is condition (3) with $n = 0$.

Theorem 3. *Every isomorphism* Φ *from* $\mathfrak{a}_{\mathfrak{B}}(\mathfrak{J})$ *into* $\mathfrak{a}_{\mathfrak{B}}(\mathfrak{J}')$ *is an elementary embedding.*

Proof. It follows from Chap. VI, § 2.3, Lemma 3 by induction on n that Φ is also an isomorphism from $\mathfrak{a}_{\mathfrak{B}\cup\{x_1,\dots,x_n\}}\left(\mathfrak{J}^{\mathfrak{r}_1\dots\mathfrak{r}_n}_{x_1\dots x_n}\right)$ into $\mathfrak{a}_{\mathfrak{B}\cup\{x_1,\dots,x_n\}}\left(\mathfrak{J}'^{\Phi(\mathfrak{r}_1)\dots\Phi(\mathfrak{r}_n)}_{x_1\ \dots\ x_n}\right)$, where $\mathfrak{r}_1,\dots,\mathfrak{r}_n$ are elements of ω. Thus these algebras are isomorphic and therefore, by the theorem in 2.2, have the same theory.

3.2 Elementary subalgebras. In analogy to 1.3, we lay down

Definition 3. $\mathfrak{a}_{\mathfrak{B}}(\mathfrak{J})$ is said to be *an elementary subalgebra* of $\mathfrak{a}_{\mathfrak{B}}(\mathfrak{J}')$ - or $\mathfrak{a}_{\mathfrak{B}}(\mathfrak{J}')$ *an elementary extension of* $\mathfrak{a}_{\mathfrak{B}}(\mathfrak{J})$ - if $\omega \subset \omega'$ and

$$(3') \qquad \Theta_{\mathfrak{B}\cup\{x_1,\dots,x_n\}}\left(\mathfrak{J}^{\mathfrak{r}_1\dots\mathfrak{r}_n}_{x_1\dots x_n}\right) = \Theta_{\mathfrak{B}\cup\{x_1,\dots,x_n\}}\left(\mathfrak{J}'^{\mathfrak{r}_1\dots\mathfrak{r}_n}_{x_1\dots x_n}\right)$$

for all $x_1,\dots,x_n$ and all $\mathfrak{r}_1,\dots,\mathfrak{r}_n$.

Remarks.

(b) If $\mathfrak{a}_{\mathfrak{B}}(\mathfrak{J})$ is an elementary subalgebra of $\mathfrak{a}_{\mathfrak{B}}(\mathfrak{J}')$, then $\mathfrak{a}_{\mathfrak{B}}(\mathfrak{J})$ is elementarily embeddable into $\mathfrak{a}_{\mathfrak{B}}(\mathfrak{J}')$.

(c) The relation "is an elementary subalgebra of" is reflexive and transitive.

(d) If $\mathfrak{a}_{\mathfrak{B}}(\mathfrak{J})$ is an elementary subalgebra of $\mathfrak{a}_{\mathfrak{B}}(\mathfrak{J}')$, then $\mathfrak{a}_{\mathfrak{B}}(\mathfrak{J})$ is a subalgebra of $\mathfrak{a}_{\mathfrak{B}}(\mathfrak{J}')$.

3.3 The union of an elementary chain of algebras. If, for each $j = 0, 1, 2, \dots$, $\mathfrak{a}_{\mathfrak{B}}(\mathfrak{J}_j)$ is an elementary subalgebra of $\mathfrak{a}_{\mathfrak{B}}(\mathfrak{J}_{j+1})$, then we say that the $\mathfrak{a}_{\mathfrak{B}}(\mathfrak{J}_j)$ form *an elementary chain of algebras*. Every elementary chain of algebras is a chain of algebras. In 1.4 we defined the union of a chain of algebras, We have the following

Theorem. *The union* $\mathfrak{a}_{\mathfrak{B}}(\mathfrak{J})$ *of an elementary chain of algebras* $\mathfrak{a}_{\mathfrak{B}}(\mathfrak{J}_j)$ *is an elementary extension of each of the* $\mathfrak{a}_{\mathfrak{B}}(\mathfrak{J}_j)$.

Proof. We prove by induction on the structure of α that, for all j, n, $x_1,\dots,x_n$ and all $\mathfrak{r}_1,\dots,\mathfrak{r}_n \in \omega$: If $\mathfrak{B}(\alpha) \subset \mathfrak{B}\cup\{x_1,\dots,x_n\}$, then $\mathrm{Mod}\,\mathfrak{J}^{\mathfrak{r}_1\dots\mathfrak{r}_n}_{x_1\dots x_n}\alpha$ iff $\mathrm{Mod}\,\mathfrak{J}_{j\,x_1\dots x_n}^{\ x_1\dots x_n}\alpha$. In doing this, we make use of the properties of $\mathfrak{a}_{\mathfrak{B}}(\mathfrak{J})$ given in 1.4.

a) If $\alpha \equiv Pt_1 \ldots t_r$ and $\mathfrak{B}(\alpha) \subset \mathfrak{B} \cup \{x_1, \ldots, x_n\}$, then we have

$$\mathrm{Mod}\, \mathfrak{J}_{j\,x_1 \ldots x_n}^{\;\mathfrak{r}_1 \ldots \mathfrak{r}_n} Pt_1 \ldots t_r$$

$$\text{iff } \mathfrak{J}_j(P) \text{ fits } \mathfrak{J}_{j\,x_1 \ldots x_n}^{\;\mathfrak{r}_1 \ldots \mathfrak{r}_n}(t_1), \ldots, \mathfrak{J}_{j\,x_1 \ldots x_n}^{\;\mathfrak{r}_1 \ldots \mathfrak{r}_n}(t_r)$$

$$\text{iff } \mathfrak{J}_j(P) \text{ fits } \mathfrak{J}_{x_1 \ldots x_n}^{\mathfrak{r}_1 \ldots \mathfrak{r}_n}(t_1), \ldots, \mathfrak{J}_{x_1 \ldots x_n}^{\mathfrak{r}_1 \ldots \mathfrak{r}_n}(t_r) \qquad \text{(by 1.3 (f))}$$

$$\text{iff } \mathfrak{J}(P) \text{ fits } \mathfrak{J}_{x_1 \ldots x_n}^{\mathfrak{r}_1 \ldots \mathfrak{r}_n}(t_1), \ldots, \mathfrak{J}_{x_1 \ldots x_n}^{\mathfrak{r}_1 \ldots \mathfrak{r}_n}(t_r) \qquad \text{(by 1.3 (1'))}$$

$$\text{iff } \mathrm{Mod}\, \mathfrak{J}_{x_1 \ldots x_n}^{\mathfrak{r}_1 \ldots \mathfrak{r}_n} Pt_1 \ldots t_r .$$

b) If $\alpha \equiv t_1 = t_2$ and if $\mathfrak{B}(\alpha) \subset \mathfrak{B} \cup \{x_1, \ldots, x_n\}$, then we have

$$\mathrm{Mod}\, \mathfrak{J}_{j\,x_1 \ldots x_n}^{\;\mathfrak{r}_1 \ldots \mathfrak{r}_n}\; t_1 = t_2$$

$$\text{iff } \mathfrak{J}_{j\,x_1 \ldots x_n}^{\;\mathfrak{r}_1 \ldots \mathfrak{r}_n}(t_1) = \mathfrak{J}_{j\,x_1 \ldots x_n}^{\;\mathfrak{r}_1 \ldots \mathfrak{r}_n}(t_2)$$

$$\text{iff } \mathfrak{J}_{x_1 \ldots x_n}^{\mathfrak{r}_1 \ldots \mathfrak{r}_n}(t_1) = \mathfrak{J}_{x_1 \ldots x_n}^{\mathfrak{r}_1 \ldots \mathfrak{r}_n}(t_2) \qquad \text{(by 1.3 (f))}$$

$$\text{iff } \mathrm{Mod}\, \mathfrak{J}_{x_1 \ldots x_n}^{\mathfrak{r}_1 \ldots \mathfrak{r}_n}\; t_1 = t_2 .$$

c) If α is a negation or a conjunction, then the assertion follows immediately from the induction hypothesis.

d) If $\alpha \equiv \bigwedge x\beta$ and $\mathfrak{B}(\alpha) \subset \mathfrak{B} \cup \{x_1, \ldots, x_n\}$, then $\mathfrak{B}(\beta) \subset \mathfrak{B} \cup \{x_1, \ldots, x_n\}$.

d_1) Let $\mathrm{Mod}\, \mathfrak{J}_{x_1 \ldots x_n}^{\mathfrak{r}_1 \ldots \mathfrak{r}_n} \bigwedge x\beta$. Then we have $\mathrm{Mod}\, \mathfrak{J}_{x_1 \ldots x_n x}^{\mathfrak{r}_1 \ldots \mathfrak{r}_n \mathfrak{r}} \beta$ for every $\mathfrak{r} \in \omega$, and hence certainly for every $\mathfrak{r} \in \omega_j$. Now it follows from the induction hypothesis that $\mathrm{Mod}\, \mathfrak{J}_{j x_1 \ldots x_n x}^{\;\mathfrak{r}_1 \ldots \mathfrak{r}_n \mathfrak{r}} \beta$ for every $\mathfrak{r} \in \omega_j$, and hence that $\mathrm{Mod}\, \mathfrak{J}_{j x_1 \ldots x_n}^{\;\mathfrak{r}_1 \ldots \mathfrak{r}_n} \bigwedge x\beta$.

d_2) Let $\mathrm{Mod}\, \mathfrak{J}_{j x_1 \ldots x_n}^{\;\mathfrak{r}_1 \ldots \mathfrak{r}_n} \bigwedge x\beta$. Let $\mathfrak{r}$ be an arbitrary element of ω. Then there is a $k \geqslant j$ such that $\mathfrak{r} \in \omega_k$. Since $\mathfrak{A}_{\mathfrak{B}}(\mathfrak{J}_j)$ form an elementary chain of algebras, it follows from (3') that $\mathrm{Mod}\, \mathfrak{J}_{k x_1 \ldots x_n}^{\;\mathfrak{r}_1 \ldots \mathfrak{r}_n} \bigwedge x\beta$, and hence that $\mathrm{Mod}\, \mathfrak{J}_{k x_1 \ldots x_n x}^{\;\mathfrak{r}_1 \ldots \mathfrak{r}_n \mathfrak{r}} \beta$. Now, by induction hypothesis, we have $\text{Mod} \mathfrak{I}_{x_1 \ldots x_n x}^{\mathfrak{r}_1 \ldots \mathfrak{r}_n \mathfrak{r}} \beta$. Since this holds for every $\mathfrak{r} \in \mathfrak{w}$, it follows that $\text{Mod} \mathfrak{I}_{x_1 \ldots x_n}^{\mathfrak{r}_1 \ldots \mathfrak{r}_n} \bigwedge x \beta$.

Exercises. **1.** Let $\mathfrak{B}$ be a set of variables and $\mathfrak{I}_1$ an interpretation over a finite domain of individuals $\mathfrak{w}_1$. Let $\mathfrak{I}_2$ be an interpretation over the domain of individuals $\mathfrak{w}_2$ and an elementary extension of $\mathfrak{I}_1$ relative to $\mathfrak{B}$. Show that: $\mathfrak{w}_1 = \mathfrak{w}_2$, and every element of $\mathfrak{B}$ has the same image under $\mathfrak{I}_1$ as under $\mathfrak{I}_2$.

2. Let K be a two-place predicate variable. Let $\mathfrak{I}_1$ be an interpretation over the set $\mathfrak{w}_1$ of the natural numbers, and $\mathfrak{I}_2$ an interpretation over $\mathfrak{w}_2 = \mathfrak{w}_1 \cup \{-1\}$. Let $\mathfrak{I}_1(K)$ and $\mathfrak{I}_2(K)$ be the $<$-relation over $\mathfrak{w}_1$ and $\mathfrak{w}_2$ respectively. Show that: $\mathfrak{I}_2$ is an extension of $\mathfrak{I}_1$ relative to $\mathfrak{B} = \{K\}$, but not an elementary extension.

3. Prove the analog of § 1, Exercise 4 for elementary extensions.

4. Let $A_{\mathfrak{B}}(\mathfrak{I})$ be a subalgebra of $A_{\mathfrak{B}}(\mathfrak{I}')$ and an elementary subalgebra of $A_{\mathfrak{B}}(\mathfrak{I}'')$. Moreover, let $A_{\mathfrak{B}}(\mathfrak{I}')$ be an elementary subalgebra of $A_{\mathfrak{B}}(\mathfrak{I}'')$. Then $A_{\mathfrak{B}}(\mathfrak{I})$ is an elementary subalgebra of $A_{\mathfrak{B}}(\mathfrak{I}')$.

5. Let $\mathfrak{B}$ be a set of variables and $\mathfrak{I}_j$ an interpretation over $\mathfrak{w}_j$ $(j = 1, 2)$. Let $\mathfrak{I}_2$ be an extension of $\mathfrak{I}_1$ relative to $\mathfrak{B}$. Show that: $\mathfrak{I}_2$ is an elementary extension of $\mathfrak{I}_1$ relative to $\mathfrak{B}$ if and only if, for all individual variables x, $x_1, \ldots, x_n$, all expressions α such that $\mathfrak{B}(\alpha) \subset \mathfrak{B} \cup \{x, x_1, \ldots, x_n\}$ and all $\mathfrak{r}_1, \ldots, \mathfrak{r}_n \in \mathfrak{w}_1$: If $\text{Mod} \mathfrak{I}_2 {}_{x_1 \ldots x_n}^{\mathfrak{r}_1 \ldots \mathfrak{r}_n} \bigvee x \alpha$, then there is an $\mathfrak{r} \in \mathfrak{w}_1$ such that $\text{Mod} \mathfrak{I}_2 {}_{x_1 \ldots x_n x}^{\mathfrak{r}_1 \ldots \mathfrak{r}_n \mathfrak{r}} \alpha$.

6. Let $\mathfrak{w}$ be a domain of individuals and $\mathfrak{R}$ a two-place predicate over $\mathfrak{w}$. Let $\mathfrak{R}$ be called a dense linear ordering over $\mathfrak{w}$ without first or last element if:

(1) $\mathfrak{R}$ is a (total) ordering $<$ over $\mathfrak{w}$ with no first or last element.

(2) For all $\mathfrak{r}, \mathfrak{y} \in \mathfrak{w}$: If $\mathfrak{R}$ fits $\mathfrak{r}, \mathfrak{y}$, then there is a $\mathfrak{z} \in \mathfrak{w}$ such that $\mathfrak{R}$ fits $\mathfrak{r}, \mathfrak{z}$ and $\mathfrak{z}, \mathfrak{y}$.

Let $\mathfrak{w}_1$ and $\mathfrak{w}_2$ be denumerable domains of individuals and $\mathfrak{R}_1$, $\mathfrak{R}_2$ be dense linear orderings over $\mathfrak{w}_1$, $\mathfrak{w}_2$ respectively without first or last elements. Show that:

(a) There is a 1-1 order-preserving mapping from $\mathfrak{w}_1$ onto $\mathfrak{w}_2$.

(b) Every 1-1 order-preserving mapping of a finite subset of $\mathfrak{w}_1$ into $\mathfrak{w}_2$ can be extended to a 1-1 order-preserving mapping from $\mathfrak{w}_1$ onto $\mathfrak{w}_2$.

(c) Let K be a two-place predicate variable and $\mathfrak{I}_j$ an interpretation over $\mathfrak{w}_j$ such that $\mathfrak{I}_j(K) = \mathfrak{R}_j$ $(j = 1, 2)$. Moreover, let $\mathfrak{w}_1 \subset \mathfrak{w}_2$. Then $\mathfrak{I}_2$ is an elementary extension of $\mathfrak{I}_1$ relative to $\mathfrak{B} = \{K\}$.

§ 4. Three lemmas about elementary embeddings

Lemmas 2' and 3' are modifications of lemmas 2 and 3, and are used in the proof in 5.1 of A. Robinson's satisfiability theorem. Lemma 1 is used to prove lemmas 2 and 3, but it is also interesting in its own right.

4.1 Lemma 1. If $\Theta_{\mathfrak{B}}(\mathfrak{I}) = \Theta_{\mathfrak{B}}(\mathfrak{I}')$ and if, for every $\mathfrak{r} \in \mathfrak{w}$, there are infinitely many variables $x \in \mathfrak{B}$ such that $\mathfrak{I}(x) = \mathfrak{r}$, then $G_{\mathfrak{B}}(\mathfrak{I})$ is elementarily embedabble into $G_{\mathfrak{B}}(\mathfrak{I}')$.

R e m a r k . If $\mathfrak{r} = \mathfrak{I}(x)$, where $x \in \mathfrak{B}$, then we can think of x as a name for the individual $\mathfrak{r}$ in the theory $\Theta_{\mathfrak{B}}(\mathfrak{I})$. The second assumption in Lemma 1 then means that every element of ω has infinitely many names in $\Theta_{\mathfrak{V}}(\mathfrak{I})$. We could weaken this "assumption about names" by requiring only that every element of ω should have at least one name in $\Theta_{\mathfrak{V}}(\mathfrak{I})$. However, we should then have to use extended substitution, whereas here ordinary substitution is sufficient. If every element of ω has at least one name in $\Theta_{\mathfrak{V}}(\mathfrak{I})$, then the set of those atomic expressions (expressions) which are in $\Theta_{\mathfrak{V}}(\mathfrak{I})$ is also called <u>a diagram</u> <u>(a complete diagram)</u> of the algebra $\mathfrak{A}_{\mathfrak{V}}(\mathfrak{I})$. This concept plays an important part in model theory.

P r o o f. We must find a 1-1 mapping Φ from ω into ω' which has the property (3).

Let $\mathfrak{r} \in \omega$. Then there is an $x \in \mathfrak{V}$ such that $\mathfrak{I}(x) = \mathfrak{r}$. We define $\Phi(\mathfrak{r}) = \mathfrak{I}'(x)$ and must show that this definition is independent of the representative x. Let us therefore suppose that $y \in \mathfrak{V}$ and $\mathfrak{I}(y) = \mathfrak{r}$. Since $\mathfrak{I}(x) = \mathfrak{I}(y)$, $\mathrm{Mod}\,\mathfrak{I}\, x{=}y$, and therefore $x{=}y \in \Theta_{\mathfrak{V}}(\mathfrak{I})$. Hence also $x{=}y \in \Theta_{\mathfrak{V}}(\mathfrak{I}')$, i.e. $\mathfrak{I}'(x) = \mathfrak{I}'(y)$.

The mapping Φ which we have just defined is 1-1: Let $\Phi(\mathfrak{r}) = \Phi(\mathfrak{y})$. Then there are elements $x, y \in \mathfrak{V}$ such that $\mathfrak{I}(x) = \mathfrak{r}$, $\mathfrak{I}(y) = \mathfrak{y}$. Now $\Phi(\mathfrak{r}) = \mathfrak{I}'(x)$, $\Phi(\mathfrak{y}) = \mathfrak{I}'(y)$, and hence $\mathfrak{I}'(x) = \mathfrak{I}'(y)$, $\mathrm{Mod}\,\mathfrak{I}'\; x{=}y$, $x{=}y \in \Theta_{\mathfrak{V}}(\mathfrak{I}')$, $x{=}y \in \Theta_{\mathfrak{V}}(\mathfrak{I})$, $\mathfrak{I}(x) = \mathfrak{I}(y)$, $\mathfrak{r} = \mathfrak{y}$.

Now we prove by induction on n that (3) holds for arbitrary $x_1, \ldots, x_n$ and for α such that $\mathfrak{V}(\alpha) \subset \mathfrak{V} \cup \{x_1, \ldots, x_n\}$.

For $n = 0$ we have

$$\begin{aligned} \mathrm{Mod}\,\mathfrak{I}\alpha \quad &\text{iff } \alpha \in \Theta_{\mathfrak{V}}(\mathfrak{I}) \\ &\text{iff } \alpha \in \Theta_{\mathfrak{V}}(\mathfrak{I}') \qquad \text{(assumption)} \\ &\text{iff } \mathrm{Mod}\,\mathfrak{I}'\alpha\,. \end{aligned}$$

Induction step: Let $\mathfrak{r}_1, \ldots, \mathfrak{r}_{n+1} \in \omega$ and α be an expression such that $\mathfrak{V}(\alpha) \subset \mathfrak{V} \cup \{x_1, \ldots, x_{n+1}\}$, for some $x_1, \ldots, x_{n+1}$. By hypothesis, there are infinitely many $x \in \mathfrak{V}$ such that $\mathfrak{I}(x) = \mathfrak{r}_{n+1}$. Thus we can find such an x which does not occur in α and is different from each of $x_1, \ldots, x_n$. Then there is an expression β such that $\mathrm{Subst}\,\alpha x_{n+1} x \beta$ (cf. Chap. II, §5.4, Theorem 6). By Theorem 7 (of Chap. II, §5.4) $\mathfrak{V}(\beta) \subset \mathfrak{V} \cup \{x_1, \ldots, x_n, x\} \subset \mathfrak{V} \cup \{x_1, \ldots, x_n\}$, since $x \in \mathfrak{V}$. Now, since $x_{n+1} = \mathfrak{I}(x) = \mathfrak{I}_{x_1 \ldots x_n}^{\mathfrak{r}_1 \ldots \mathfrak{r}_n}(x)$ and $\Phi(\mathfrak{r}_{n+1}) = \mathfrak{I}'(x) = \mathfrak{I}'^{\,\mathfrak{r}_1 \ldots \mathfrak{r}_n}_{\,x_1 \ldots x_n}(x)$ (because $x \neq x_1, \ldots, x_n$), we have

$$\begin{aligned} \mathrm{Mod}\,\mathfrak{I}_{x_1 \ldots x_n x_{n+1}}^{\mathfrak{r}_1 \ldots \mathfrak{r}_n \mathfrak{r}_{n+1}}\alpha \quad &\text{iff } \mathrm{Mod}\,\mathfrak{I}_{x_1 \ldots x_n}^{\mathfrak{r}_1 \ldots \mathfrak{r}_n}\beta && \text{(substitution theorem)} \\ &\text{iff } \mathrm{Mod}\,\mathfrak{I}'^{\,\Phi(\mathfrak{r}_1) \ldots \Phi(\mathfrak{r}_n)}_{\,x_1 \;\ldots\; x_n}\beta && \text{(induction hypothesis)} \\ &\text{iff } \mathrm{Mod}\,\mathfrak{I}'^{\,\Phi(\mathfrak{r}_1) \ldots \Phi(\mathfrak{r}_n) \Phi(\mathfrak{r}_{n+1})}_{\,x_1 \;\ldots\; x_n \;\; x_{n+1}}\alpha && \text{(substitution theorem).} \end{aligned}$$

<u>4.2 Lemma 2. Let</u> $\Theta_{\mathfrak{B}}(\mathfrak{J}) \cup \{\alpha\}$ <u>be satisfiable</u>, $\mathfrak{B}$ <u>be finite and</u> ω <u>at most denumerable. Then there is an interpretation</u> $\mathfrak{J}'$ <u>over a finite or denumerable domain of individuals</u> ω', <u>such that</u> Mod $\mathfrak{J}'\alpha$ <u>and such that</u> $\mathfrak{A}_{\mathfrak{B}}(\mathfrak{J})$ <u>is elementarily embeddable into</u> $\mathfrak{A}_{\mathfrak{B}}(\mathfrak{J}')$.

Proof. (1) We want to apply Lemma 1. In order to do this, we shall first of all extend $\mathfrak{B}$ by adding to it a set $\mathfrak{W}$ of individual variables, so that every element of ω has infinitely many "names" in $\mathfrak{B} \cup \mathfrak{W}$.

Let $\mathfrak{W}$ be an infinite set of individual variables which is disjoint from $\mathfrak{B} \cup \mathfrak{B}(\alpha)$. For the hypothesis and the assertion of Lemma 2 it is irrelevant how $\mathfrak{J}$ is defined for the variables in $\mathfrak{W}$. Thus we can assume w.l.o.g. that $\mathfrak{J}$ is defined on $\mathfrak{W}$ in such a way that every element of the (at most denumerable) domain of individuals ω appears infinitely often as the image under $\mathfrak{J}$ of a variable in $\mathfrak{W}$.

(2) In (3) we shall show that $\Theta_{\mathfrak{B}\cup\mathfrak{W}}(\mathfrak{J}) \cup \{\alpha\}$ is satisfiable. Thus, by the Löwenheim-Skolem theorem, there is an interpretation $\mathfrak{J}'$ over a finite or denumerable domain ω' such that Mod $\mathfrak{J}'\,\Theta_{\mathfrak{B}\cup\mathfrak{W}}(\mathfrak{J}) \cup \{\alpha\}$. It follows that Mod $\mathfrak{J}'\alpha$. Since Mod $\mathfrak{J}'\,\Theta_{\mathfrak{B}\cup\mathfrak{W}}(\mathfrak{J})$, it also follows from 2.3, Lemma 1 that $\Theta_{\mathfrak{B}\cup\mathfrak{W}}(\mathfrak{J}) = \Theta_{\mathfrak{B}\cup\mathfrak{W}}(\mathfrak{J}')$. Now, by Lemma 1, which we have just proved, $\mathfrak{A}_{\mathfrak{B}\cup\mathfrak{W}}(\mathfrak{J})$ is elementarily embedabble into $\mathfrak{A}_{\mathfrak{B}\cup\mathfrak{W}}(\mathfrak{J}')$. But then, trivially, $\mathfrak{A}_{\mathfrak{B}}(\mathfrak{J})$ is elementarily embeddable into $\mathfrak{A}_{\mathfrak{B}}(\mathfrak{J}')$.

(3) It remains to be shown that $\Theta_{\mathfrak{B}\cup\mathfrak{W}}(\mathfrak{J}) \cup \{\alpha\}$ is satisfiable. If this were not the case, then we should have $\Theta_{\mathfrak{B}\cup\mathfrak{W}}(\mathfrak{J}) \models \neg\alpha$. Then, by 2.3, Lemma 2, there would be a $\delta \in \Theta_{\mathfrak{B}\cup\mathfrak{W}}(\mathfrak{J})$ such that $\delta \models \neg\alpha$. Let $\bigvee\delta$ stand for an expression which is obtained from δ by applying the existential quantifier to all variables from $\mathfrak{W}$ which occur free in δ ($\bigvee\delta \equiv \delta$ if none of the variables in $\mathfrak{W}$ occurs free in δ). It follows from Mod $\mathfrak{J}\delta$, $\delta \models \bigvee\delta$, $\mathfrak{B}(\bigvee\delta) \subset \mathfrak{B}$ that $\bigvee\delta \in \Theta_{\mathfrak{B}}(\mathfrak{J})$. Thus, by hypothesis, $\{\bigvee\delta, \alpha\}$ is satisfiable. But this contradicts the fact that $\bigvee\delta \models \neg\alpha$, which follows from $\delta \models \neg\alpha$ by (possibly repeated) application of the rule (P_x'), making use of the fact that $\mathfrak{W}$ and $\mathfrak{B}(\alpha)$ are disjoint.

<u>Lemma 2'. Let</u> $\Theta_{\mathfrak{B}}(\mathfrak{J}) \cup \{\alpha\}$ <u>be satisfiable. Let</u> $\mathfrak{B}$ <u>be finite and</u> ω <u>finite or denumerable. Then there is an interpretation</u> $\mathfrak{J}'$ <u>over a finite or denumerable domain of individuals</u> ω', <u>such that</u> Mod $\mathfrak{J}'\alpha$ <u>and such that</u> $\mathfrak{A}_{\mathfrak{B}}(\mathfrak{J})$ <u>is an elementary subalgebra of</u> $\mathfrak{A}_{\mathfrak{B}}(\mathfrak{J}')$.

Proof. By Lemma 2, there is an interpretation $\mathfrak{J}^*$ over a finite or denumerable domain of individuals ω^*, such that Mod $\mathfrak{J}^*\alpha$ and such that $\mathfrak{A}_{\mathfrak{B}}(\mathfrak{J})$ is elementarily embeddable into $\mathfrak{A}_{\mathfrak{B}}(\mathfrak{J}^*)$. Let Φ be one of these elementary embeddings, and let $\omega_0^* = \Phi(\omega)$, $\omega_1^* = \omega^* - \omega_0^*$. Let Φ_1 be a 1-1 mapping from ω_1^* onto a set ω_1 which is disjoint from ω. Let $\omega' = \omega \cup \omega_1$. ω^* can be mapped onto ω' in a 1-1 way by a mapping ψ which is identical with Φ^{-1} on ω_0^* and with Φ_1 on ω_1^*. By Lemma 1 of Chap. VI, §2.2, there is an interpretation $\mathfrak{J}'$ over ω' such that ψ is an isomorphism from $\mathfrak{A}_{\mathfrak{B}\cup\mathfrak{B}(\alpha)}(\mathfrak{J}^*)$ into $\mathfrak{A}_{\mathfrak{B}\cup\mathfrak{B}(\alpha)}(\mathfrak{J}')$. Then $\Psi \circ \Phi$ is an elmentary embedding of $\mathfrak{A}_{\mathfrak{B}}(\mathfrak{J})$ into $\mathfrak{A}_{\mathfrak{B}}(\mathfrak{J}')$. By construction, $\Psi \circ \Phi$ is the identity, and hence $\mathfrak{G}_{\mathfrak{B}}(\mathfrak{I})$ is an elementary subalgebra of $\mathfrak{G}_{\mathfrak{B}}(\mathfrak{I}')$. Mod $\mathfrak{I}'\alpha$ follows from the fact that Mod $\mathfrak{I}^*\alpha$ and that $\mathfrak{G}_{\mathfrak{B}\cup\mathfrak{B}(\alpha)}(\mathfrak{I}^*)$ and $\mathfrak{G}_{\mathfrak{B}\cup\mathfrak{B}(\alpha)}(\mathfrak{I}')$ are isomorphic.

4.3 Lemma 3. Let $\mathfrak{G}_{\mathfrak{B}}(\mathfrak{I})$ be an elementary subalgebra of $\mathfrak{G}_{\mathfrak{B}}(\mathfrak{I}')$, $\mathfrak{B}$ be finite and ω finite or denumerable. Then there is an interpretation $\mathfrak{I}''$ over a finite or denumerable domain of individuals ω'' and (1) an elementary embedding Φ' from $\mathfrak{G}_{\mathfrak{B}}(\mathfrak{I}')$ into $\mathfrak{G}_{\mathfrak{B}}(\mathfrak{I}'')$ and (2) an elementary embedding Φ from $\mathfrak{G}_{\mathfrak{B}\cup\mathfrak{B}(\alpha)}(\mathfrak{I})$ into $\mathfrak{G}_{\mathfrak{B}\cup\mathfrak{B}(\alpha)}(\mathfrak{I}'')$. Moreover, Φ and Φ can be chosen in such a way that Φ is the restriction of Φ' to ω.

Proof. (1) We proceed in a way similar to the proof of Lemma 2. Let $\mathfrak{W}'$ be an infinite set of individual variables which is disjoint from $\mathfrak{B}\cup\mathfrak{B}(\alpha)$. We can assume that $\mathfrak{I}'$ is defined on $\mathfrak{W}'$ in such a way that every element of ω' appears infinitely often as the image under $\mathfrak{I}'$ of a variable in $\mathfrak{W}'$.

(2) Let $\mathfrak{W}$ be the set of those elements $x \in \mathfrak{W}'$ such that $\mathfrak{I}'(x) \in \omega$ (note that $\omega \subset \omega'$). We can assume that $\mathfrak{I}$ is defined on $\mathfrak{W}$ in such a way that $\mathfrak{I}(x) = \mathfrak{I}'(x)$ for each $x \in \mathfrak{W}$. Then every element of ω appears infinitely often as the image under $\mathfrak{I}$ of a variable in $\mathfrak{W}$ (for, if $\mathfrak{r} \in \omega$, $\mathfrak{r}$ appears infinitely often as the image under $\mathfrak{I}'$ of a variable in $\mathfrak{W}'$; but every such variable lies in $\mathfrak{W}$, and $\mathfrak{I}$ and $\mathfrak{I}'$ are identical on W).

(3) In (4) we shall show that $\Theta_{\mathfrak{B}\cup\mathfrak{B}(\alpha)\cup\mathfrak{W}}(\mathfrak{I}) \cup \Theta_{\mathfrak{B}\cup\mathfrak{W}'}(\mathfrak{I}')$ is satisfiable. Thus, by the Löwenheim-Skolem theorem, there is an interpretation $\mathfrak{I}''$ over a finite or denumerable domain ω'', such that Mod $\mathfrak{I}''\Theta_{\mathfrak{B}\cup\mathfrak{B}(\alpha)\cup\mathfrak{W}}(\mathfrak{I}) \cup \Theta_{\mathfrak{B}\cup\mathfrak{W}'}(\mathfrak{I}')$. It follows that $\Theta_{\mathfrak{B}\cup\mathfrak{B}(\alpha)\cup\mathfrak{W}}(\mathfrak{I}) = \Theta_{\mathfrak{B}\cup\mathfrak{B}(\alpha)\cup\mathfrak{W}}(\mathfrak{I}'')$ and $\Theta_{\mathfrak{B}\cup\mathfrak{W}'}(\mathfrak{I}') = \Theta_{\mathfrak{B}\cup\mathfrak{W}'}(\mathfrak{I}'')$. Now Lemma 1 shows that $\mathfrak{G}_{\mathfrak{B}\cup\mathfrak{B}(\alpha)}(\mathfrak{I})$ is embeddable into $\mathfrak{G}_{\mathfrak{B}\cup\mathfrak{B}(\alpha)}(\mathfrak{I}'')$ and $\mathfrak{G}_{\mathfrak{B}}(\mathfrak{I}')$ into $\mathfrak{G}_{\mathfrak{B}}(\mathfrak{I}'')$. In Lemma 1, the embedding Φ' of $\mathfrak{G}_{\mathfrak{B}}(\mathfrak{I}')$ into $\mathfrak{G}_{\mathfrak{B}}(\mathfrak{I}'')$ was so constructed that $\Phi'(\mathfrak{r}) = \mathfrak{I}''(x)$, where x is any variable such that $x \in \mathfrak{W}'$ and $\mathfrak{I}'(x) = \mathfrak{r}$. Correspondingly, the embedding Φ of $\mathfrak{G}_{\mathfrak{B}\cup\mathfrak{B}(\alpha)}(\mathfrak{I})$ into $\mathfrak{G}_{\mathfrak{B}\cup\mathfrak{B}(\alpha)}(\mathfrak{I}'')$ was so constructed that $\Phi(\mathfrak{r}) = \mathfrak{I}''(x)$, where x is any variable such that $x \in \mathfrak{W}$ and $\mathfrak{I}'(x) = \mathfrak{r}$. Now suppose that $\mathfrak{r} \in \omega$, $x \in \mathfrak{W}$ and $\mathfrak{I}(x) = \mathfrak{r}$, Then $x \in \mathfrak{W}'$ and $\mathfrak{I}'(x) = \mathfrak{r}$, and hence $\Phi(\mathfrak{r}) = \Phi'(\mathfrak{r})$. Thus, all the assertions have been proved.

(4) It remains to be shown that $\Theta_{\mathfrak{B}\cup\mathfrak{B}(\alpha)\cup\mathfrak{W}}(\mathfrak{I}) \cup \Theta_{\mathfrak{B}\cup\mathfrak{W}'}(\mathfrak{I}')$ is satisfiable. If this were not the case, then, by 2.3, Lemma 3, there would be expressions $\delta \in \Theta_{\mathfrak{B}\cup\mathfrak{B}(\alpha)\cup\mathfrak{W}}(\mathfrak{I})$ and $\delta' \in \Theta_{\mathfrak{B}\cup\mathfrak{W}'}(\mathfrak{I}')$ such that $\{\delta, \delta'\}$ is not satisfiable. Let $\bigvee\delta'$ be obtained from δ' by applying the existential quantifier to all the variables from $\mathfrak{W}' - \mathfrak{W}$ which occur free in δ'. If there were a model of $\{\delta, \bigvee\delta'\}$, then it could be altered to obtain a model of $\{\delta, \delta'\}$, since the variables which are quantified in obtaining $\bigvee\delta'$ from δ' do not occur free in δ. Hence $\{\delta, \bigvee\delta'\}$ is not satisfiable. On the other hand, Mod $\mathfrak{I}\delta$ and Mod $\mathfrak{I}\bigvee\delta'$. The former is trivial. The latter follows from the fact that Mod $\mathfrak{I}'\delta'$, and hence that Mod $\mathfrak{I}'\bigvee\delta'$, as follows: Let $x_1, \ldots, x_n$ be those variables in $\mathfrak{W}$ which occur free in $\bigvee\delta'$. Then we have $\mathfrak{B}(\bigvee\delta') \subset \mathfrak{B} \cup \{x_1, \ldots, x_n\}$. For the following, note that $\mathfrak{I}'(x_j) \in \omega$ and that

$\mathfrak{J}(x_j) = \mathfrak{J}'(x_j)$, since $x_j \in \mathfrak{W}$; and, moreover, that, by hypothesis, $\mathfrak{A}_{\mathfrak{B}}(\mathfrak{J})$ is an elementary subalgebra of $\mathfrak{A}_{\mathfrak{B}}(\mathfrak{J}')$. We have

$$\begin{aligned}
\mathrm{Mod}\,\mathfrak{J}' \vee \delta' \quad &\text{iff } \mathrm{Mod}\,\mathfrak{J}'\,\frac{\mathfrak{J}'(x_1)\ldots\mathfrak{J}'(x_n)}{x_1\ \ldots\ x_n} \vee \delta' \\
&\text{iff } \mathrm{Mod}\,\mathfrak{J}\,\frac{\mathfrak{J}'(x_1)\ldots\mathfrak{J}'(x_n)}{x_1\ \ldots\ x_n} \vee \delta' \\
&\text{iff } \mathrm{Mod}\,\mathfrak{J}\,\frac{\mathfrak{J}(x_1)\ldots\mathfrak{J}(x_n)}{x_1\ \ldots\ x_n} \vee \delta' \\
&\text{iff } \mathrm{Mod}\,\mathfrak{J} \vee \delta' ,
\end{aligned}$$

and hence $\mathrm{Mod}\,\mathfrak{J} \vee \delta'$, q.e.d.

Lemma 3'. Let $\mathfrak{A}_{\mathfrak{B}}(\mathfrak{J})$ be an elementary subalgebra of $\mathfrak{A}_{\mathfrak{B}}(\mathfrak{J}')$, $\mathfrak{B}$ be finite and ω' finite or denumerable. Moreover, suppose that $\mathrm{Mod}\,\mathfrak{J}\alpha$. Then there is an interpretation $\mathfrak{J}''$ over a finite or denumerable domain ω'', such that (1) $\mathfrak{A}_{\mathfrak{B}}(\mathfrak{J}')$ is an elementary subalgebra of $\mathfrak{A}_{\mathfrak{B}}(\mathfrak{J}'')$ and (2) $\mathfrak{A}_{\mathfrak{B}\cup\mathfrak{B}(\alpha)}(\mathfrak{J})$ is an elementary subalgebra of $\mathfrak{A}_{\mathfrak{B}\cup\mathfrak{B}(\alpha)}(\mathfrak{J}'')$.

Proof. By Lemma 3, there is an interpretation $\mathfrak{J}^*$ over a finite or denumerable domain ω^* and (1) an elementary embedding Φ' of $\mathfrak{A}_{\mathfrak{B}}(\mathfrak{J}')$ into $\mathfrak{A}_{\mathfrak{B}}(\mathfrak{J}^*)$ and (2) an elementary embedding Φ of $\mathfrak{A}_{\mathfrak{B}\cup\mathfrak{B}(\alpha)}(\mathfrak{J})$ into $\mathfrak{A}_{\mathfrak{B}\cup\mathfrak{B}(\alpha)}(\mathfrak{J}^*)$ such that Φ is the restriction of Φ' to ω. Now, as in the proof of Lemma 2', we replace the interpretation $\mathfrak{J}^*$ by an interpretation $\mathfrak{J}''$ such that $\mathfrak{A}_{\mathfrak{B}\cup\mathfrak{B}(\alpha)}(\mathfrak{J})$ is isomorphic to $\mathfrak{A}_{\mathfrak{B}\cup\mathfrak{B}(\alpha)}(\mathfrak{J}'')$, where the isomorphism Ψ is chosen in such a way that $\Psi\circ\Phi'$ is the identity on ω'. Then $\Psi\circ\Phi$ is the identity on ω. The other assertions can be proved in a way similar to the proof of Lemma 2'.

Exercises. 1. Let $\Theta_{\mathfrak{B}}(\mathfrak{J})$ and $\Theta_{\mathfrak{B}}(\mathfrak{J}')$ contain the same quantifier-free expressions. For every $\mathfrak{r} \in \omega$ let there be infinitely many $x \in \mathfrak{B}$ such that $\mathfrak{J}(x) = \mathfrak{r}$. Then $\mathfrak{A}_{\mathfrak{B}}(\mathfrak{J})$ is embeddable into $\mathfrak{A}_{\mathfrak{B}}(\mathfrak{J}')$.

2. Show that Lemma 2, 2', 3, 3' also hold if $\mathfrak{B}$ is infinite.

3. Let $\mathfrak{A}_{\mathfrak{B}}(\mathfrak{J})$ have a denumerable underlying set. Show that there is a proper elementary extension of $\mathfrak{A}_{\mathfrak{B}}(\mathfrak{J})$.

§5. The theorems of A. Robinson and Craig

5.1 A. Robinson's satisfiability theorem. Let the sets of expressions $\Theta_{\mathfrak{B}}(\mathfrak{J}) \cup \{\alpha\}$ and $\Theta_{\mathfrak{B}}(\mathfrak{J}) \cup \{\alpha'\}$ be satisfiable, $\mathfrak{B}$ finite and $\mathfrak{B}(\alpha) \cap \mathfrak{B}(\alpha') \subset \mathfrak{B}$. Then $\{\alpha, \alpha'\}$ is satisfiable.

Proof. (1) For $j = 0, 1, 2, \ldots$ we put $\alpha_j \equiv \alpha$ or $\alpha_j \equiv \alpha'$ according to whether j is even or odd. In (2) we shall show that, for every j, there is an interpretation $\mathfrak{J}_j$ over a finite or denumerable domain of individuals ω_j, such that $\mathrm{Mod}\,\mathfrak{J}_j\alpha_j$ and $\mathfrak{A}_{\mathfrak{B}}(\mathfrak{J}_j)$ is an elementary subalgebra of $\mathfrak{A}_{\mathfrak{B}}(\mathfrak{I}_{j+1})$ and $\mathfrak{A}_{\mathfrak{B}\cup\mathfrak{B}(\alpha_j)}(\mathfrak{I}_j)$ is an elementary subalgebra of $\mathfrak{A}_{\mathfrak{B}\cup\mathfrak{B}(\alpha_{j+2})}(\mathfrak{I}_{j+2})$. Thus, we have here two elementary chains of algebras, namely

$$\mathfrak{A}_{\mathfrak{B}\cup\mathfrak{B}(\alpha)}(\mathfrak{I}_0),\ \mathfrak{A}_{\mathfrak{B}\cup\mathfrak{B}(\alpha)}(\mathfrak{I}_2),\ \mathfrak{A}_{\mathfrak{B}\cup\mathfrak{B}(\alpha)}(\mathfrak{I}_4),\ldots$$

and

$$\mathfrak{A}_{\mathfrak{B}\cup\mathfrak{B}(\alpha')}(\mathfrak{I}_1),\ \mathfrak{A}_{\mathfrak{B}\cup\mathfrak{B}(\alpha')}(\mathfrak{I}_3),\ \mathfrak{A}_{\mathfrak{B}\cup\mathfrak{B}(\alpha')}(\mathfrak{I}_5),\ldots .$$

As in 3.3, we form the unions $\mathfrak{A}_{\mathfrak{B}\cup\mathfrak{B}(\alpha)}(\mathfrak{I}^0)$ and $\mathfrak{A}_{\mathfrak{B}\cup\mathfrak{B}(\alpha')}(\mathfrak{I}^1)$ of these two chains. Now, for each j, $\omega_j \subset \omega_{j+1}$, since $\mathfrak{A}_{\mathfrak{B}}(\mathfrak{I}_j)$ is a subalgebra of $\mathfrak{A}_{\mathfrak{B}}(\mathfrak{I}_{j+1})$. Hence the domain of individuals $\omega_0 \cup \omega_2 \cup \omega_4 \cup \ldots$ belonging to $\mathfrak{I}^0$ is identical with the domain of individuals $\omega_1 \cup \omega_3 \cup \omega_5 \cup \ldots$ belonging to $\mathfrak{I}^1$. Thus $\mathfrak{I}^0$ and $\mathfrak{I}^1$ are interpretations with the same domain of individuals $\omega^* = \omega_0 \cup \omega_1 \cup \omega_2 \cup \ldots$. Now we introduce an interpretation $\mathfrak{I}^*$ over ω^* by requiring that $\mathfrak{I}^*_{\mathfrak{B}\cup\mathfrak{B}(\alpha)} = \mathfrak{I}^0$ and $\mathfrak{I}^*_{\mathfrak{B}\cup\mathfrak{B}(\alpha')}\mathfrak{I}^1$. Thus, $\mathfrak{I}^*$ is fixed only for the variables in $\mathfrak{B}\cup\mathfrak{B}(\alpha)\cup\mathfrak{B}(\alpha')$; but this is sufficient for the following. However, we need to show that our two requirements do not contradict each other. A variable which lies in both $\mathfrak{B}\cup\mathfrak{B}(\alpha)$ and $\mathfrak{B}\cup\mathfrak{B}(\alpha')$ must lie in $\mathfrak{B}$, since, by our hypothesis, $\mathfrak{B}(\alpha)\cap\mathfrak{B}(\alpha')\subset\mathfrak{B}$. Thus, we must show that $\mathfrak{I}^0$ and $\mathfrak{I}^1$ are identical for the variables in $\mathfrak{B}$. We confine ourselves to the case of a predicate variable $P \in \mathfrak{B}$, since the proof for a function variable can be carried out in just the same way. Let P be r-place and $\mathfrak{x}_1,\ldots,\mathfrak{x}_r$ be arbitrary elements of ω^*. Then there is a j such that $\mathfrak{x}_1,\ldots,\mathfrak{x}_r \in \omega_{2j}$, and we have

$\mathfrak{I}^0(P)$ fits $\mathfrak{x}_1,\ldots,\mathfrak{x}_r$

iff $\mathfrak{I}_{2j}(P)$ fits $\mathfrak{x}_1,\ldots,\mathfrak{x}_r$ ($\mathfrak{A}_{\mathfrak{B}}(\mathfrak{I}_{2j})$ is a subalgebra of $\mathfrak{A}_{\mathfrak{B}}(\mathfrak{I}^0)$)

iff $\mathfrak{I}_{2j+1}(P)$ fits $\mathfrak{x}_1,\ldots,\mathfrak{x}_r$ ($\mathfrak{A}_{\mathfrak{B}}(\mathfrak{I}_{2j})$ is a subalgebra of $\mathfrak{A}_{\mathfrak{B}}(\mathfrak{I}_{2j+1})$)

iff $\mathfrak{I}^1(P)$ fits $\mathfrak{x}_1,\ldots,\mathfrak{x}_r$ ($\mathfrak{A}_{\mathfrak{B}}(\mathfrak{I}_{2j+1})$ is a subalgebra of $\mathfrak{A}_{\mathfrak{B}}(\mathfrak{I}^1)$) .

Since $\mathfrak{A}_{\mathfrak{B}\cup\mathfrak{B}(\alpha)}(\mathfrak{I}_{2j})$ is an elementary subalgebra of $\mathfrak{A}_{\mathfrak{B}\cup\mathfrak{B}(\alpha)}(\mathfrak{I}^0)$, it follows from $\mathrm{Mod}\,\mathfrak{I}_{2j}\alpha$ that $\mathrm{Mod}\,\mathfrak{I}^0\alpha$ and hence, by the coincidence theorem, that $\mathrm{Mod}\,\mathfrak{I}^*\alpha$. Analogously, it follows from $\mathrm{Mod}\,\mathfrak{I}_{2j+1}\alpha'$ that $\mathrm{Mod}\,\mathfrak{I}^*\alpha'$. Thus, $\{\alpha, \alpha'\}$ is satisfiable.

(2) We define interpretations $\mathfrak{I}_j$ by induction on j and show simultaneously that they have the properties given in (1):

(a) By hypothesis, $\Theta_{\mathfrak{B}}(\mathfrak{I}) \cup \{\alpha\}$ is satisfiable. Let $\mathfrak{I}_0$ be an interpretation over a denumerable domain of individuals ω_0, such that $\mathrm{Mod}\,\mathfrak{I}_0\Theta_{\mathfrak{B}}(\mathfrak{I}) \cup \{\alpha\}$.

(b) <u>Definition of</u> $\mathfrak{I}_1$: By hypothesis, $\Theta_{\mathfrak{B}}(\mathfrak{I}) \cup \{\alpha'\}$ is satisfiable. Since $\mathrm{Mod}\,\mathfrak{I}_0\Theta_{\mathfrak{B}}(\mathfrak{I})$ and by 2.3, Lemma 1, $\Theta_{\mathfrak{B}}(\mathfrak{I}) = \Theta_{\mathfrak{B}}(\mathfrak{I}_0)$. Hence, $\Theta_{\mathfrak{B}}(\mathfrak{I}_0) \cup \{\alpha'\}$ is satisfiable. We can now apply 4.2, Lemma 2'. Thus, there is an interpretation $\mathfrak{I}_1$ over a finite or denumerable domain of individuals ω_1, such that $\mathrm{Mod}\,\mathfrak{I}_1\alpha'$ and such that $\mathfrak{A}_{\mathfrak{B}}(\mathfrak{I}_0)$ is an elementary subalgebra of $\mathfrak{A}_{\mathfrak{B}}(\mathfrak{I}_1)$.

(c) <u>Definition of</u> $\mathfrak{J}_{j+2}$ (from $\mathfrak{J}_j$ and $\mathfrak{J}_{j+1}$): By induction hypothesis, $\operatorname{Mod}\mathfrak{J}_j\alpha_j$, ω_{j+1} is finite or denumerable and $\mathfrak{A}_{\mathfrak{V}}(\mathfrak{J}_j)$ is an elementary subalgebra of $\mathfrak{A}_{\mathfrak{V}}(\mathfrak{J}_{j+1})$. Now we can apply 4.3, Lemma 3'; thus, there is an interpretation $\mathfrak{J}_{j+2}$ over a finite or denumerable domain of individuals ω_{j+2} such that $\mathfrak{A}_{\mathfrak{V}}(\mathfrak{J}_{j+1})$ is an elementary subalgebra of $\mathfrak{A}_{\mathfrak{V}}(\mathfrak{J}_{j+2})$ and $\mathfrak{A}_{\mathfrak{V}\cup\mathfrak{V}(\alpha_j)}(\mathfrak{J}_j)$ is an elementary subalgebra of $\mathfrak{A}_{\mathfrak{V}\cup\mathfrak{V}(\alpha_j)}(\mathfrak{J}_{j+2}) = \mathfrak{A}_{\mathfrak{V}\cup\mathfrak{V}(\alpha_{j+2})}(\mathfrak{J}_{j+2})$.

<u>5.2 Craig's interpolation lemma. Let</u> α, γ <u>be expressions such that</u> $\alpha \models \gamma$. <u>Then there is an expression</u> β <u>such that</u> $\mathfrak{V}(\beta) \subset \mathfrak{V}(\alpha) \cap \mathfrak{V}(\gamma)$ <u>and</u> $\alpha \models \beta$ <u>and</u> $\beta \models \gamma$.

Proof. (1) Let $\mathfrak{V} = \mathfrak{V}(\alpha) \cap \mathfrak{V}(\gamma)$. $\mathfrak{V}$ is finite. Let $\mathfrak{C}$ be the set of those expressions ρ such that $\mathfrak{V}(\rho) \subset \mathfrak{V}$ and $\alpha \models \rho$. $\mathfrak{C}$ is nonempty since, for example, $\bigwedge x\, x{=}x \in \mathfrak{V}$.

(2) <u>We want to show that</u> $\mathfrak{C} \cup \{\neg\gamma\}$ <u>is not satisfiable</u>. Otherwise, there would be an interpretation $\mathfrak{J}$ such that $\operatorname{Mod}\mathfrak{J}\mathfrak{C}\cup\{\neg\gamma\}$. (a) $\Theta_{\mathfrak{V}}(\mathfrak{J}) \cup \{\alpha\}$ is satisfiable. For, if it were not, we should have $\Theta_{\mathfrak{V}}(\mathfrak{J}) \models \neg\alpha$. Then, by 2.3, Lemma 2, there would be a $\delta \in \Theta_{\mathfrak{V}}(\mathfrak{J})$ such that $\delta \models \neg\alpha$, i.e. $\alpha \models \neg\alpha$. It would follow that $\neg\delta \in \mathfrak{C}$ and hence that $\operatorname{Mod}\mathfrak{J}\neg\delta$ (since $\operatorname{Mod}\mathfrak{J}\mathfrak{C}$), which would contradict $\delta \in \Theta_{\mathfrak{V}}(\mathfrak{J})$. (b) $\Theta_{\mathfrak{V}}(\mathfrak{J}) \cup \{\neg\gamma\}$ is satisfiable, since $\mathfrak{J}$ is a model of this set of expressions. Moreover, (c) $\mathfrak{V}$ is finite and (d) $\mathfrak{V}(\alpha) \cap \mathfrak{V}(\neg\gamma) \subset \mathfrak{V}$. Because of (a), ..., (d) we can apply Robinson's satisfiability theorem and conclude that $\{\alpha, \neg\gamma\}$ is satisfiable. However, this contradicts $\alpha \models \gamma$.
(3) We have just seen that $\mathfrak{C} \cup \{\neg\gamma\}$ is not satisfiable. Hence, $\mathfrak{C} \models \gamma$. Now the conjunction of a finite number of elements of $\mathfrak{C}$ is also an element of $\mathfrak{C}$; hence there is an element β of $\mathfrak{C}$ such that $\beta \models \gamma$. β clearly has all the required properties.

<u>Exercises. 1.</u> Using A. Robinson's satisfiability theorem and §4, Exercise 2, prove the following, more general form of the satisfiability theorem (also due to A. Robinson):

Let $\mathfrak{V}$ be a set of variables. Let $\mathfrak{M}, \mathfrak{M}_0, \mathfrak{M}_1$ be sets of expressions which satisfy the following conditions:

(1) For all $\alpha \in \mathfrak{M}$, $\mathfrak{V}(\alpha) \subset \mathfrak{V}$.

(2) For all α such that $\mathfrak{V}(\alpha) \subset \mathfrak{V}$, $\mathfrak{M} \models \alpha$ or $\mathfrak{M} \models \neg\alpha$.

(3) For all $\alpha_0 \in \mathfrak{M}_0$ and $\alpha_1 \in \mathfrak{M}_1$, $\mathfrak{V}(\alpha_0) \cap \mathfrak{V}(\alpha_1) \subset \mathfrak{V}$.

(4) $\mathfrak{M} \cup \mathfrak{M}_j$ is satisfiable $(j = 1, 2)$.

Then $\mathfrak{M}_0 \cup \mathfrak{M}_1$ is satisfiable.

<u>2.</u> If $\mathfrak{M}_1 \cup \mathfrak{M}_2$ is not satisfiable, then there is an α such that $\mathfrak{V}(\alpha) \subset \mathfrak{V}(\mathfrak{M}_1) \cap \mathfrak{V}(\mathfrak{M}_2)$ and $\mathfrak{M}_1 \models \alpha$, $\mathfrak{M}_2 \models \neg\alpha$.

<u>3.</u> Let P and Q be two distinct two-place predicate variables, f a one-place function variable and x, y, z pairwise distinct individual variables.

(a) Show that: $\bigwedge x\, Pfyx \models \bigvee x \bigvee y \bigwedge z((Pxy \wedge \neg Qxy) \vee Qxz)$.

(b) Find an interpolation formula.

<u>4.</u> Using Craig's interpolation lemma, prove A. Robinson's satisfiability theorem. (Hint: Prove the theorem indirectly.)

§ 6. Beth's definability theorem

6.1 The notion of definability. It often happens that, in a mathematical axiom system $\mathfrak{M}$, some of the primitive notions of $\mathfrak{M}$ are "superfluous" in the sense that they can be defined by means of the other primitive notions of $\mathfrak{M}$. This situation gives rise to the following definitions:

Definition 1. α defines the r-place predicate variable P relative to $\mathfrak{M}$ by means of the variables $x_1, \ldots, x_r$ if and only if

(a) $x_1, \ldots, x_r$ are pairwise distinct,

(b) $\mathfrak{B}(\alpha) \subset \mathfrak{B}(\mathfrak{M}) \cup \{x_1, \ldots, x_r\}$,

(c) $P \notin \mathfrak{B}(\alpha)$,

(d) $\mathfrak{M} \models \bigwedge x_1 \ldots \bigwedge x_r (Px_1 \ldots x_r \leftrightarrow \alpha)$.

P is said to be definable relative to $\mathfrak{M}$ if there are variables $x_1, \ldots, x_r$ and an expression α which defines P relative to $\mathfrak{M}$ by means of the variables $x_1, \ldots, x_r$.

Definition 2. α defines the r-place function variable f relative to $\mathfrak{M}$ by means of the variables $x_1, \ldots, x_r$, x if and only if

(a') $x_1, \ldots, x_r$, x are pairwise distinct,

(b') $\mathfrak{B}(\alpha) \subset \mathfrak{B}(\mathfrak{M}) \cup \{x_1, \ldots, x_r, x\}$,

(c') $f \notin \mathfrak{B}(\alpha)$,

(d') $\mathfrak{M} \models \bigwedge x_1 \ldots \bigwedge x_r \bigwedge x (fx_1 \ldots x_r = x \leftrightarrow \alpha)$.

f is said to be definable relative to $\mathfrak{M}$ if there are variables $x_1, \ldots x_r$, x and an expression α which defines f relative to $\mathfrak{M}$ by means of the variables $x_1, \ldots, x_r$, x.

6.2 Beth's definability theorem for a predicate variable. In the following we shall assume (for the sake of simplicity) that $\mathfrak{B}(\mathfrak{M})$ is finite. Let P be an r-place predicate variable. Let P' be r-place, $P' \neq P$ and $P' \notin \mathfrak{B}(\mathfrak{M})$. We can assign to every expression α an expression α' such that Subst αP P' α'. (This form of substitution can be defined analogously to the original form - substitution for individual variables - and there are no complications, since predicate variables are not quantified. α' is obtained from α by replacing the variable P by P' everywhere in α.) Let $\mathfrak{M}'$ be the set of those μ' such that $\mu \in \mathfrak{M}$. Under these assumptions we assert that the three following statements are equivalent:

(*) P is definable relative to $\mathfrak{M}$.

(**) If two models of $\mathfrak{M}$ with the same domain of individuals are identical on $\mathfrak{B}(\mathfrak{M}) - P$, then they also have the same value at P.

(***) $\mathfrak{M} \cup \mathfrak{M}' \models \bigwedge x_1 \dots \bigwedge x_r(Px_1 \dots x_r \leftrightarrow P'x_1 \dots x_r)$, where $x_1, \dots, x_r$ are pairwise distinct variables.

(1) If (*), then (**). Let α be a definition of P by menas of the variables $x_1, \dots, x_r$. Let $\mathfrak{I}$, $\mathfrak{I}'$ be interpretations over the same domain of individuals ω such that $\mathrm{Mod}\,\mathfrak{I}\mathfrak{M}$, $\mathrm{Mod}\,\mathfrak{I}'\mathfrak{M}$ and $\mathfrak{I} \underset{\mathfrak{B}(\mathfrak{M})-P}{=} \mathfrak{I}'$. By (d), $\mathfrak{I}$ and $\mathfrak{I}'$ are models of $\bigwedge x_1 \dots \bigwedge x_r(Px_1 \dots x_r \leftrightarrow \alpha)$. Now if $\mathfrak{r}_1, \dots, \mathfrak{r}_r$ are arbitrary elements of ω, then we have

$$\mathfrak{I}(P) \text{ fits } \mathfrak{r}_1, \dots, \mathfrak{r}_r$$

$$\text{iff } \mathrm{Mod}\,\mathfrak{I}^{\mathfrak{r}_1 \dots \mathfrak{r}_r}_{x_1 \dots x_r} Px_1 \dots x_r$$

$$\text{iff } \mathrm{Mod}\,\mathfrak{I}^{\mathfrak{r}_1 \dots \mathfrak{r}_r}_{x_1 \dots x_r} \alpha$$

$$\text{iff } \mathrm{Mod}\,\mathfrak{I}'^{\mathfrak{r}_1 \dots \mathfrak{r}_r}_{x_1 \dots x_r} \alpha \qquad \text{(coincidence theorem)}$$

$$\text{iff } \mathrm{Mod}\,\mathfrak{I}'^{\mathfrak{r}_1 \dots \mathfrak{r}_r}_{x_1 \dots x_r} Px_1 \dots x_r$$

$$\text{iff } \mathfrak{I}'(P) \text{ fits } \mathfrak{r}_1, \dots, \mathfrak{r}_r .$$

(2) If (**), then (***). Let $\mathrm{Mod}\,\mathfrak{I}\mathfrak{M}$ and $\mathrm{Mod}\,\mathfrak{I}\mathfrak{M}'$. We need to show that $\mathrm{Mod}\,\mathfrak{I} \bigwedge x_1 \dots \bigwedge x_r(Px_1 \dots x_r \leftrightarrow P'x_1 \dots x_r)$, i.e. that $\mathfrak{I}(P) = \mathfrak{I}'(P)$. We have $\mathrm{Subst}\,\mathfrak{M}PP'\mathfrak{M}'$. Now, by the substitution theorem (which can be proved for this sort of substitution in the same way as for ordinary substitution).

$$\mathrm{Mod}\,\mathfrak{I}_P^{\mathfrak{I}(P')}\mathfrak{M} \quad \text{if and only if} \quad \mathrm{Mod}\,\mathfrak{I}\mathfrak{M}' ,$$

and hence $\mathrm{Mod}\,\mathfrak{I}_P^{\mathfrak{I}(P')}\mathfrak{M}$. Now we can apply (**) to the two interpretations $\mathfrak{I}$ and $\mathfrak{I}_P^{\mathfrak{I}(P')}$ and obtain $\mathfrak{I}(P) = \mathfrak{I}_P^{\mathfrak{I}(P')}(P)$, i.e. $\mathfrak{I}(P) = \mathfrak{I}(P')$.

(3) If (***), then (*). We shall assume that none of the variables $x_1, \dots, x_r$ mentioned in (***) lies in $\mathfrak{B}(\mathfrak{M})$ (otherwise we would replace $x_1, \dots, x_r$ by suitable $y_1, \dots, y_r$ and apply the rule $(\mathrm{ReG}_{x,y})$). Moreover, we shall assume that $\mathfrak{M}$ is nonempty (we leave it to the reader to reformulate the following proof for the case that $\mathfrak{M}$ is empty). By (***), there are finitely many elements $\mu_1, \dots, \mu_s$ of $\mathfrak{M}$ such that

$$\vdash \mu_1 \dots \mu_s \mu_1' \dots \mu_s' \bigwedge x_1 \dots \bigwedge x_r(Px_1 \dots x_r \leftrightarrow P'x_1 \dots x_r) .$$

Let $\mu \equiv \mu_1 \wedge \dots \wedge \mu_s$. Then we have

$$\vdash \mu\mu' \bigwedge x_1 \dots \bigwedge x_r(Px_1 \dots x_r \leftrightarrow P'x_1 \dots x_r)$$

$$\vdash \mu\mu' Px_1 \dots x_r \leftrightarrow P'x_1 \dots x_r$$

$$\vdash \mu\mu' Px_1 \dots x_r \rightarrow P'x_1 \dots x_r$$

$$\vdash \mu\,\mu' P x_1 \ldots x_r \;\; P' x_1 \ldots x_r$$

$$\vdash \mu \wedge P x_1 \ldots x_r\, \mu' \rightarrow P' x_1 \ldots x_r .$$

In this last sequent, P' does not occur in the antecedent and P does not occur in the succedent. We can now apply Craig's interpolation lemma. Thus there is an expression α such that $\mathfrak{V}(\alpha) \subset \mathfrak{V}(\mu \wedge P x_1 \ldots x_r) \cap \mathfrak{V}(\mu' \rightarrow P' x_1 \ldots x_r)$. i.e. such that $V(\alpha) \subset \mathfrak{V}(\mathfrak{M}) \cup \{x_1, \ldots, x_r\}$ and $P \notin \mathfrak{V}(\alpha)$ and

(1) $$\mu \wedge P x_1 \ldots x_r \models \alpha$$

and

(2) $$\alpha \models \mu' \rightarrow P' x_1 \ldots x_r .$$

If follows from (1) that

(3) $$\vdash \mu_1 \ldots \mu_r P x_1 \ldots x_r \rightarrow \alpha .$$

From (2) we obtain

(4) $$\vdash \mu_1' \ldots \mu_r' \alpha \rightarrow P' x_1 \ldots x_r .$$

If, in any derivation of (4), we replace P' everywhere by P, then, clearly, we obtain another derivation (cf. the proof of the first part of Theorem 3 in Chap. V, § 1.3). Thus we have

(5) $$\vdash \mu_1 \ldots \mu_r \alpha \rightarrow P x_1 \ldots x_r .$$

From (3) and (5) we obtain

(6) $$\vdash \mu_1 \ldots \mu_r P x_1 \ldots x_r \leftrightarrow \alpha .$$

Since $x_1, \ldots, x_r$ do not occur free in $\mathfrak{M}$, we may generalise over these variables in the right-hand expression. Thus, finally, we obtain

(7) $$\vdash \mu_1 \ldots \mu_r \bigwedge x_1 \ldots \bigwedge x_r (P x_1 \ldots x_r \leftrightarrow \alpha),$$

which proves (*).

6.3 Beth's definability theorem for a function variable. Once again, let $\mathfrak{V}(\mathfrak{M})$ be finite. Let f and f' be r-place, $f \not\equiv f'$ and $f' \notin \mathfrak{V}(\mathfrak{M})$. Let $\mathfrak{M}'$ be defined analogously to the $\mathfrak{M}'$ in 6.2. The following three statements are equivalent:

(*) f is definable relative to $\mathfrak{M}$.

(**) If two models of $\mathfrak{M}$ with the same domain of individuals are identical on $\mathfrak{V}(\mathfrak{M})$ - f, then they also have the same value at f.

(***) $\mathfrak{M} \cup \mathfrak{M}' \models \bigwedge x_1 \dots \bigwedge x_r (fx_1 \dots x_r = f'x_1 \dots x_r)$, <u>where</u> $x_1, \dots, x_r$ <u>are pairwise distinct variables</u>.

Proof. As in 6.2. In proving (*) from (***), the separation of f from f' can be achieved as follows: First of all, we have

$$\vdash \mu\mu' \; fx_1 \dots x_r = f'x_1 \dots x_r.$$

Now if x is a new individual variable, then by (ET):

$$\vdash fx_1 \dots x_r = f'x_1 \dots x_r \;\; f'x_1 \dots x_r = x \;\; fx_1 \dots x_r = x.$$

It follows that

$$\vdash \mu\mu' \; f'x_1 \dots x_r = x \;\; fx_1 \dots x_r = x,$$

and thus the separation of f and f' can be carried out as in 6.2.

<u>Exercises. 1.</u> In the expression $(\neg(\alpha \wedge \neg(\neg\beta \vee \bigwedge x\gamma)) \wedge \delta)$, δ is not in the domain of influence of a negator, $(\alpha \wedge \neg(\neg\beta \vee \bigwedge x\gamma))$, α and $\neg(\neg\beta \vee \bigwedge x\gamma)$ are in the domain of influence of the first negator, $(\neg\beta \vee \bigwedge x\gamma)$, $\neg\beta$, $\bigwedge x\gamma$ and γ are in the domain of influence of the first two negators and β is in the domain of influence of all three negators. In the following, let P be an n-place predicate variable and α not contain $\rightarrow$, $\leftrightarrow$. Let $\mathrm{Pos}\,P\alpha$ mean that every expression of the form $Pt_1 \dots t_n$ which occurs in α at some point occurs there in the domain of influence of an even number of negators. Let $\mathrm{Neg}\,P\alpha$ mean that every expression of the form $Pt_1 \dots t_n$ which occurs in α at some point occurs there in the domain of influence of an uneven number of negators.

Lay down the rules of a calculus in which rows of symbols of the form $+P\alpha$ and $-P\alpha$ are derivable; $+P\alpha$ if and only if $\mathrm{Pos}\,P\alpha$, and $-P\alpha$ if and only if $\mathrm{Neg}\,P\alpha$. (As an example cf. the calculus of free occurrence in Chap. II, §4.2.)

<u>2.</u> Let α, P and P' fulfil the hypotheses of Beth's definability theorem. Suppose that α does not contain either of $\rightarrow$, $\leftrightarrow$. For each of the cases $\mathrm{Pos}\,P\alpha$ and $\mathrm{Neg}\,P\alpha$, find an expression β which defines P in the sense of Beth's theorem. (Hint: Put α into a prenex normal form whose kernel is a disjunctive normal form.)

<u>3.</u> Let $\mathfrak{C}_{\mathfrak{B}}(\mathfrak{I})$ be an at most denumerable structure over the domain of individuals ω. Let $\omega_1 \subset \omega$ and $\omega_2 \subset \omega$ and $\mathfrak{B}^* \subset \mathfrak{B}$. Let Φ be a partial $\mathfrak{B}^*$-isomorphism in the extended sense (i.e.s.) from ω_1 onto ω_2; i.e. let Φ be a 1-1 mapping from ω_1 onto ω_2 with the property: For all $n \geqslant 0$, all pairwise distinct $x_1, \dots, x_n$, all α such that $\mathfrak{B}(\alpha) \subset \mathfrak{N}^* \cup \{x_1, \dots, x_n\}$ and all $\mathfrak{r}_1, \dots, \mathfrak{r}_n \in \omega_1$,

$$\mathrm{Mod}\,\mathfrak{I}^{\mathfrak{r}_1 \dots \mathfrak{r}_n}_{x_1 \dots x_n}\,\alpha \quad \text{if and only if} \quad \mathrm{Mod}\,\mathfrak{I}^{\Phi(\mathfrak{r}_1) \dots \Phi(\mathfrak{r}_n)}_{x_1 \quad \dots \quad x_n}\,\alpha.$$

Show that:

a) There is an at most denumerable elementary extension $\mathfrak{C}_{\mathfrak{B}}(\mathfrak{I}')$ of $\mathfrak{C}_{\mathfrak{B}}(\mathfrak{I})$ with a domain of individuals ω' and a partial $\mathfrak{B}^*$-isomorphism i.e.s. from ω onto a subset of ω' which is an extension of Φ. (Cf. the proof of §4, Lemma 3.)

b) Assertion a) with Φ^{-1} instead of Φ.

c) There is an at most denumerable elementary extension $\mathfrak{C}_{\mathfrak{B}}(\mathfrak{I}'')$ of $\mathfrak{C}_{\mathfrak{B}}(\mathfrak{I})$ and an isomorphism from $\mathfrak{C}_{\mathfrak{B}}(\mathfrak{I}'')$ onto itself which is an extension of Φ. (Use a) and b). Cf. the proof of §4, Lemma 3.)

4. Let $\mathfrak{B} = \mathfrak{B}^* \cup \{P\}$, $P \notin \mathfrak{B}^*$. Let $\mathfrak{A}_{\mathfrak{B}}(\mathfrak{J})$, $\mathfrak{A}_{\mathfrak{B}}(\mathfrak{J}')$ be different structures, both at most denumerable, such that $\mathfrak{A}_{\mathfrak{B}^*}(\mathfrak{J}) = \mathfrak{A}_{\mathfrak{B}^*}(\mathfrak{J}')$, $\Theta_{\mathfrak{B}}(\mathfrak{J}) = \Theta_{\mathfrak{B}}(\mathfrak{J}')$. Then there is a structure $\mathfrak{A}_{\mathfrak{B}}(\mathfrak{J}'')$ which is at most denumerable and satisfies:

a) $\Theta_{\mathfrak{B}}(\mathfrak{J}'') = \Theta_{\mathfrak{B}}(\mathfrak{J}')$.

b) There is an isomorphism from $\mathfrak{A}_{\mathfrak{B}^*}(\mathfrak{J})$ onto itself which maps onto itself the set of those tuples which $\mathfrak{J}''(P)$ fits.

(Suggestion: We may assume that there are pairwise distinct individual variables y_1, $y_2, \ldots, y_1', y_2', \ldots$ which do not occur in $\mathfrak{B}$ and satisfy $\mathfrak{J}(y_1) = \mathfrak{J}'(y_1')$, $\mathfrak{J}(y_2) = \mathfrak{J}'(y_2'), \ldots$, and that the $\mathfrak{J}(y_i)$ run trough the whole of the underlying set of $\mathfrak{A}_{\mathfrak{B}}(\mathfrak{J})$. Show that $\Theta_{\mathfrak{B} \cup \{y_1, \ldots\}}(\mathfrak{J}) \cup \Theta_{\mathfrak{B} \cup \{y_1', \ldots\}}(\mathfrak{J}')$ is satisfiable and apply the result of the previous exercise.)

5. Let $\mathfrak{B} = \mathfrak{B}^* \cup \{P\}$, $P \notin \mathfrak{B}^*$. Let $\mathfrak{M}$ be a set of expressions such that $\mathfrak{B}(\mathfrak{M}) \subset \mathfrak{B}$. Suppose that all $\mathfrak{A}_{\mathfrak{B}}(\mathfrak{J})$, $\mathfrak{A}_{\mathfrak{B}}(\mathfrak{J}')$ which are models of $\mathfrak{M}$ satisfy: If $\mathfrak{A}_{\mathfrak{B}^*}(\mathfrak{J}) = \mathfrak{A}_{\mathfrak{B}}(\mathfrak{J}')$ and $\mathfrak{A}_{\mathfrak{B}}(\mathfrak{J})$ is isomorphic to $\mathfrak{A}_{\mathfrak{B}}(\mathfrak{J}')$, then $\mathfrak{J}(P) = \mathfrak{J}'(P)$. Show that:

a) If $\mathfrak{A}_{\mathfrak{B}}(\mathfrak{J})$ is a model of $\mathfrak{M}$, then P is definable relative to $\Theta_{\mathfrak{B}}(\mathfrak{J})$. (Use the preceding example and Beth's Theorem.)

b) If $\mathfrak{A}_{\mathfrak{B}}(\mathfrak{J})$ is a model of $\mathfrak{M}$, then there are pairwise distinct $x_1, \ldots, x_n$ and an α such that $\mathfrak{B}(\alpha) \subset \mathfrak{B}^* \cup \{x_1, \ldots, x_n\}$ and such that $\mathfrak{A}_{\mathfrak{B}}(\mathfrak{J})$ is a model of $\bigwedge x_1 \ldots \bigwedge x_n (Px_1 \ldots x_n \leftrightarrow \alpha)$.

c) (Svenonius' Theorem (1959)) There are pairwise distinct $x_1, \ldots, x_n$ and $\alpha_1, \ldots, \alpha_m$ such that, for each $i \leq m$, $\mathfrak{B}(\alpha_i) \subset \mathfrak{B}^* \cup \{x_1, \ldots, x_n\}$ and such that $\mathfrak{M} \models \bigwedge x_1 \ldots \bigwedge x_n (Px_1 \ldots x_n \leftrightarrow \alpha_1) \vee \ldots \vee \bigwedge x_1 \ldots \bigwedge x_n (Px_1 \ldots x_n \leftrightarrow \alpha_m)$.

IX. Miscellaneous

§ 1. Extended substitution

1.1 Introduction. In Chap. II, § 5 we defined the relation of substitution Subst $\alpha x t \beta$. For each α, x, t there is at most one β such that Subst $\alpha x t \beta$. However, as we have seen, not every α, x, t possesses a β such that Subst $\alpha x t \beta$. In the following, we shall deal with such exceptions. We want to associate with each α, x, t a unique expression $\alpha\frac{t}{x}$ which, we shall say, is obtained from α by extended substitution of t for x. The expression "extended substitution" is justified by the fact that $\alpha\frac{t}{x} \equiv \beta$ whenever Subst $\alpha x t \beta$.

$\alpha\frac{t}{x}$ is defined as follows: First of all, we associate with α an "equivalent" expression α_x^t, depending also on x and t, where $\alpha_x^t \equiv \alpha$ if and only if there is a β such that Subst $\alpha x t \beta$. α_x^t is chosen in such a way that α_x^t, x, t possess a β such that Subst $\alpha_x^t x t \alpha$. We take this β, which is uniquely determined by x and t, as the required expression $\alpha\frac{t}{x}$.

The "equivalence" of α and α_x^t is characterised by the fact that the same variables occur free in α as in α_x^t and that $\alpha \dashv\vdash \alpha_x^t$.

If Subst $\alpha x t \beta$, then β is obtained from α by replacing the individual variable x by the term t at all the places where x occurs free in α. In this sense, we can speak of "substitution by replacement". In the definition of substitution in Chap. II, § 5, the reason why we did not assign a β such that Subst $\alpha x t \beta$ to every α, x, t was that we wanted to prevent the situation in which a variable x occurs free in an expression α, whilst a variable which appears in the same position after the substitution of t for x is not free in this position. This reason may at first seem somewhat unimportant; be that as it may, we can now give a second reason: If we were to define substitution as substitution by replacement with no restrictions, then the substitution rule (S_x^t), which is one of the rules of predicate logic (Chap. IV, § 2.3), would be unsound (Chap. IV, § 3).

In order to show this, we consider the following "derivation", in which substitution by replacement is used to pass to the third line in such a way that the situation described above is created (x occurs free in $\neg\bigwedge yPxy$; we replace x here by y, with the result that the variable y which we have introduced in this position is bound by the universal quantifier):

1) $\bigwedge x \neg \bigwedge y Pxy : \bigwedge x \neg \bigwedge y Pxy$		(A)
2) $\bigwedge x \neg \bigwedge y Pxy : \neg \bigwedge y Pxy$		(G)
3) $\bigwedge x \neg \bigwedge y Pxy : \neg \bigwedge y Pyy$		"(S_x^y)"

Whereas the second row, which is obtained by (A) and (G), is sound, this is not true of the third row. For otherwise we should have

$$\bigwedge x \neg \bigwedge y Pxy \models \neg \bigwedge y Pyy.$$

But this is false: Let us consider the domain of individuals ω of the natural numbers and an interpretation $\mathfrak{J}$ over it such that $\mathfrak{J}(P)$ is the equality predicate (i.e. the predicate which fits two natural numbers $\mathfrak{x}$, $\mathfrak{y}$ if and only if $\mathfrak{x} = \mathfrak{y}$). $\mathrm{Mod}_\omega \mathfrak{J} \bigwedge x \neg \bigwedge y Pxy$ and $\mathrm{Mod}_\omega \mathfrak{J} \bigwedge y Pyy$, which contradicts the relation of consequence given above.

Thus, in $\alpha \equiv \neg \bigwedge y Pxy$ we cannot replace x by y. The reason for this is that y occurs in the component $\bigwedge y$. Now an obvious idea is to consider, instead of α, the expression $\alpha_x^y \equiv \neg \bigwedge z Pxz$, where z is a uniquely determined individual variable which is different from x and y. Clearly,

(1) the same individual variables occur free in α as in α_x^y,

(2) $\alpha \dashv\vdash \alpha_x^y$,

(3) Subst $\alpha_x^y xy \neg \bigwedge z Pyz$.

Thus, we shall say that the expression $\neg \bigwedge z Pyz$ is obtained from $\neg \bigwedge y Pxy$ by extended substitution of y for x, i.e. that $[\neg \bigwedge y Pxy]_x^y \equiv \neg \bigwedge z Pyz$. We could say that the expression α_x^y is obtained from α by "renaming the bound variables". In general, we shall use this method of renaming bound variables in the cases in which normal substitution cannot be carried out (see 6.2, case 2b).

In order that the expressions α_x^t (and hence also $\alpha\frac{t}{x}$) should be uniquely determined for a given α, x, t it is necessary to characterise uniquely the new variables which are introduced by the process of renaming bound variables. In doing this, we make use of the fact that there are countably many individual variables. We assume that we have (in a way that is arbitrary but is kept fixed during the following) numbered the individual variables. The number which is assigned to the individual variable x (which will be one of the natural numbers 0, 1, 2,...) is to be called the <u>index of x</u>, denoted by Ind(x).

The new variables introduced by the process of renaming bound variables, which we have just mentioned, satisfy certain conditions (for example, we required above that z should be different from x and y), which, however, are not sufficient to determine them uniquely. However, we want to determine these variables uniquely by requiring that, each time, the variable chosen should be the one with the smallest index among those which satisfy the given conditions.

1.2 The operations α^t_x and $\alpha\frac{t}{x}$. In Chap. II, § 1.2, we assigned to every expression α a rank $R(\alpha)$, which was defined as the total number of logical sumbols occuring in α. Now we shall define α^t_x and $\alpha\frac{t}{x}$ simultaneously under the assumption that, for all expression β such that $R(\beta) < R(\alpha)$, we have already defined the operations $\beta^{t'}_{x'}$ and $\beta\frac{t'}{x'}$ for arbitrary x' and t'. At the same time we shall show that the following assertions hold:

(1) The same variables occur free in α as in α^t_x.

(2) $\alpha \dashv\vdash \alpha^t_x$.

(3) Subst $\alpha^t_x \, x \, t \, \alpha\frac{t}{x}$.

(4) If there is a β such that Subst $\alpha x t \beta$, then $\alpha^t_x \equiv \alpha$ and $\alpha\frac{t}{x} \equiv \beta$.

Here, we can assume that the corresponding assertions hold for every β, x' and t' such that $R(\beta) < R(\alpha)$. It can be seen immediately that the operations α^t_x and $\alpha\frac{t}{x}$ are effective, i.e. that these expressions can be found effectively for given α, x, t (where we can assume that this is true for β, x', t').

In the definition of α^t_x and $\alpha\frac{t}{x}$ we distinguish between the following cases:

(a) α is atomic. We put $\alpha^t_x \equiv \alpha$. By Chap. II, § 5.4, Theorem 2, there is a uniquely determined β such that Subst $\alpha x t \beta$. We put $\alpha\frac{t}{x} \equiv \beta$. The assertions (1), ..., (4) can be verified immediately.

(b) α is a negation. Then there is an expression α_1 such that $\alpha \equiv \neg \alpha_1$. $R(\alpha_1) < R(\alpha)$. We put

$$\alpha^t_x \equiv \neg \alpha^t_{1x}, \qquad \alpha\frac{t}{x} \equiv \neg \alpha_1\frac{t}{x}.$$

Proof of (1), ..., (4):

to (1): The same variables occur free in α^t_x as in α^t_{1x}. By hypothesis, the same variables occur free in α^t_{1x} as in α_1. The same variables occur free in α_1 as in α. Hence the same variables occur free in α^t_x as in α.

to (2): Our assumption gives us $\alpha_1 \dashv\vdash \alpha^t_{1x}$. From this, by (CaPo) (Chap. IV, § 4.2), we obtain $\neg \alpha_1 \dashv\vdash \neg \alpha^t_{1x}$, i.e. $\alpha \dashv\vdash \alpha^t_x$.

to (3): By our assumption Subst $\alpha^t_{1x} \, x \, t \, \alpha^t_{1x}$, and by using the definition of substitution, we obtain Subst $\neg\alpha^t_{1x} \, x \, t \, \neg\alpha_1\frac{t}{x}$, i.e. Subst $\alpha^t_x \, x \, t \, \alpha\frac{t}{x}$.

to (4): Let there be a β such that Subst $\alpha x t \beta$. Then we can see from the substitution calculus (by inversion) that there is a β_1 such that Subst $\alpha_1 x t \beta_1$ and $\beta \equiv \neg \beta_1$. Then, by our assumption, $\alpha^t_{1x} \equiv \alpha_1$ and $\alpha_1\frac{t}{x} \equiv \beta_1$, from which it is clear that $\alpha^t_x \equiv \alpha$ and $\alpha\frac{t}{x} \equiv \beta$.

(c) α is a conjunction. Then there are expressions α_1 and α_2 such that $\alpha \equiv (\alpha_1 \wedge \alpha_2)$. Also, $R(\alpha_1) < R(\alpha)$ and $R(\alpha_2) < R(\alpha)$. We put:

$$\alpha_x^t \equiv (\alpha_{1x}^{\ t} \wedge \alpha_{2x}^{\ t}), \qquad \alpha_{\overline{x}}^t \equiv (\alpha_{1\overline{x}}^{\ t} \wedge \alpha_{2\overline{x}}^{\ t}).$$

We can prove the assertions (1), ..., (4) as in (b), calling upon (UU) (Chap. IV, § 4.3) in the proof of (2).

(d) α is a generalisation. Then there is a variable u and an expression α_1 such that $\alpha \equiv \bigwedge u\alpha_1$. Also, $R(\alpha_1) < R(\alpha)$. We distinguish two cases, according to whether or not x occurs free in α.

Case 1. x does not occur free in α. Then we put $\alpha_x^t \equiv \alpha_{\overline{x}}^t \equiv \alpha$. It is easy to verify the assertions (1), ..., (4).

Case 2. x occurs free in α. We distinguish two cases, according to whether or not u occurs in t.

Subcase 2a. u does not occur in t. We put

$$\alpha_x^t \equiv \bigwedge u\alpha_{1x}^{\ t}, \qquad \alpha_{\overline{x}}^t \equiv \bigwedge u\alpha_{1\overline{x}}^{\ t}.$$

Proof of (1), ..., (4):

to (1): Clearly α and α_x^t contain the same more-than-one-place function variables and the same predicate variables. Thus it is sufficient to show that the same individual variables occur free in α as in α_x^t. Now, for an arbitrary individual variable y,

y occurs free in α_x^t iff y occurs free in $\alpha_{1x}^{\ t}$ and $y \not\equiv u$
iff y occurs free in α_1 and $y \not\equiv u$
iff y occurs free in $\bigwedge u\alpha_1$, i.e. in α.

to (2): By our assumption, $\alpha_1 \dashv\vdash \alpha_{1x}^{\ t}$. Hence, by Chap. II, § 5.2 (14), $\bigwedge u\alpha_1 \dashv\vdash \bigwedge u\alpha_{1x}^{\ t}$, i.e. $\alpha \dashv\vdash \alpha_x^t$.

to (3): By our assumption, Subst $\alpha_{1x}^{\ t}$ x t $\alpha_{1\overline{x}}^{\ t}$. x occurs free in α_x^t, since x occurs free in α and the same variables occur free in α as in α_x^t, as we have shown above in proving (1). Since u does not occur in t, we have Subst $\bigwedge u\alpha_{1x}^{\ t}$ x t $\bigwedge u\alpha_{1\overline{x}}^{\ t}$, i.e. Subst α_x^t x t $\alpha_{\overline{x}}^t$.

to (4): Let there be a β such that Subst α x t β. Since x occurs free in α, there is a β_1 such that Subst α_1 x t β_1 and $\beta \equiv \bigwedge u\beta_1$. By our assumption it follows from Subst α_1 x t β_1 that $\alpha_{1x}^{\ t} \equiv \alpha_1$ and $\alpha_{1\overline{x}}^{\ t} \equiv \beta_1$. Hence, $\alpha_x^t \equiv \bigwedge u\alpha_{1x}^{\ t} \equiv \bigwedge u\alpha_1 \equiv \alpha$ and $\alpha_{\overline{x}}^t \equiv \bigwedge u\alpha_{1\overline{x}}^{\ t} \equiv \bigwedge u\beta_1 \equiv \beta$.

Subcase 2b: u occurs in t. Let v be the individual variable with the smallest index such that:

(*) $v \not\equiv u$,

(**) v does not occur in α_1,

(***) v does not occur in t.

Since v does not occur in t there is, by Chap. II, § 5.4, Theorem 6, an expression α_1' such that

$$\text{Subst}\,\alpha_1\,u\,v\,\alpha_1', \qquad \text{Subst}\,\alpha_1'\,v\,u\,\alpha_1 .$$

By Chap. II, § 5.6, Theorem 1, $R(\alpha_1') = R(\alpha_1)$ and hence $R(\alpha_1') < R(\alpha)$. Put:

$$\alpha_x^t \equiv \bigwedge v\alpha_{1x}'^{\,t}, \qquad \alpha_{\bar{x}}^{t} \equiv \bigwedge v\alpha_{1\bar{x}}'^{\,t} .$$

Proof of (1),...(4):

<u>to</u> (1): As before, we need only consider the individual variables.

If z occurs free in α, then z occurs free in α_1 and $z \not\equiv u$. Since $z \not\equiv u$ and $\text{Subst}\,\alpha_1\,uv\,\alpha_1'$, by Chap. II, § 5.4, Theorem 7 (i) z occurs free in α_1'. Since, by our assumption, the same variables occur free in α_1' as in $\alpha_{1x}'^{\,t}$, z occurs free in $\alpha_{1x}'^{\,t}$. In order to prove that z also occurs free in $\bigwedge v\,\alpha_{1x}'^{\,t}$, i.e. in α_x^t, we need only show that $z \not\equiv v$. If $z \equiv v$ then, since $\text{Subst}\,\alpha_1'\,v\,u\,\alpha_1$ and by Chap. II, § 5.4, Theorem 8, the variable z would not occur free in α_1 (contrary to what we showed above), since, by (*), $v \not\equiv u$. Thus, we have shown that every variable which occurs free in α also occurs free in α_x^t. <u>Conversely:</u>

If z occurs free in α_x^t, then z occurs free in $\alpha_{1x}'^{\,t}$ and $z \not\equiv v$. Since, by our assumption, the same variables occur free in $\alpha_{1x}'^{\,t}$ as in α_1', z occurs free in α_1'. Since $\text{Subst}\,\alpha_1'\,v\,u\,\alpha_1$ and $z \not\equiv v$, z occurs free in α_1 [Chap. II, § 5.4, Theorem 7 (i)]. The assertion that z occurs free in α, i.e. in $\bigwedge u\alpha_1$, will follow as soon as we have shown that $z \not\equiv u$. But this follows from the fact that z occurs free in α_1' whereas u does not, since $\text{Subst}\,\alpha_1\,u\,v\,\alpha_1'$ and $u \not\equiv v$ (Chap. II, § 5.4, Theorem 8).

<u>to</u> (2): We prove the assertion by means of the following two derivations:

$\bigwedge u\,\alpha_1$	$:\ \bigwedge u\,\alpha_1$	(A)
$\bigwedge u\,\alpha_1$	$:\ \alpha_1$	(G)
$\bigwedge u\,\alpha_1$	$:\ \alpha_1'$	(S_u^v)
α_1'	$:\ \alpha_{1x}'^{\,t}$	(assumption)
$\bigwedge u\,\alpha_1$	$:\ \alpha_{1x}'^{\,t}$	(CuRu)
$\bigwedge u\,\alpha_1$	$:\ \bigwedge v\,\alpha_{1x}'^{\,t}$	(G_v)

The application of the critical rule (G_v) is a legitimate one, since, by (**), v does not occur in α_1, and hence it does not occur free in $\bigwedge u\alpha_1$. The last line of the derivation shows that $\alpha \vdash \alpha^t_x$.

$$\begin{array}{ll} \bigwedge v\,\alpha'^{\,t}_{1x} : \bigwedge v\,\alpha'^{\,t}_{1x} & (A) \\ \bigwedge v\,\alpha'^{\,t}_{1x} : \alpha'^{\,t}_{1x} & (G) \\ \alpha'^{\,t}_{1x} : \alpha'_1 & \text{(assumption)} \\ \bigwedge v\,\alpha'^{\,t}_{1x} : \alpha'_1 & \text{(CuRu)} \\ \bigwedge v\,\alpha'^{\,t}_{1x} : \alpha_1 & (S^u_v) \\ \bigwedge v\,\alpha'^{\,t}_{1x} : \bigwedge u\,\alpha_1 & (G_u) \end{array}$$

The application of the critical rule (G_u) in the last line is legitimate, since u does not occur free in $\bigwedge v\alpha'^{\,t}_{1x}$. For, if u occured free in this expression, then it would also occur free in $\alpha'^{\,t}_{1x}$ and therefore also in α'_1, since, by our assumption, the same variables occur free in the two last-named expressions. But u cannot occur free in α'_1, since

$$\text{Subst}\,\alpha_1\,u\,v\,\alpha'_1 \quad \text{and, by } (*), \quad u \not\equiv v$$

(cf. Chap. II, § 5.4, Theorem 8).

<u>to</u> (3): By our assumption, Subst $\alpha'^{\,t}_{1x}\,x\,t\,\alpha'^{\,t}_{1\bar{x}}$. We assert that

$$\text{Subst}\ \bigwedge v\alpha'^{\,t}_{1x}\,x\,t\ \bigwedge v\alpha'^{\,t}_{1\bar{x}}, \qquad \text{i.e.} \qquad \text{Subst}\,\alpha^t_x\,x\,t\,\alpha^t_{\bar{x}}\,.$$

It is sufficient to show that x occurs free in $\bigwedge v\,\alpha'^{\,t}_{1x}$ and that v does not occur in t. The latter holds because of (***). The first assertion follows from the fact that x occurs free in α, and therefore also in α^t_x ($\equiv \bigwedge v\,\alpha'^{\,t}_{1x}$), since the same variables occur free in α as in α^t_x, as we showed in proving (1).

<u>to</u> (4): Since x occurs free in α ($\equiv \bigwedge u\,\alpha_1$) and u occurs in t, there is no β such that Subst $\alpha\,x\,t\,\beta$, and hence there is nothing to prove.

1.3 Applications

Extended substitution rule

$$\frac{\alpha_1 \ldots \alpha_n : \alpha}{\alpha_{1\bar{x}}^{\,t} \ldots \alpha_{n\bar{x}}^{\,t} : \alpha^t_{\bar{x}}} \qquad (\text{ExS}^t_x)$$

Justification:

$$\begin{array}{rcll} \alpha_1\ldots\alpha_n & : & \alpha & \\ \alpha & : & \alpha_x^t & (1.2(2)) \\ \alpha_1\ldots\alpha_n & : & \alpha_x^t & (\text{CuRu}) \\ \alpha_{1x}^{\;t} & : & \alpha_1 & (1.2(2)) \\ \alpha_{1x}^{\;t}\ldots\alpha_n & : & \alpha_x^t & (\text{CuRu}) \\ \ldots\ldots & & \ldots & \ldots\ldots \\ \alpha_{nx}^{\;t} & : & \alpha_n & (1.2(2)) \\ \alpha_{1x}^{\;t}\ldots\alpha_{nx}^{\;t} & : & \alpha_x^t & (\text{CuRu}) \\ \alpha\frac{t}{1x}\ldots\alpha\frac{t}{nx} & : & \alpha\frac{t}{x} & (S_x^t) \end{array}$$

In obtaining the last line, we make use of the fact that, for any β, $\text{Subst}\,\beta_x^t\,x\,t\,\beta\frac{t}{x}$, which we proved in 1.2(3).

<u>Theorem.</u> $\bigwedge x\,\alpha \vdash \alpha\frac{t}{x}$.

Proof.

$$\begin{array}{rcll} \bigwedge x\,\alpha & : & \bigwedge x\,\alpha & (A) \\ \bigwedge x\,\alpha & : & \alpha & (G) \\ [\bigwedge x\,\alpha]_x^t & : & \alpha\frac{t}{x} & (\text{ExS}_x^t) \end{array}$$

But $[\bigwedge x\,\alpha]\frac{t}{x} \equiv \bigwedge x\,\alpha$ by 6.2 (4), since $\text{Subst}\,\bigwedge x\alpha\, x\, t\, \bigwedge x\alpha$.

<u>Extended substitution theorem.</u> $\text{Mod}\,\mathfrak{J}_x^{\mathfrak{J}(t)}\,\alpha$ <u>iff</u> $\text{Mod}\,\mathfrak{J}\,\alpha\frac{t}{x}$.

Proof. By 1.2(2) we have $\text{Mod}\,\mathfrak{J}_x^{\mathfrak{J}(t)}\,\alpha$ iff $\text{Mod}\,\mathfrak{J}_x^{\mathfrak{J}(t)}\,\alpha_x^t$. Together with 1.2(3) the Substitution theorem (Chap. III, § 3.2) gives us

$$\text{Mod}\,\mathfrak{J}_x^{\mathfrak{J}(t)}\,\alpha_x^t \text{ iff } \text{Mod}\,\mathfrak{J}\,\alpha\frac{t}{x}\,.$$

<u>Exercises. 1.</u> (In the following let $x_1, x_2, \ldots$ be the first, second, ... individual variable.) Carry out the following extended substitution effectively:

(a) $(Px_1 \wedge Qx_2)\frac{fx_1x_2}{x_3}$ (b) $(Px_1 \wedge Qx_2)\frac{fx_1x_2}{x_1}$

(c) $\bigwedge x_1(Px_1 \wedge Qx_2)\frac{fx_2}{x_1}$ (d) $\bigwedge x_2(Px_1 \wedge Qx_2)\frac{fx_2}{x_1}$

(e) $(\bigwedge x_1(fx_1 = ghx_1 \wedge Px_1) \wedge \bigwedge x_2 \neg Qx_1x_2)\frac{fx_2}{x_1}$.

<u>2.</u> Show that:

a) x occurs free in $\alpha\frac{t}{x}$ if and only if x occurs free in t and in α.

b) y occurs free in $\alpha\frac{t}{x}$ $(y \not\equiv x)$ if and only if either y occurs free in α or y occurs in t and x occurs free in α.

c) $R(\alpha) = R(\alpha_x^t) = R(\alpha\frac{t}{x})$.

3. Show that:

a) $\dfrac{\Sigma : \alpha}{\Sigma\ x = t : \alpha\frac{t}{x}}$ (extended second equality rule) is a derivable rule.

b) $\bigwedge x\alpha \vdash \bigwedge x\alpha\frac{t}{x}$.

§2. The admissability of definitions

2.1 Statement of the problem. Let us consider a theory which is based on an axiom system $\mathfrak{A}$ formulated in the language of predicate logic. For the sake of simplicity we shall assume that the axiom system is finite; let it be the set of members of the sequent σ.

When building up a theory, it is usual to introduce definitions. The definitions, formally speaking, act as additional axioms; thus, with them, more theorems can be proved than before. It is therefore natural to ask whether a consistent theory can be made inconsistent by adding definitions. We shall show that this is not so.

We shall restrict ourselves to definitions δ for predicate symbols (cf. Exercise 3, however). Such a definition has the form:

(1) $$\delta \equiv \bigwedge x_1 \dots \bigwedge x_r (Px_1 \dots x_r \leftrightarrow \rho),$$

where ρ is an expression such that $\mathfrak{B}(\rho) \subset \{x_1, \dots, x_r\}$. P is a "new" predicate symbol, i.e. one which does not occur in σ. Now we say that the definition (1) is admissable if:

(2) If $\vdash \sigma\delta\alpha$ and $\mathfrak{B}(\alpha) \subset \mathfrak{B}(\sigma)$, then $\vdash \sigma\alpha$.

From (2) it follows immediately that, if $\mathfrak{A}$ is consistent, so is $\mathfrak{A} \cup \{\delta\}$.

The assertion (2) can easily be proved by semantic means: We assume that the assumptions in (2) hold and that Mod $\mathfrak{J}\sigma$; and we have to show that Mod $\mathfrak{J}\alpha$. We define an r-place predicate $\mathfrak{P}$ over the domain of individuals ω which belongs to $\mathfrak{J}$, as follows:

(3) $$\mathfrak{P}\mathfrak{r}_1 \dots \mathfrak{r}_r \text{ if and only if } \mathrm{Mod}\, \mathfrak{J}_{x_1 \dots x_r}^{\mathfrak{r}_1 \dots \mathfrak{r}_r}\, \rho$$

for all $\mathfrak{r}_1, \dots, \mathfrak{r}_r \in \omega$. From (3) it follows that Mod $\mathfrak{J}_P^{\mathfrak{P}}\delta$. Then, by the coincidence theorem, Mod $\mathfrak{J}_P^{\mathfrak{P}}\sigma$. From $\vdash \sigma\delta\alpha$ it follows that Mod $\mathfrak{J}_P^{\mathfrak{P}}\alpha$ and from this, again by the coincidence theorem, that Mod $\mathfrak{J}\alpha$.

We may ask whether this semantic justification for the admissability of a definition can be replaced by a purely syntactic one; this question is motivated by the attempts (initiated by Hilbert) to show by purely constructive methods that mathematical theories are consistent. It is a characteristic of the constructive approach that infinite sets, in so far as they are seen as given totalities, are not recognised. Thus the use of semantic methods, which use infinite domains of individuals, is barred.

In this section we want, among other things, to show the truth of statement (2) by purely syntactic means. Typically, this is considerably more troublesome than our demonstration above of the truth of the semantic equivalent.

2.2 Simultaneous extended substitution. In the following let $x_1, \ldots, x_r$ be distinct individual variables. To every expression α and all terms $t_1, \ldots, t_r$ we want to assign an expression $\alpha\frac{t_1 \ldots t_r}{x_1 \ldots x_r}$; we shall say that $\alpha\frac{t_1 \ldots t_r}{x_1 \ldots x_r}$ is obtained from α by simultaneous extended substitution. For $r = 1$, $\alpha\frac{t_1}{x_1}$ is different from the expression with the same notation which was introduced in § 1, but equivalent to it (cf. Exercise 1). The way of defining $\alpha\frac{t_1 \ldots t_r}{x_1 \ldots x_r}$ which we use here is due to Tarski.

Definition. Let $y_1, \ldots, y_r$ be pairwise distinct and distinct from all the $x_1, \ldots, x_r$. Suppose also that none of the y_i occurs in α or in any of the terms $t_1, \ldots, t_r$. We abbreviate these conditions by saying: $y_1, \ldots, y_r$ are new individual variables (i.e. "new" as far as the expression α and the terms $x_1, \ldots, x_r$, $t_1, \ldots, t_r$ are concerned). (In the following, whenever we speak of "new" individual variables, we mean "new" in this sense.) We now put:

$$\alpha \begin{smallmatrix} t_1 \ldots t_r \\ y_1 \ldots y_r \\ x_1 \ldots x_r \end{smallmatrix} \equiv \bigwedge y_1 \ldots \bigwedge y_r (y_1 = t_1 \wedge \ldots \wedge y_r = t_r$$

$$\rightarrow \bigwedge x_1 \ldots \bigwedge x_r (x_1 = y_1 \wedge \ldots \wedge x_r = y_r \rightarrow \alpha));$$

$$\alpha\frac{t_1 \ldots t_r}{x_1 \ldots x_r} \equiv \alpha \begin{smallmatrix} t_1 \ldots t_r \\ y_1 \ldots y_r \\ x_1 \ldots x_r \end{smallmatrix}$$

where $y_1, \ldots, y_r$ are, in that order, the variables with the smallest index which satisfy the conditions given above.

For the sake of brevity, we shall often write $\alpha \begin{smallmatrix} t_1 \ldots \\ y_1 \ldots \\ x_1 \ldots \end{smallmatrix}$ instead of $\alpha \begin{smallmatrix} t_1 \ldots t_r \\ y_1 \ldots y_r \\ x_1 \ldots x_r \end{smallmatrix}$.

<u>Lemma 1.</u> $\vdash \alpha \begin{smallmatrix} t_1\ldots \\ y_1\ldots \\ x_1\ldots \end{smallmatrix} \leftrightarrow \alpha \begin{smallmatrix} t_1\ldots \\ z_1\ldots \\ x_1\ldots \end{smallmatrix}$, <u>and</u>, <u>in particular</u>, $\vdash \alpha \begin{smallmatrix} t_1\ldots \\ y_1\ldots \\ x_1\ldots \end{smallmatrix} \leftrightarrow \alpha \dfrac{t_1 \ldots t_r}{x_1 \ldots x_r}$.

Proof. Let $u_1, \ldots, u_r$ be "new" individual variables. Then the following sequents are derivable:

$$\alpha \begin{smallmatrix} t_1\ldots \\ y_1\ldots \\ x_1\ldots \end{smallmatrix} \quad y_1 = t_1 \ldots x_1 = y_1 \ldots : \alpha$$

$$- \text{"} - \; u_1 = t_1 \ldots x_1 = u_1 \ldots : \alpha$$

$$- \text{"} - \; z_1 = t_1 \ldots x_1 = z_1 \ldots : \alpha$$

$$- \text{"} - \; : \alpha \begin{smallmatrix} t_1\ldots \\ z_1\ldots \\ x_1\ldots \end{smallmatrix} ,$$

and so is the last sequent with the position of its members exchanged; the assertion then follows easily.

<u>Lemma 2.</u> $\vdash x_1 = t_1 \ldots x_r = t_r \; \alpha \begin{smallmatrix} t_1\ldots \\ y_1\ldots \\ x_1\ldots \end{smallmatrix} \; \alpha$.

Proof by means of a derivation whose main steps are:

$$\alpha \begin{smallmatrix} t_1\ldots \\ y_1\ldots \\ x_1\ldots \end{smallmatrix} y_1 = t_1 \ldots x_1 \ldots : \alpha$$

$$\alpha \begin{smallmatrix} t_1\ldots \\ y_1\ldots \\ x_1\ldots \end{smallmatrix} x_1 = t_1 \ldots : \alpha \quad \text{using} \quad (S^{t_1}_{y_1}), \ldots .$$

<u>Lemma 3.</u> $\vdash x_1 = t_1 \ldots x_r = t_r \alpha \; \alpha \begin{smallmatrix} t_1\ldots \\ y_1\ldots \\ x_1\ldots \end{smallmatrix}$.

Proof. Let $u_1, \ldots, u_r$ be new individual variables. Then there are expressions $\alpha_1, \ldots, \alpha_{r-1}, \beta$ such that

$$\text{Subst}\, \alpha x_1 u_1 \alpha_1, \; \text{Subst}\, \alpha_1 x_2 u_2 \alpha_2, \ldots, \; \text{Subst}\, \alpha_{r-1} x_r u_r \beta ,$$

$$\text{Subst}\, \alpha_1 u_1 x_1 \alpha, \; \text{Subst}\, \alpha_2 u_2 x_2 \alpha_1, \ldots, \; \text{Subst}\, \beta u_r x_r \alpha_{r-1} .$$

We prove the lemma by means of a derivation whose main steps are:

$$\begin{array}{rcll}
 & \alpha : \alpha & & \\
 & \alpha\, x_1 = u_1 : \alpha_1 & & (E^{u_1}_{x_1}) \\
 & \cdots\cdots\cdots & & \\
 & \alpha_{r-1}\, x_r = u_r : \beta & & (E^{u_r}_{x_r}) \\
 & \alpha\, x_1 = u_1 \ldots x_r = u_r : \beta & & \\
 & x_1 = t_1\; y_1 = t_1\; u_1 = y_1 : x_1 = u_1 & & \\
 & \cdots\cdots\cdots\cdots\cdots\cdots & &
\end{array}$$

$$\begin{array}{rl}
\alpha x_1 = t_1 \ldots y_1 = t_1 \ldots u_1 = y_1 \ldots & : \beta \\
\alpha\, x_1 = t_1 \ldots y_1 = t_1 \ldots & : \bigwedge u_1 \ldots (u_1 = y_1 \wedge \ldots \to \beta) \\
\bigwedge u_1 \ldots (u_1 = y_1 \wedge \ldots \to \beta) & : \bigwedge x_1 \ldots (x_1 = y_1 \wedge \ldots \to \alpha) \qquad (\mathrm{ReG}_{u_1, x_1}) \\
\alpha\, x_1 = t_1 \ldots y_1 = t_1 \ldots & : \bigwedge x_1 \ldots (x_1 = y_1 \wedge \ldots \to \alpha) \\
\alpha\, x_1 = t_1 \ldots & : \bigwedge y_1 \ldots (y_1 = t_1 \wedge \ldots \\
 & \quad \to \bigwedge x_1 \ldots (x_1 = y_1 \wedge \ldots \to \alpha).
\end{array}$$

Lemma 4. $\vdash x_1 = t_1 \ldots x_r = t_r\; \alpha \frac{t_1 \ldots t_r}{x_1 \ldots x_r} \leftrightarrow \alpha.$

Proof by Lemmas 2, 3.

Lemma 5. $\vdash \alpha \frac{x_1 \ldots x_r}{x_1 \ldots x_r} \leftrightarrow \alpha.$

Proof by Lemma 4.

Lemma 6. z occurs free in $\alpha \begin{smallmatrix} t_1 \ldots \\ y_1 \ldots \\ x_1 \ldots \end{smallmatrix}$ if and only if z occurs free in one of the terms t_i or (z occurs free in α and $z \notin \{x_1, \ldots, x_r\}$). (Compare this statement for $r = 1$ with the corresponding one for extended substitution in § 1, Exercise 2.)

Proof. First of all, it follows immediately from the definition that z is free in $\alpha \begin{smallmatrix} t_1 \ldots \\ y_1 \ldots \\ x_1 \ldots \end{smallmatrix}$ if and only if

(4) $z \notin \{y_1, \ldots, y_r\}$ and $[(z \equiv y_1$ or z is in $t_1)$ or ... or $(z \notin \{x_1, \ldots, x_r\}$ and $[(z \equiv x_1$ or $z \equiv y_1)$ or ... or z is free in $\alpha]))]$.

Now, if we write

a for $z \notin \{y_1, \ldots, y_r\}$, b for $z \notin \{x_1, \ldots, x_r\}$,
c_1 for $z \equiv y_1$, d_1 for z occurs in t_1,
e_1 for $z \equiv x_1$, f for z occurs free in α,

then, (if $\wedge$, $\vee$, $\neg$ are taken as metalinguistic symbols) (4) has the form

(4') $$a \wedge [(c_1 \vee d_1) \vee \ldots \vee (b \wedge [(e_1 \vee c_1) \vee \ldots \vee f])].$$

Because $\neg(\alpha \wedge c_1)$, this is equivalent to

(4'') $$(a \wedge d_1) \vee \ldots \vee (a \wedge b \wedge [(e_1 \vee c_1) \vee \ldots \vee f]);$$

and, because $\neg(b \wedge e_1)$, this is equivalent to

(4''') $$(a \wedge d_1) \vee \ldots \vee (a \wedge b \wedge f)$$

which, finally, because $f \to a$ and $d_1 \to a$, is equivalent to

(4'''') $$d_1 \vee \ldots \vee (b \wedge f).$$

Lemma 7. Let α, x, t, $x_1,\ldots,x_r$, $t_1,\ldots,t_r$ be fixed and $y_1,\ldots,y_r$ be new variables. If $x \in \mathfrak{V}(\alpha)$, then let $x \in \{x_1,\ldots,x_r\}$. Then

$$\text{Subst}\; \alpha\, {}^{t_1\ldots}_{\substack{y_1\ldots\\ x_1\ldots}}\; x\, t\; \alpha\, {}^{\Delta^t_x t_1\ldots}_{\substack{y_1\ldots\\ x_1\ldots}}$$ (For Δ^t_x cf. Chap. II, § 5.2.)

Proof. (a) If x is not free in $\alpha\, {}^{t_1\ldots}_{\substack{y_1\ldots\\ x_1\ldots}}$, then x does not occur in any of the t_i, so that $\Delta^t_x t_i \equiv t_i$. Then the assertion follows from Chap. II, § 5.4, Theorem 4.

(b) If x occurs free in $\alpha\, {}^{t_1\ldots}_{\substack{y_1\ldots\\ x_1\ldots}}$, then, by Rule 4b of the substitution calculus (Chap. II, § 5.3), it suffices to show that:

i) None of the variables y_i occurs in t (this holds by the choice of the y_i);

ii) $y_1 = t_1 \wedge \ldots \to \bigwedge x_1 \ldots (x_1 = y_1 \wedge \ldots \to \alpha)$ is transformed by substitution of t for x into $y_1 = \Delta^t_x t_1 \wedge \ldots \wedge \bigwedge x_1 \ldots (x_1 = y_1 \wedge \ldots \to \alpha)$. This holds if, as in our assumption, x does not occur free in $\bigwedge x_1 \ldots (x_1 = y_1 \wedge \ldots \to \alpha)$.

Theorem 1. If $\vdash \sigma$, then $\vdash \sigma \dfrac{t_1 \ldots t_r}{x_1 \ldots x_r}$.

Proof. (We shall write $\sigma \dfrac{t_1\ldots}{x_1\ldots}$ instead of $\sigma \dfrac{t_1\ldots t_r}{x_1\ldots x_r}$.) Let $u_1,\ldots,u_r$ be new individual variables. Let s_i be obtained from t_i by replacing the $x_1,\ldots,x_r$ by the corresponding $u_1,\ldots,u_r$. Let $\sigma \equiv \alpha_1 \ldots \alpha_n \alpha$. By Lemma 4 we have

$$\vdash x_1 = s_1 \ldots \alpha_i \frac{s_1\ldots}{x_1\ldots} \leftrightarrow \alpha_i \qquad (i = 1,\ldots,r)$$

and
$$\vdash x_1 = s_1 \ldots \alpha \frac{s_1\ldots}{x_1\ldots} \leftrightarrow \alpha\,,$$

and hence, because $\vdash \sigma$,

$$\vdash x_1 = s_1 \ldots \sigma \frac{s_1\ldots}{x_1\ldots}\,.$$

By construction, none of the x_i occurs in any s_j and none of the x_i occurs free in $\sigma\frac{s_1\ldots}{x_1\ldots}$. Thus we have

$$\vdash \bigvee x_1\, x_1 = s_1 \ldots \sigma\frac{s_1\ldots}{x_1\ldots} \qquad (P'_{x_1}),\ldots$$

$$\vdash \sigma\frac{s_1\ldots}{x_1\ldots}$$

(since $\bigvee x_1\, x_1 = s_1$ can be derived from $s_1 = s_1$ by (ExP_{x_1, s_1}))

$$\vdash \sigma\frac{t_1\ldots}{x_1\ldots} \qquad (S_{u_1}^{x_1}),\ldots\ ; \quad \text{cf. Lemma 7.}$$

<u>Lemma 8.</u> If none of the variables $x_1,\ldots,x_r$ occurs free in α, then $\vdash \alpha\,\alpha\frac{t_1\ldots t_r}{x_1\ldots x_r}$.

Proof by means of a derivation whose main steps are:

$$\alpha : x_1 = y_1 \wedge \ldots \rightarrow \alpha$$

$$\alpha : \bigwedge x_1 \ldots (x_1 = y_1 \wedge \ldots \rightarrow \alpha)$$

$$\alpha : y_1 = t_1 \wedge \ldots \rightarrow \bigwedge x_1 \ldots (x_1 = y_1 \wedge \ldots \rightarrow \alpha)$$

$$\alpha : \bigwedge y_1 \ldots (y_1 = t_1 \wedge \ldots \rightarrow \bigwedge x_1 \ldots (x_1 = y_1 \wedge \ldots \rightarrow \alpha)).$$

<u>Lemma 9.</u> $\vdash Px_1\ldots x_r\frac{t_1\ldots t_r}{x_1\ldots x_r}\quad Pt_1\ldots t_r$.

Proof. Let $u_1,\ldots,u_r$ be new variables. There is a derivation whose main steps are:

$$Px_1\ldots x_r\frac{t_1\ldots t_r}{x_1\ldots x_r}\, y_1 = t_1 \ldots x_1 = y_1 \ldots : Px_1\ldots x_r$$

$$Px_1\ldots x_r\frac{t_1\ldots t_r}{x_1\ldots x_r}\, y_1 = t_1 \ldots z_1 = y_1 \ldots : Pz_1\ldots z_r$$

$$Px_1\ldots x_r\frac{t_1\ldots t_r}{x_1\ldots x_r}\, z_1 = t_1 \ldots : Pz_1\ldots z_r \qquad (S_{y_1}^{t_1}),\ldots$$

$$Px_1\ldots x_r\frac{t_1\ldots t_r}{x_1\ldots x_r} : Pt_1\ldots t_r \qquad (S_{z_1}^{t_1}),\ldots$$

<u>Lemma 10.</u> $\vdash Pt_1\ldots t_r\quad Px_1\ldots x_r\frac{t_1\ldots t_r}{x_1\ldots x_r}$.

Proof by means of a derivation whose main steps are:

$$
\begin{array}{rll}
Py_1\ldots y_r & : Py_1\ldots y_r & \\
Py_1\ldots y_r\, y_1 = x_1\ldots & : Px_1\ldots x_r & (E^{x_1}_{y_1}) \\
Py_1\ldots y_r & : \bigwedge x_1\ldots (x_1 = y_1 \wedge \ldots \rightarrow Px_1\ldots x_r) & \\
\neg\bigwedge x_1\ldots (x_1 = y_1 \wedge \ldots \rightarrow Px_1\ldots x_r) & : \neg Py_1\ldots y_r & \\
\neg\bigwedge x_1\ldots (x_1 = y_1 \wedge \ldots \rightarrow Px_1\ldots x_r) y_1 = t_1\ldots & : \neg Pt_1\ldots t_r & (E^{t_1}_{y_1}) \\
Pt_1\ldots t_r & : \bigwedge y_1\ldots (y_1 = t_1 \wedge \ldots \rightarrow \bigwedge x_1\ldots & \\
& \quad (x_1 = y_1 \wedge \ldots \rightarrow Px_1\ldots x_r)). &
\end{array}
$$

2.3 Substitution for a predicate variable. In this subsection, let P be a fixed predicate variable, $x_1, \ldots, x_r$ be the first r individual variables and ρ be an expression in which at most $x_1, \ldots, x_r$ occur free. To every expression α we assign an expression α^* which, we say, is obtained from α by substituting ρ for P. The inductive definition of α^* reads as follows:

$$[Pt_1\ldots t_r]^* \equiv \rho\,\frac{t_1\ldots t_r}{x_1\ldots x_r}$$

$\alpha^* \equiv \alpha$ for every other atomic expression

$$[\neg\alpha]^* \equiv \neg\alpha^*$$

$$(\alpha \wedge \beta)^* \equiv (\alpha^* \wedge \beta^*)$$

$$[\bigwedge x\alpha]^* \equiv \bigwedge x\alpha^*.$$

If σ is a sequent, then let σ^* be the sequent whose members are the $*$-images of the members of σ.

Lemma 11. If P does not occur in α, then $\alpha^* \equiv \alpha$.

Lemma 12. $\mathfrak{B}(\alpha) = \mathfrak{B}(\alpha^*)$.

P r o o f by induction on the structure of α. The only case worth mentioning explicitly is $\alpha \equiv Pt_1\ldots t_r$. By Lemma 6, z occurs free in α if and only if (z occurs in t_1 or ... or z occurs in t_r) or (z occurs free in ρ and $z \notin \{x_1, \ldots, x_r\}$). But the latter member of the disjunction is false because of the condition on ρ; and (z occurs in t_1 or ... or z occurs in t_r) if and only if z occurs free in α.

Lemma 13. $\vdash \delta^*$. (For δ cf. (1) in §2.1.)

P r o o f. $\delta \equiv \bigwedge x_1\ldots\bigwedge x_r\left(\rho\,\frac{x_1\ldots}{x_1\ldots} \leftrightarrow \rho\right)$. Apply Lemma 5.

Lemma 14. Let Subst $\alpha x t \beta$. Then there are expressions α', β' such that

(i) $\mathfrak{B}(\alpha') = \mathfrak{B}(\alpha^*)$, $\vdash \alpha' \leftrightarrow \alpha^*$,

(ii) $\mathfrak{B}(\beta') = \mathfrak{B}(\beta^*)$, $\vdash \beta' \leftrightarrow \beta^*$,

(iii) Subst $\alpha' x t \beta'$.

P r o o f by induction on the structure of α.

1a) $\alpha \equiv Pt_1 \ldots t_r$, and thus $\beta \equiv P \Delta_x^t t_1 \ldots \Delta_x^t t_r$. We have

$$\alpha^* \equiv \rho \frac{t_1 \ldots}{x_1 \ldots} , \qquad \beta^* \equiv \rho \frac{\Delta_x^t t_1 \ldots}{x_1 \ldots} .$$

Let $y_1, \ldots, y_r$ be new individual variables, and put

$$\alpha' \equiv \rho \begin{smallmatrix} t_1 \ldots \\ y_1 \ldots \\ x_1 \ldots \end{smallmatrix} , \qquad \beta' \equiv \rho \begin{smallmatrix} \Delta_x^t t_1 \ldots \\ y_1 \ldots \\ x_1 \ldots \end{smallmatrix} .$$

(i) and (ii) follow by Lemmas 1 and 6. Because at most $x_1, \ldots, x_r$ occur free in ρ, (iii) follows by Lemma 7.

1b) If α is any other atomic expression, then the required result is achieved by putting $\alpha' \equiv \alpha$, $\beta' \equiv \beta$.

2) Let Subst $\neg \alpha x t \neg \beta$, and hence Subst $\alpha x t \beta$. By induction hypothesis, there are expressions α', β' with the properties (i), (ii), (iii). Put $[\neg \alpha]' \equiv \neg \alpha'$, $[\neg \beta]' \equiv \neg \beta'$.

3) If α is a conjunction, we prove the Lemma in a way analogous to 2).

4) Let Subst $\bigwedge z\alpha x t \bigwedge z\beta$.

C a s e 1: x does not occur free in $\bigwedge z\alpha$. Then $\beta \equiv \alpha$. By Lemma 12, x does not occur free in $[\bigwedge z\alpha]^*$. It follows that Subst $[\bigwedge z\alpha]^* x t [\bigwedge z\alpha]^*$. Put $[\bigwedge z\alpha]' \equiv [\bigwedge z\beta]' \equiv$ $\equiv [\bigwedge z\alpha]^*$.

C a s e 2: x occurs free in $\bigwedge z\alpha$, z does not occur in t, Subst $\alpha x t \beta$. By induction hypothesis, there are expressions α', β' with the required properties. In particular, Subst $\bigwedge z\alpha' x t \bigwedge z\beta'$. Put $[\bigwedge z\alpha]' \equiv \bigwedge z\alpha'$, $[\bigwedge z\beta]' \equiv \bigwedge z\beta'$. It follows that $\mathfrak{V}([\bigwedge z\alpha]') = \mathfrak{V}(\alpha') - \{z\} = \mathfrak{V}(\alpha^*) - \{z\} = \mathfrak{V}([\bigwedge z\alpha]^*)$, and that $\vdash [\bigwedge z\alpha]' \leftrightarrow [\bigwedge z\alpha]^*$, since $\vdash \alpha' \leftrightarrow \alpha^*$. Analogously, β also has the required properties.

<u>Theorem 2.</u> <u>If</u> $\vdash \sigma$, <u>then</u> $\vdash \sigma^*$.

P r o o f by induction on the length of the derivation of σ. We consider the last rule which was applied in the derivation of σ and assume, by induction hypothesis, that Theorem 2 holds for all sequents with a shorter derivation. We have the following cases, according to which rule was applied last in the derivation of σ:

(A). Then $\sigma \equiv \alpha\alpha$, $\sigma^* \equiv \alpha^*\alpha^*$, and hence $\vdash \sigma^*$.

(E). Analogous to (A).

(C). $\sigma \equiv \sigma_{12}(\alpha \wedge \beta)$ is obtained from $\sigma_1\alpha$ and $\sigma_2\beta$. By induction hypothesis, we have $\vdash \sigma_1^*\alpha^*$ and $\vdash \sigma_2^*\beta^*$. It follows that $\vdash \sigma_{12}^*(\alpha^* \wedge \beta^*)$, i.e. $\sigma_{12}^*(\alpha \wedge \beta)^*$, i.e. $\vdash \sigma^*$.

(C'), (C''), (R), (X), (G). Analogously to (C).

(G_x). $\sigma \equiv \tau \bigwedge x\alpha$ is obtained from $\tau\alpha$, where x does not occur free in τ. By induction hypothesis, we have $\vdash \tau^*\alpha^*$. By Lemma 12, x does not occur free in τ^*. Hence $\vdash \tau^* \bigwedge x\alpha^*$, i.e. $\vdash \tau^*[\bigwedge x\alpha]^*$, i.e. $\vdash \sigma^*$.

(E_x^t). $\sigma \equiv \tau\, x{=}t\ \beta$ is obtained from $\tau\alpha$, where Subst $\alpha x t \beta$. By induction hypothesis, we have $\vdash \tau^*\alpha^*$. By Lemma 14, there are expressions α', β' with the properties given there. Because $\vdash \alpha' \leftrightarrow \alpha^*$ we have $\vdash \tau^*\alpha'$ and hence, by (E_x^t), $\vdash \tau^* x{=}t\ \beta'$. Because $\vdash \beta' \leftrightarrow \beta^*$ it follows that $\vdash \tau^* x = t\,\beta^*$, i.e. $\vdash \sigma^*$.

(S_x^t). (DdRu) and (DdRu') (Chap. IV, § 4.3) show that it is sufficient to prove (S_x^t) only for cases in which σ consists of a single member. Let $\sigma \equiv \beta$. β is obtained from α, where Subst $\alpha x t \beta$. Using Lemma 14, we carry out the proof as in the case (E_x^t).

<u>Theorem 3.</u> <u>If</u> $\vdash \delta\sigma$ <u>and if P does not occur in</u> σ, <u>then</u> $\vdash \sigma$. (For δ cf. (1) in § 2.1.)

P r o o f . If $\vdash \delta\sigma$ then, by Theorem 2, $\vdash \delta^*\sigma^*$. We then apply Lemmas 11 and 13.

<u>Theorem 4.</u> <u>If</u> $\mathfrak{A}$ <u>is consistent and P does not occur in</u> $\mathfrak{A}$, <u>then</u> $\mathfrak{A} \cup \{\delta\}$ <u>is also consistent.</u>

P r o o f . If $\mathfrak{A} \cup \{\delta\}$ were inconsistent, then there would be a sequent σ whose members were elements of $\mathfrak{A}$ and an expression α which did not contain P, such that $\vdash \delta\sigma(\alpha \wedge \neg\alpha)$. We now apply Theorem 3.

<u>Theorem 5.</u> $\vdash \delta \rightarrow (\alpha \leftrightarrow \alpha^*)$.

P r o o f by induction on the structure of α. We restrict ourselves to the only nontrivial case, $\alpha \equiv Pt_1 \ldots t_r$. Now $\vdash \delta\delta$, i.e. $\vdash \delta(Px_1 \ldots x_r \leftrightarrow \rho)$, and hence $\vdash \delta Px_1 \ldots x_r\,\rho$ and $\vdash \delta\,\rho\,Px_1 \ldots x_r$. We now apply Theorem 1 and, by Lemmas 8, 9, 10, obtain

$$\vdash \delta Pt_1 \ldots t_r\ \rho\frac{t_1 \ldots}{x_1 \ldots} \quad \text{and} \quad \vdash \delta\,\rho\frac{t_1 \ldots}{x_1 \ldots}\ Pt_1 \ldots t_r ,$$

from which it follows that $\vdash \delta Pt_1 \ldots t_r \leftrightarrow [Pt_1 \ldots t_r]^*$.

<u>Exercises. 1.</u> To every α, x, t we assigned an expression α_1 in this section and an expression α_2 in § 1, both of which we denoted by $\alpha\frac{t}{x}$. Show that $\alpha_1 \dashv\vdash \alpha_2$.

<u>2.</u> Show that, in general, it is not true that $\alpha\frac{t_1 t_2}{x_1 x_2} \dashv\vdash [\alpha\frac{t_1}{x_1}]\frac{t_2}{x_2}$.

<u>3.</u> In analogy to (1) in § 2.1, we can regard a formula

$$(*) \qquad \bigwedge x_1 \ldots \bigwedge x_r \bigwedge y(fx_1 \ldots x_r = y \leftrightarrow \rho)$$

as a definition of a function symbol f. In analogy to the definition of α^* given at the beginning of § 2.3, give a definition of an expression α^0 (for each expression α) in which f is eliminated with the aid of (*), and prove analogs to Theorems 2, 3, 4. In order to do this, it will be necessary to stipulate that ρ satisfies a condition which guarantees that an f which is introduced by (*) always has the nature of a function.

§ 3. The process of relativisation

3.1 Introduction. In this section, let R be a fixed one-place predicate variable. To every expression α we shall assign effectively an expression α^R, the (R-) relativisation of α, by means of the following inductive

Definition 1.

$$(1)\quad \begin{cases} \alpha^R \equiv \alpha \quad \text{for atomic } \alpha \\ [\neg\alpha]^R \equiv \neg\alpha^R \\ (\alpha \wedge \beta)^R \equiv (\alpha^R \wedge \beta^R) \\ [\bigwedge x\alpha]^R \equiv \bigwedge x(Rx \to \alpha^R). \end{cases}$$

The process of relativisation is often applied in relative consistency proofs (cf. the remark after Theorem 4). In view of these applications, we shall prove the following theorems constructively (see also the deliberations in § 2.1). For the semantic significance of relativisation cf. Exercise 4.

If σ is a sequent, then let σ^R be the sequent whose members are the R-relativisations of the members of σ.

3.2 Theorems about relativisation. The following two lemmas can easily be proved by induction on the structure of α.

Lemma 1. $\mathfrak{B}(\alpha^R) = \mathfrak{B}(\alpha) \cup \{R\}$.

Lemma 2. If $\mathrm{Subst}\,\alpha x t \beta$, then $\mathrm{Subst}\,\alpha^R x t \beta^R$.

Definition 2. Let s be an arbitrary term, and $v_1, \ldots, v_r$ be the individual variables which occur in s, taken in a standard order. Then let

$$(2)\qquad R^s \equiv \bigwedge v_1 \ldots \bigwedge v_r(Rv_1 \wedge \ldots \wedge Rv_r \to Rs)\ .$$

Note that no individual variable occurs free in R^s.

Theorem 1. Let a derivation of the sequent σ be given. Then we can effectively find finitely many individual variables $u_1, \ldots, u_n$ $(n \geqslant 0)$ and finitely many terms $s_1, \ldots, s_m$ $(m \geqslant 0)$ and a derivation of the sequent

$$Ru_1 \ldots Ru_n R^{s_1} \ldots R^{s_m} \sigma^R.$$

P r o o f by induction on the length of the given derivation of σ. We have the following cases, according to which rule was applied last in the derivation:

(A): This rule gives a sequent $\alpha\alpha.\ \vdash \alpha^R\alpha^R$, i.e. our assertion with $n = m = 0$, holds.

(E): Similar to (A).

(C): This rule gives a sequent $\sigma_{12}(\alpha_1 \wedge \alpha_2)$, where shorter derivations for the sequents $\sigma_1\alpha_1$ and $\sigma_2\alpha_2$ are already given. By induction hypothesis, we can find $u_1, \ldots, s_1, \ldots$ and $v_1, \ldots, t_1, \ldots$ such that $\vdash Ru_1 \ldots R^{s_1} \sigma_1^R$ and $\vdash Rv_1 \ldots R^{t_1} \ldots \sigma_2^R \alpha_2^R$. It follows from this that $\vdash Ru_1 \ldots Rv_1 \ldots R^{s_1} \ldots R^{t_1} \ldots \sigma_{12}^R (\alpha_1^R \wedge \alpha_2^R)$, and $(\alpha_1^R \wedge \alpha_2^R) \equiv (\alpha_1 \wedge \alpha_2)^R$.

(C'), (C''), (R), (X): Analogous to (C).

(E_x^t): This rule gives a sequent $\tau\ \ x{=}t\ \beta$, where a shorter derivation for $\tau\alpha$ is already given and Subst $\alpha x t \beta$. By induction hypothesis, we can find $u_1, \ldots, s_1, \ldots$ such that $\vdash Ru_1 \ldots R^{s_1} \ldots \tau^R \alpha^R$. By (E_x^t) this gives $\vdash Ru_1 \ldots R^{s_1} \ldots \tau^R x = t\, \beta^R$ (cf. Lemma 2), and $x = t \equiv [x = t]^R$.

(S_x^t): This rule gives a sequent $\tau'\alpha'$, where a shorter derivation for $\tau\alpha$ is already given and Subst $\tau\alpha x t \tau'\alpha'$. By induction hypothesis we can find $u_1, \ldots, s_1, \ldots$ such that $\vdash Ru_1 \ldots R^{s_1} \ldots \tau^R \alpha^R$. By adding Rx to the antecedent (provided none of the u_i is x) we obtain $\vdash Ru_1 \ldots Rx R^{s_1} \ldots \tau^R \alpha^R$, where we may now assume that all $u_i \not\equiv x$. By (S_x^t) this gives us $\vdash Ru_1 \ldots Rt R^{s_1} \ldots \tau'^R \alpha'^R$. If $v_1, \ldots, v_r$ are the individual variables which occur in t in the standard order then, by (2), we have $\vdash R^t Rv_1 \ldots Rv_r Rt$ and hence, finally, $\vdash Ru_1 \ldots Rv_1 \ldots R^t R^{s_1} \ldots \tau'^R \alpha'^R$.

(G): This rule gives a sequent $\tau\alpha$, where a shorter derivation of a sequent $\tau \bigwedge x\alpha$ is already given. By induction hypothesis we can find $u_1, \ldots, s_1 \ldots$ such that $\vdash Ru_1 \ldots R^{s_1} \ldots \tau^R \bigwedge x(Rx \to \alpha^R)$. From this we obtain $\vdash Ru_1 \ldots Rx R^{s_1} \ldots \tau^R \alpha^R$.

(G_x): This rule gives a sequent $\tau \bigwedge x\alpha$, where a shorter derivation of the sequent $\tau\alpha$ is already given and x does not occur free in τ. By induction hypothesis we can find $u_1, \ldots, s_1, \ldots$ such that $\vdash Ru_1 \ldots R^{s_1} \ldots \tau^R \alpha^R$. As in the case (S_x^t), we can then obtain $\vdash Ru_1 \ldots Rx R^{s_1} \ldots \tau^R \alpha^R$, where we may assume that all the $u_i \not\equiv x$. With the aid of (G_x), we now obtain $\vdash Ru_1 \ldots R^{s_1} \ldots \tau^R \bigwedge x(Rx \to \alpha^R)$ (note that, by Lemma 1, x does not occur free in τ^R).

Theorem 2. Let a derivation of the sequent σ be given. Suppose also that σ contains no occurrences of genuine (i.e. $\geqslant$ 1-place) function variables and no free occurrences of individual variables. Then we can effectively find a derivation of the sequent $\bigvee x Rx \sigma^R$.

Proof. We can effectively transform the given derivation of σ into one in which the rule (S_x^t) (and, incidentally, also the rule (E_x^t)) is applied only when t is an individual variable. This simplifies the proof of Theorem 1 and gives us a derivation of a sequent $Ru_1 \ldots \sigma^R$. Since, by hypothesis, no individual variable occurs free in σ, we can particularise the Ru_i.

Theorem 3a. Let a derivation of $\sigma(P \wedge \neg P)$ be given and also, for every member α of σ, a derivation of $\tau\alpha^R$, and a derivation of $\tau \bigvee x Rx$. Suppose also that σ contains no occurrences of genuine function variables and no free occurrences of individual variables. Then we can effectively find a derivation of the sequent $\tau(P \wedge \neg P)$.

Proof. By Theorem 2, we can find a derivation of $\bigvee x Rx \sigma^R (P \wedge \neg P)$. The assertion follows by applications of the cut rule.

From the constructive formulation of Theorem 3a we at once obtain the following purely existential statement:

Theorem 3b. Let $\vdash \sigma(P \wedge \neg P)$, $\vdash \tau\alpha^R$ for every member α of σ and $\vdash \tau \bigvee x Rx$. Suppose also that σ contains no genuine function variables and no free occurrences of individual variables. Then $\vdash \tau(P \wedge \neg P)$.

Theorem 4. Let $\mathfrak{A}$ be a consistent set of expressions, and suppose that no genuine function variables and no individual variables occur free in the elements of the set of expressions $\mathfrak{B}$. Also, let $\mathfrak{A} \vdash \beta^R$ for every $\beta \in \mathfrak{B}$ and $\mathfrak{A} \vdash \bigvee x Rx$. Then $\mathfrak{B}$ is also consistent.

Proof. If $\mathfrak{B}$ were inconsistent, then there would be a sequent σ whose members were elements of $\mathfrak{B}$, such that $\vdash \sigma(P \wedge \neg P)$. For every element α of σ there is, by hypothesis, a sequent τ_α whose members are elements of $\mathfrak{A}$, such that $\vdash \tau_\alpha \alpha^R$. Moreover, there is a sequent τ_R whose members are elements of $\mathfrak{A}$, such that $\vdash \tau_R \bigvee x Rx$. Let τ be a sequent whose members are the members of the τ_α and of τ_R. Then $\vdash \tau\alpha^R$ for every element α of σ, and $\vdash \tau \bigvee x Rx$. We now apply Theorem 3b.

Remark. Theorem 4 allows us to deduce the consistency of a set of expressions $\mathfrak{B}$ from that of a set $\mathfrak{A}$, i.e. it gives us so-called relative consistency proofs. If a derivation of $\bigvee x Rx$ from $\mathfrak{A}$ is given effectively and if a method is known by means of which, for every $\beta \in \mathfrak{B}$, a derivation of β from $\mathfrak{A}$ can be obtained, then the relative consistency proof given in Theorem 4 can be carried out constructively. Relative consistency proofs are most widely used in axiomatic set theory.

Exercises. 1. Prove the first assertion in the proof of Theorem 2. (Suggestion: Replace the genuine terms which occur in the given derivation of σ in a suitable way.)

2. In Theorems 2 to 4 we require that only predicate variables should occur free in σ. Find a way of dispensing with this requirement; e.g. prove the following generalisation of Theorem 2: Let a derivation of the sequent σ be given. As i varies, let f_i (with n_i places) run through all the ($\geqslant$ 0-place) function variables which occur free in σ. Then we can effectively find a derivation of the sequent

$$\bigwedge x_1 \ldots \bigwedge x_{n_1}(Rx_1 \wedge \ldots \wedge Rx_{n_1} \rightarrow Rf_1x_1\ldots x_{n_1}) \ldots \bigvee xRx\sigma^R.$$

3. Find an axiom system $\mathfrak{A}$ for group theory with the primitive notions e (unit element), f (inverse function), g (group operation). Show constructively that: If $\mathfrak{A}$ is consistent, then so is $\mathfrak{A} \cup \{\bigwedge x \bigwedge y\, x = y\}$. (Suggestion: Introduce a predicate R by means of the definition $\bigwedge x(Rx \leftrightarrow x = e)$, and relativise with respect to R. See also Exercise 2.)

4. Let $\mathfrak{J}$ be an arbitrary interpretation over ω. Let Mod$\mathfrak{J}$Rx. Moreover, suppose that, for every ($\geqslant$ 0-place) function variable f: If $\mathfrak{J}(R)$ fits every one of the individuals $\mathfrak{x}_j$ $(j = 1,\ldots,r)$, then $\mathfrak{J}(R)$ also fits $\mathfrak{J}(f)(\mathfrak{x}_1,\ldots,\mathfrak{x}_r)$. Under these assumptions we can define an interpretation $\mathfrak{J}^R$ over the domain ω_R of those elements of ω which $\mathfrak{J}(R)$ fits, as follows: $\mathfrak{J}^R(P)$ fits $\mathfrak{x}_1,\ldots,\mathfrak{x}_r$ if and only if $\mathfrak{J}(P)$ fits $\mathfrak{x}_1,\ldots,\mathfrak{x}_r$; $\mathfrak{J}^R(f)(\mathfrak{x}_1,\ldots,\mathfrak{x}_r) = \mathfrak{J}(f)(\mathfrak{x}_1,\ldots,\mathfrak{x}_r)$. Show that Mod$\mathfrak{J}\alpha^R$ iff Mod$\mathfrak{J}^R\alpha$.

5. With the aid of the previous exercise, give a semantic proof of Theorem 1.

6. Let $\mathfrak{B}$ be a given set of variables. Let the interpretation $\mathfrak{J}$ satisfy the conditions given in Exercise 4 (where the requirement on the function variables can be restricted to the elements of $\mathfrak{B}$). If $\mathfrak{a}_{\mathfrak{B}}(\mathfrak{J}^R)$ is defined as in Chap. VI, § 2.1 (for $\mathfrak{J}^R$ cf. Exercise 4 above), then show that:

a) $\mathfrak{a}_{\mathfrak{B}}(\mathfrak{J}^R)$ is a subalgebra of $\mathfrak{a}_{\mathfrak{B}}(\mathfrak{J})$.

b) For all α such that $\mathfrak{B}(\alpha) \subset \mathfrak{B}$: $\mathfrak{a}_{\mathfrak{B}}(\mathfrak{J}^R)$ is a model of α if and only if $\mathfrak{a}_{\mathfrak{B} \cup \{R\}}(\mathfrak{J})$ is a model of α^R.

7. Let $\mathfrak{B}$ and $\mathfrak{J}$ satisfy the conditions given in Exercise 6, and suppose also that $R \notin \mathfrak{B}$. Show that: $\mathfrak{a}_{\mathfrak{B} \cup \{R\}}(\mathfrak{J}^R)$ is an elementary subalgebra of $\mathfrak{a}_{\mathfrak{B} \cup \{R\}}(\mathfrak{J})$ if and only if, for each n, the algebra $\mathfrak{a}_{\mathfrak{B} \cup \{R\}}(\mathfrak{J})$ is a model for all expressions $\bigwedge x_1 \ldots \bigwedge x_n(Rx_1 \wedge \ldots \wedge Rx_n \rightarrow (\alpha \leftrightarrow \alpha^R))$ for which $\mathfrak{B}(\alpha) \subset \mathfrak{B} \cup \{x_1,\ldots,x_n\}$.

Further Reading

A) Logic textbooks (a selection)

Carnap, R.: The Logical Syntax of Language, London 1964.

Church, A.: Introduction to Mathematical Logic, Vol. I. Princeton 1956.

Curry, H.B.: Foundations of Mathematical Logic, New York - San Fransisco - Toronto - London 1963.

Hilbert, D., Ackermann, W.: Principles of Mathematical Logic, New York 1950.

Kalish, D., Montague, R.: Logic Techniques of Formal Reasoning, New York - Burlingame 1964.

Kleene, S.C.: Introduction to Metamathematics, Amsterdam 1964.

Kleene, S.C.: Mathematical Logic, New York - London - Sydney 1967.

Kreisel, G., Krivine, J.L.: Elements of Mathematical Logic (Model Theory), Amsterdam 1967.

Lorenzen, P.: Formale Logik, Berlin 1967.

Mendelson, E.: Introduction to Mathematical Logic, Princeton, N.J. - Toronto - New York - London 1964.

Novikov, P.S.: Elements of Mathematical Logic, Edinburgh - London 1964.

Quine, W.V.: Mathematical Logic, 1951, new edition 1962.

Rosser, J.B.: Logic for Mathematicians, New York - Toronto - London 1953.

Schmidt, A.: Mathematische Gesetze der Logik I, Vorlesungen über Aussagenlogik, Berlin - Göttingen - Heidelberg 1960.

Shoenfield, J.R.: Mathematical Logic, Reading/Mass. 1967.

Smullyan, R.M.: First-Order Logic, Berlin - Heidelberg - New York 1968.

Tarski, A.: Introduction to Logic and to the Methodology of Deductive Sciences, New York 1965.

B) On symbolisation of everyday statements consult also

Carnap, R.: Introduction to Symbolic Logic and its Applications, New York 1958.

Mates, B.: Elementary Logic, Oxford 1965.

C) On Peano's axiom system and model theory c.f.

Bell, J.L., Slomson, A.B.: Models and Ultraproducts, Amsterdam 1969.

Henkin, L., Smith, W.N., Varineau, V.J., Walsh, M.J.: Retracing Elementary Mathematics, London - New York 1962.

Robinson, A.: Introduction to Model Theory and to the Metamathematics of Algebra, Amsterdam 1963.

D) The concept of decidability is treated by

Davis, M.: Computability and Unsolvability, New York - Toronto - London 1958.

Davis, M.: The Undecidable. Basic Papers on Undecidable Propositions, Unsolvable Problems and Computable Functions, New York: Hewlett 1965.

Hermes, H.: Enumerability, Decidability, Computability, New York 1969.

Kleene, S.C.: Introduction to Metamathematics, Amsterdam 1964.

Péter, R.: Recursive Functions, New York 1967.

Rogers, H., Jr.: Theory of Recursive Functions and Effective Computability, New York 1967.

Smullyan, R.M.: Theory of Formal Systems, 1971.

E) On the decision problem in predicate logic see

Ackermann, W.: Solvable Cases of the Decision Problem, Amsterdam 1954.

Surányi, J.: Reduktionstheorie des Entscheidungsproblems im Prädikatenkalkül der ersten Stufe, Budapest 1959.

Tarski, A., Mostowski, A., Robinson, R.M.: Undecidable Theories, Amsterdam 1953.

F) Non-classical views of logic are presented in

Beth, E.W.: The Foundations of Mathematics, Amsterdam 1959.

Fitting, M.C.: Intuitionistic Logic, Model Theory and Forcing, Amsterdam 1969.

Heyting, A.: Intuitionism. An Introduction, Amsterdam 1966.

Lorenzen, P.: Einführung in die operative Logik und Mathematik, Berlin - Göttingen - Heidelberg 1955.

Lorenzen, P.: Metamathematik, Mannheim 1962.

Schütte, K.: Beweistheorie, Berlin - Göttingen - Heidelberg 1960.

Schütte, K.: Vollständige Systeme modaler und intuitionistischer Logik, Berlin - Heidelberg - New York 1968.

G) Some publications of historical importance for the development of predicate logic

Bolzano, B.: Wissenschaftslehre, 4 vols. 1837, New edition Leipzig 1929 - 1931.

Frege, G.: Begriffsschrift, Eine der arithmetischen nachgebildete Formelsprache des reinen Denkens. Halle a. S. 1879. English translation in: J. van Heijenoort: From Frege to Gödel, Cambridge/Mass. 1967.

Whitehead, A.N., Russell, B.: Principia Mathematica, 3 vols, Cambridge 1910 - 1913, 1925 - 1927. New edition 1950.

Gödel, K.: Die Vollständigkeit der Axiome des logischen Funktionenkalküls, Mh. Math. Phys. 37, 349-360 (1930). English translation in: J. van Heijenoort: From Frege to Gödel, Cambridge/Mass. 1967.

Tarski, A.: Der Wahrheitsbegriff in den formalisierten Sprachen, Studia Philosophica, Lemberg 1, 261-405 (1936). English translation in: Logic, Semantics, Metamathematics. Papers from 1923 to 1938 by A. Tarski. Oxford 1956.

Heijenoort, J. van: From Frege to Gödel. A Source Book in Mathematical Logic, 1979 - 1931. Cambridge/Mass. 1967.

H) On the history of logic

Bochenski, J.M.: A History of Formal Logic, Notre Dame/Ind. 1961.

Kneale, W., Kneale, M.: The Development of Logic, Oxford 1964.

I) Summaries

Klibansky, R.: Ed. Contemporary Philosophy, Florence 1968 ff.

Mostowski, A.: Thirty Years of Foundational Studies, Helsinki 1965.

K) The most significant periodical in the field of mathematical logic

The Journal of Symbolic Logic. Published by the Association of Symbolic Logic since 1936. Vol. 1 contains a bibliography of symbolic logic. The following volumes contain, among other things, reviews of all the more important logical publications which have appeared since then.

Index of Abbreviations for Defining and Derived Rules

1. Defining rules

2. Derived rules

The abbreviations are chosen so that each capital letter (or capital letter followed by a lower-case letter) always stands for the same word; e.g.: A for "assumption", An for "antecedent", As for "associative", At for "assertion", etc.

Notation

Isomorphism of expressions

α^{Φ}, $\mathfrak{M}^{\Phi}$	122

Conjunctive and disjunctive normal form

$\alpha_{\wedge}$, $\alpha_{\vee}$	181

Iterated conjunction and disjunction

$\bigwedge_{i=1}^{n}$	181
$\bigvee_{i=1}^{n}$	182

Inverse prefix

$\underline{P}^{-1}$	184

Algebras and theories

$\mathfrak{A}_{\mathfrak{b}}(\mathfrak{J})$	149
$\theta_{\mathfrak{b}}(\mathfrak{J})$	192

Index of individual variables

Ind (x)	210

Relativisation

α^{R}	225

Name and Subject Index